The Work of Language in Multicultural Classrooms

"This book focuses on the linguistic challenges of the academic language through which science is constructed and taught, but also recognizes the role of community and home/cultural discourses in positioning students as science learners, and addresses issues of student identity. With all of these elements, it is a valuable bringing-together of current state-of-the-art research in one volume that is likely to be widely used in teacher education."

Mary Schleppegrell, University of Michigan

"... an important read for all those working with children and students in our increasingly multicultural societies with their increasingly hybridized and hybridizing cultures. This book charts new waters in raising educators' awareness of the many different situations and facets at the intersection of cultures and languages."

Wolff-Michael Roth, University of Victoria

"... offers a much needed look into the language practices that shape science teaching and learning, and how these practices both facilitate and constrain access to and success in science for all students."

Angela Calabrese Barton, Michigan State University

"The need for a scientifically literate citizenry has never been greater and the scholars contributing to this volume provide insights into the potential of language to afford knowing science in ways that shape identity and contribute to an improved social life."

Kenneth Tobin, The Graduate Center of the City University of New York

How does language comprise the implicit or explicit curriculum of teaching and learning in multicultural science settings? Building on a growing interest in how language and literacy practices interact with science teaching and learning to facilitate or obstruct successful student outcomes, this book

- contributes to scholarship on the role of language in developing classroom scientific communities of practice,
- expands that work by highlighting the challenges faced specifically by ethnic- and linguistic- "minority" students and their teachers in joining those communities, and
- showcases exemplary teaching and research initiatives for helping to meet these challenges.

Offering lenses through which readers can critically consider the myriad settings, approaches, materials, and topics involved in what it means to teach science while pointedly addressing concerns about equity of educational opportunity, this volume serves as a powerful resource for linking theory and practice. End-of-chapter reflection questions and engagement activities facilitate discussion and provide rich opportunities to consider implications and to apply insights developed in real-world science teaching and learning contexts.

Katherine Richardson Bruna is Assistant Professor of Multicultural and International Curriculum Studies in the Department of Curriculum and Instruction, College of Human Sciences, Iowa State University.

Kimberley Gomez is Assistant Professor of Literacy, Language and Culture and Learning Sciences at the University of Illinois at Chicago.

Language, Culture, and Teaching
Sonia Nieto, Series Editor

The Work of Language in Multicultural Classrooms

Talking Science, Writing Science

Edited by
Katherine Richardson Bruna and
Kimberley Gomez

Routledge
Taylor & Francis Group

NEW YORK AND LONDON

First published 2009
by Routledge
605 Third Avenue, New York, NY 10017
4 Park Square, Milton Park, Abingdon, Oxon OX14 4RN

Routledge is an imprint of the Taylor & Francis Group, an informa business

© 2009 Taylor and Francis

Typeset in Minion by Wearset Ltd, Boldon, Tyne and Wear
Printed and bound in the United States of America on acid-free paper by
Sheridan Books, Inc

Library of Congress Cataloging in Publication Data
The work of language in multicultural classrooms: talking science, writing
science/Katherine Richardson Bruna and Kimberley Gomez (Eds.).

p. cm.—(Language, culture, and teaching)
Includes bibliographical references and index.
1. Science–Study and teaching–United States. 2. Multicultural
education–United States. 3. Language arts. I. Bruna, Katherine Richardson.
II. Gomez, Kimberley, 1958–
LB1585.3.W67 2008
507.1–dc22   2008003035

ISBN13: 978-0-8058-6427-4 (hbk)
ISBN13: 978-0-8058-6428-1 (pbk)
ISBN13: 978-1-4106-1860-3 (ebk)

Contents

PART II
Science Learning Funds of Knowledge 115

PART III
The Development of a Science Learner Identity 191

Foreword

Okhee Lee

When I started in the early 1990s my line of research on issues of diversity and equity with culturally and linguistically diverse students in science education, the field was at an infant stage and I was barely beginning my academic career. This book gives me a chance to look back on the field (and my own work) over the last 15 years. The literature has been emerging and growing, whereas the education system for the non-dominant (ethnic- and linguistic-"minority") student populations in science education has remained stagnant. This book painfully attests to this inconstancy.

As forcefully and eloquently articulated in the book, the crisis of science education for non-dominant student populations still exists. Despite the slogan of "science for all" and educational initiatives to achieve this goal throughout the latest science education reform since late 1980s, disparities in educational opportunities and achievement gaps remain between the haves and have-nots. The premise of this book is that such inequities will continue unless the education system and its players recognize students' linguistic and cultural resources and better understand how to teach them well, with the goal of promoting both their science learner identity and their own cultural and linguistic identity.

While lamenting the failure of the education system, the book celebrates the emerging research in many aspects. First of all, the literature has matured enough so that the book can be "dedicated exclusively to examining language and literacy issues in multicultural science settings" (p. 3 in Introduction). When the literature started to emerge since the late 1980s, the majority of the work was either advocacy for the need and the importance of issues of diversity and equity or conceptual grounding for empirical research. Over the years, research literature emerged focusing on language and literacy in science education, science education in multicultural classrooms, and science education for English language learners (ELLs) (as summarized by Stevens, Jeffries, Brisk, and Kaczmarek in Chapter 13). In this book, we see the merging of these research areas with a specific focus on language and literacy in multicultural and multilingual science classrooms.

Not only is the book dedicated to the merging of language and literacy in multicultural and multilingual science classrooms, but also it is based on contemporary research in the field. It is exciting to witness a collection of studies that are grounded on alternative, but complementary, theoretical or conceptual frameworks and were conducted using multiple research approaches. Additionally, it is encouraging that there is sufficient research literature that can be further grouped into thematic clusters, including *the language of science schooling, science learning funds of knowledge,* and *development of a science learner identity.*

For the merging of multiple areas of research to happen, researchers need to go beyond the comforts of their own areas of expertise into new areas where they have less (or limited or little) expertise. It is heartening that more researchers, both "novice" and "highly-regarded veteran researchers" (p. xix in Preface), move beyond the boundaries of their own areas into this emerging field. The two editors are examples of this new group of researchers. "While neither of our professional paths started out in science education, our paths have brought us here now" because of their long-term commitments to issues of culture, language, identity, power, and equity with non-dominant students in multicultural and multilingual settings (p. 3 in Introduction). All the authors from multiple disciplinary backgrounds have made such a transition (as an illustration, see Torres-Guzmán and Howe in Chapter 13). These researchers will serve as models for others (hopefully many including both veteran and novice researchers) to follow in coming years.

Taking a decidedly forward-looking outlook, this book is aimed at prospective science educators to develop understandings of culture, language, diversity, and equity. The editors express their hope, "We are proud to be experiencing what we consider a very exciting and much-awaited historical moment in science education and urge you to take to heart the crucial role that you, yourselves, can play" (p. 3 in Introduction). This presents challenges to the chapter authors, editors, and readers. To address these challenges, the authors and editors take innovative approaches. The authors, who typically write scholarly articles for academic journals, present their research findings for laypeople, especially prospective science educators. And the authors succeed in translating scholarly language to laypeople's language. The editors take additional steps to take theory to practice for prospective science educators. Each chapter starts with a preface highlighting key issues in the chapter. Then each chapter is followed by a common set of questions to ensure the readers' understanding of key ideas in the chapter in relation to culturally responsive instruction as described by Villegas, and Lucas (2002) and the role of language in a multicultural science context. Finally, each chapter presents to the readers a set of reflection questions and engagement activities that are specifically designed for each chapter. These questions and activities ask the readers to relate key research findings presented in each chapter to classroom practices. The four-part structure of each chapter (introduction, text, questions to ensure understanding of common themes of culture and language across all the chapters, and application of research findings presented in each chapter to classroom practice) provides a seamless integration of the 13 chapters which, otherwise, would stand separate from one another.

The seamless integration of the chapters is also enhanced by core bodies of literature in this emerging area. Across the chapters, common themes appear consistently, including identity, cultural and linguistic resources that students bring to science classrooms, community of practice in the classroom, and connection of classrooms to larger contexts of home/community and society. Additionally, common citations appear across the chapters, especially discourse in science classrooms by Lemke (1990), scientific register (Halliday & Martin, 1993), and funds of knowledge by Moll (1992). Other citations that focus specifically on science education with culturally and linguistically diverse students also appear, including the work by Rosebery and Warren (Rosebery, Warren, & Conant, 1992; Warren, Ballenger, Ogonowski, Rosebery, & Hudicourt-Barnes, 2001) and my own work (Lee, 2002; Lee & Fradd, 1998).

As a result of the conscientious efforts of the authors and editors in representing this emerging field of research, the targeted audience of prospective science educators has ample opportunities to analyze and reflect on research findings and apply these findings to classroom practices. First, the book expects the readers to understand theoretical or conceptual frameworks guiding research studies and to analyze research findings. Then, the book provides "richly descriptive and analytic examples of science teaching learning in multicultural settings within a variety of classroom and other learning contexts" (p. xviii in Preface). The examples illustrate both effective and ineffective practices across science topics and across grade levels. Finally, the book asks the readers to apply their understandings of theory and research findings to multicultural and multilingual contexts. The approach taken in this book is a stark contrast to teaching methods textbooks which typically present prepared lists of teaching strategies or tips.

Over the past 15 years of my academic career in the field, I have seen the literature grow, while the education system continues to fail the non-dominant student populations dismally. For the 15 years to come, I hope the growing body of literature provides important insights to promote these students to succeed academically in science classrooms and to remain connected to their cultural and linguistic roots in their home and community. This book contributes to such efforts.

References

Halliday, M. A. K., & Martin, J. R. (1993). *Writing science: Literacy and discursive power.* Pittsburgh: University of Pittsburgh Press.

Lee, O. (2002). Promoting scientific enquiry with elementary students from diverse cultures and languages. In W. G. Secada (Ed.), *Review of Research in Education, Vol. 26* (pp. 23–69). Washington, DC: American Educational Research Association.

Lee, O., & Fradd, S. H. (1998). Science for all, including students from non-English-language backgrounds. *Educational Researcher, 27*(4), 12–21.

Lemke, J. (1990). *Talking science: Language, learning and values.* Norwood, NJ: Ablex.

Moll, L. D. (1992). Bilingual classroom studies and community analysis: Some recent trends. *Educational Researcher, 21*(2), 20–24.

Rosebery, A. S., Warren, B., & Conant, F. R. (1992). Appropriating scientific discourse: Findings from language minority classrooms. *Journal of the Learning Sciences, 21,* 61–94.

Villegas, A. M., & Lucas, T. (2002). *Educating culturally responsive teachers: A coherent approach*. Albany, NY: State University of New York Press.

Warren, B., Ballenger, C., Ogonowski, M., Rosebery, A. S., & Hudicourt-Barnes, J. (2001). Rethinking diversity in learning science: The logic of everyday sense-making. *Journal of Research in Science Teaching, 38*, 529–552.

Preface

Much has been written in the last decade about the importance of science in facilitating new discoveries, advancing theoretical understandings about the natural world, and in creating an informed and analytic citizenry. As the editors of this volume, we wholeheartedly support these ideals. More particularly, like many others who have explored science teaching and learning in K-12 settings, we believe the latter goal, that of creating an informed and analytic citizenry, is one of the cornerstones of a democratic society and, as such, should be well supported by approaches to science curriculum and instruction. An important implication of this aim is that *all* children have the right to be exposed to, understand, and engage in the inquiry and analytic work of science. The chapters in this book seek to provide richly descriptive and analytic examples of science teaching learning in multicultural settings within a variety of classroom and other learning contexts. The intended message, we hope, is one of inclusivity and hope. That is, not only can all children learn science but also there are a variety of contexts and pedagogies that can support and extend science learning within multicultural settings.

This project builds off a growing interest, advanced in other volumes, on the ways in which language and literacy practices interact with science teaching and learning to facilitate or obstruct successful student outcomes (see, for example, Lemke, 1990; Halliday & Martin, 1993; Kress, Jewitt, Ogborn, & Tsatsarelis, 2001; Saul, 2004; Roth, 2005; Yerrick & Roth, 2005. Yet, what is missing, and what nudged, spurred, and finally compelled us toward the creation of *The Work of Language in Multicultural Classrooms: Talking Science, Writing Science* is a recognition of the need to specifically examine the intersection of language and science in the learning experiences of non-dominant (ethnic- and linguistic- "minority") populations. Because we know that the population of linguistically and ethnically diverse students comprise increasingly larger numbers within general U.S. K-12 enrollment, we recognize that what is needed, in the field of science education and multicultural education, are readings that offer teacher practitioners, and researchers a means through which they can critically consider the myriad of classroom settings, instructional approaches, curricular materials, and scientific topics that are part of what it means to teach science while pointedly addressing concerns about equity of educational opportunity. In this volume, we have assembled a collection of 13 chapters to achieve this goal. We hope to incite a lively and critical conversation among educators and researchers about the centrality of language in science teaching and learning, and how to recognize the value of, and leverage, the language resources of diverse students in science.

The Work of Language in Multicultural Classrooms: Talking Science, Writing Science is a compilation of original research articles that explore the ways in which language comprises the implicit or explicit curriculum of teaching and learning in multicultural science settings. Because it seeks to develop understandings of language, diversity, and equity issues particularly among prospective science educators, all the chapters in the volume are followed by a set of reflection questions and engagement activities that facilitate focused discussion around these issues. These reflections provide rich opportunities for the reader to consider the implications of each chapter's findings for science instruction and research and, moreover, to apply insights developed by each chapter in a real-world science context. In this regard, as a resource for teacher educators trying to promote such awareness, this volume seeks to build capacity among the science education community and serves as a powerful resource for those who wish to draw a bright line linking theory and practice. *The Work of Language in Multicultural Classrooms: Talking Science, Writing Science* showcases the current state of the field with respect to language-oriented approaches to science education theory and practice and does so with the intentionality of building capacity among future generations of science teachers to attend to issues of language in the multicultural context. It contributes, not only to the growing body of work on the role of language in developing classroom scientific communities of practice, but also by highlighting the challenges faced specifically by diverse learners and their teachers in joining and nurturing those communities.

Additionally, this volume is notable in a number of other important ways. First, there is the collective expertise of the contributing authors. While the book spotlights innovative research done by novice science education researchers, among the contributors are some of the most highly-regarded, veteran researchers of the day. Second, the collective expertise represented in the book cuts across grade-level, science domain, and sub-specialty foci. For example, there are chapters detailing research done at the elementary, middle, high school, and university levels; there are chapters taking up both the formal and informal domains of science learning; and there are chapters specializing in, as we have divided them into sections, The Language of Science Schooling, Science Learning Funds of Knowledge, Development of a Science Learner Identity, and Promising Practices for Responsive Instruction. It is this variety, as well as the quality, of the research, all united by the goal of enhancing science education for ethnically- and linguistically-non-dominant populations, that gives this volume its distinctive place among the current language-oriented science education literature. The chapters in this volume make compellingly clear the implications of the research for classroom practice, and student achievement and engagement. They will undoubtedly serve to heighten awareness about the multifaceted, language-rich nature of science teaching and learning, and clearly, cumulatively and collectively, articulate that providing a quality science education for all students, while understanding of the work of language in multicultural classrooms, is, in fact, the work of *all* science education professionals.

The material in the book is written for individuals in a number of positions with respect to science education. In one regard, much of the content—analytic descriptions of the use of language-in-and-of-science, inquiry-oriented science curricular materials, and reform-oriented science pedagogy, for example—will seem familiar to those who conduct research in science classrooms. Likewise, for preservice and inservice teachers, the rhythms of classroom life and student-teacher interaction will surely be recognizable. However, both classroom science researchers *and* preservice teachers will find that the content of the volume invites an even more nuanced understanding of the subject matter and context of science instruction. The authors in *The Work of Language in Multicultural Classrooms: Talking Science, Writing Science* ask that the reader consider how issues of language interact with students' and teachers' formal and informal experiences and pedagogies to generatively inform the link between theory and practice. We believe it is this link between theory and practice in everyday teaching and learning that holds the potential to make possible excellence in science education for all students.

Acknowledgments

Undertaking this project has been a wonderful journey in which we both have expanded our intellectual horizons and our scholarly community. We wish to thank our series editor, Sonia Nieto, for her support of this work and hope that the volume will be a worthy addition to the outstanding Language, Culture and Teaching series. We extend deep gratitude to our editor at Routledge, Naomi Silverman, for encouraging us in the earliest stages and for making each step of the process crystal clear. Your guidance and support is the unwritten text of this book.

We also wish to thank two reviewers, Wolff-Michael Roth and Mary Schleppegrell, who provided helpful comments and suggestions about the structure and content of the volume, as well as a number of other reviewers who gave initial feedback to chapter authors. Thank you, in particular, to our colleagues at Iowa State University and the University of Illinois at Chicago who provided encouragement all along the way and to Rachel Wilkes for her work related to this project in her undergraduate research assistant role. Finally, a special thank you to our significant others (Louis Gomez and Greg Bruna) and to our children (Felicia, Joi, Michelle, Celia, Michael and Satchel and Polly) for tolerating month after month of our dutiful distraction. There are no words to express how valuable and cherished you are.

1 Introduction

In remarks delivered at the National Education Summit on High Schools in February of 2005, Bill Gates observed that high schools, "even when they're working exactly as designed," are obsolete. The fact that high schools fail to teach U.S. youth "what they need to know" isn't, Gates says, "an accident or a flaw in the system." Rather, he says, "it *is* the system." The development of this book on science education rests on the very same premise. Science education, as traditionally conceived, fails to take account of students' cultural and linguistic backgrounds. This is not a simple oversight of a few ill-prepared science teachers, but a sincere reflection of the history of scientific thinking and science education as a whole. Since science was conceived as an undertaking of the disembodied observer, the "trappings" of culture and language were considered irrelevant to its idealized neutrality and objectivity (Rampel, 1992). Just as improving high school education will require transforming how we understand high schools, so too will improving science education require transforming how we understand science. The purpose of this volume is to inform and contribute to this transformation.

Gates' remarks, it is important to note, were included in the National Science Board's 2006 report, *America's Pressing Challenge—Building a Stronger Foundation*. This report documents the "clear and urgent" need to improve science education in the Nation (National Science Board, 2006a, p. 1), outlining, first, the mediocrity of outcomes associated with science education and professionalization in general, before, turning, next, to the disparities that exist in science education and professionalization between cultural and linguistic groups. These disparities, as attested to in the National Science Foundation's 2006 Science and Engineering Indicators report, consist of gaps in achievement, course taking, and AP test passing rates, and immediate college enrollment between white and Asian/Pacific Islander students, on the one hand, and black, Hispanic, and American Indian/Alaskan Native students, on the other (National Science Foundation, 2006b). The Indicators report points to teacher quality, broadly, as something to be considered in taking stock of what improvements are needed in science education, particularly eliminating out-of-field teaching or a mismatch between teachers' academic training and certification and their teaching assignments.

Given that 83% of all public school teachers come from the dominant white group (National Center for Education Statistics, 2006a) and that there are substantial growth projections in the non-dominant student populations who are currently underserved in science education (27% for blacks, 42% for Hispanics, and 30% for American Indian/Alaskan Native) (National Center for Education Statistics, 2006b), it is time to consider science teachers' need for preparation to work with culturally- and linguistically-non-dominant students as another way in which they, indeed, may be teaching "out-of-field."

A look at the data available on the professional preparation of science teachers for work with students for whom English is a second or additional language indicates that of the science teachers who instruct more than 50% of such students in their classes, 71% report receiving "some training." While this number may, at first glance, seem promising, consider the experience of those students in the classrooms of the 29% of science teachers who have not received "some training" and consider the fact, that "some training" may be a single course on multicultural education, in which the education of language-minority students was the topic of just one session. More disturbing is that science teachers who instruct fewer numbers of these students report having received "some training" in lower numbers. Thus, culturally- and linguistically-responsive science education for language-minority students is far from ensured. While it appears that teaching in geographic areas where the presence of these students is more strongly felt tends to predict teacher exposure to training, the quality of that training is highly variable. This is evidence of how the current system is configured to predispose such youth toward negative outcomes in science. This is the system we want to transform. We hope that, as you read this volume on the work of language in multicultural science classrooms, you will become part of this transformation by taking advantage of this and other opportunities in your professional preparation to equip yourself with the knowledge and skills to be an effective teacher for all of your students.

Put simply, our aim in this volume is to centralize language and culture in the arena of science and science education. For too long, these issues have been marginalized with the result that science learning for non-dominant youth has itself been a marginalizing force, and, in fact, a barrier in the school lives of these students, a site of their learning that contributes to unsuccessful academic outcomes. We seek to make science education, instead, a site of cultural and linguistic responsiveness, a place of learning in which students' experiences are not considered "trappings" that get in the way, nor quaint "extras," but are regarded as significant, substantial, and serious science-learning resources, and, as such, are acknowledged and responded to in the planning and delivery of science instruction. Ensuring this kind of responsive and, indeed, responsible science instruction will require moving beyond traditional understandings of the nature of science and science teaching.

As editors of and contributors to this volume, each of us is committed to this science-transforming movement, albeit in different ways. Just as we want science students to see the relevance of their life experiences to their science learning, we know that we bring our life experiences to the conceptualization and production of this text. Our personal journeys are part of our professional work as science education researchers and we hope that, by the end of your reading, you will have a strong sense that this text has developed a unique "voice" because it is, indeed, a unique collection. While some books emphasize, for example, the development of scientific literacy and others emphasize teaching science in multicultural classrooms, this book is dedicated exclusively to examining language and literacy issues in multicultural science settings. Its title, *The Work of Language in Multicultural Classrooms: Talking Science, Writing Science*, is meant to both acknowledge the legacy of language-oriented research in science education (Lemke, 1990; Halliday & Martin, 1993; Roth, 2005), while signaling a distinct and strategic turn in that focus toward multicultural settings and, of course, the attendant questions and concerns about equitable educational experiences and outcomes for all science learners.

As a multicultural teacher educator who explores issues of difference and disparity within the context of science education, for Katherine, these questions and concerns about educational, and, ultimately, social equity have always been central. So have they been, as well, for Kimberley, who, in her role as a literacy teacher educator, explores the gatekeeping function of language in science teaching and learning. While neither of our professional paths started out in science education, our paths have brought us here now, just, in fact, when it seems the science education community, in general, is beginning to take seriously the forces of culture, language, identity, and power that each of us has been engaging with for so long. We are proud to be experiencing what we consider a very exciting and much-awaited historical moment in science education and urge you to take to heart the crucial role that you, yourselves, can play.

In order to accomplish this volume's goal of transforming understandings of science education by placing cultural and linguistic concerns at the center of science education theory and practice, we want to highlight here, and throughout the text, what Villegas and Lucas (2002) present as essential features of culturally-responsive instruction. In their book, *Educating Culturally Responsive Teachers: A Coherent Approach*, Villegas & Lucas describe an approach to teacher preparation that is guided by these essential features. We feel these features are just as relevant to science teacher preparation and are, indeed, essential to achieving our vision of a culturally-responsive science practice. We summarize these features now to orient you to the overarching themes of the volume and will ask that you revisit them after each chapter as a way of consolidating your learning. Drawing on Villegas and Lucas' (2002) framework, we view the essential features of culturally-responsive science instruction as:

1 **Sociocultural Consciousness:** When science instruction is informed by sociocultural consciousness it is aware of and reflective about the teacher and students' social identities and how dimensions of those identities, such as race/ethnicity, class, gender, language, and citizenship status, among others, shape experience and perception in the classroom, community, and society. Sociocultural consciousness is characterized by enduring cognizance, not only of *differences* between social identities, but of historically-construed *disparities* between social identities. That is, with sociocultural consciousness comes the understanding that "power is differentially

distributed in society and that social institutions, including the educational system, are typically organized to advantage the more powerful" (Villegas & Lucas, 2002, p. 33). Socioculturally-conscious instruction is critical of, and attempts to interrupt, the traditional organization of advantage.

2   **Cultural and Linguistic Affirmation:** Affirming science instruction acknowledges and validates ways of being, believing, and behaving that vary from those of the dominant white, middle-class, monolingual English-speaking culture. This stance of affirmation is generated out of a profound recognition (acquired through sociocultural consciousness) that all cultures' practices are essentially valid. It is not inherent, only power-acquired, superiority that attributes greater status to the cultural practices of the dominant group. Affirming science instruction seeks to extend to all students the status that comes from participation in those dominant cultural practices while taking seriously how "all students—not just those who conform to the dominant cultural norms—have experiences, knowledge, and skills that can be used as resources to help them learn even more" (Villegas & Lucas, 2002, p. 36).

3   **A Commitment to Change:** Culturally-responsive science instruction is change-oriented science instruction; it resists "business-as-usual" thinking. This "business-as-usual" thinking in science education would have us place the "trappings" of culture and language on the margins and thus continue to devalue the learning resources that students from non-dominant communities bring to the classroom, community, and society. Instead, change-oriented science instruction links efforts of culture- and language-centered educational transformation with those of social transformation more broadly, articulating how the movement to "make room" for culture and language in the science classroom is inextricably connected to the movement to "make room" for culturally- and linguistically-non-dominant communities in the U.S. social, economic, and political order. Therefore, science instruction that is committed to change is science instruction that is much more than instructional activity alone. Change-oriented science is an ethical endeavor in and by which "teachers are participants in a larger struggle to promote equity in society" (Villegas & Lucas, 2002, p. 54).

4   **Constructivist Understandings:** An under-utilized association is that between constructivist understandings in science instruction and the constructivist underpinnings of culturally-responsive teaching. While science educators may easily accept that knowledge is "always filtered through knowers' frames of reference" and that learning is "an active process by which students give meaning to new input based on their preexisting knowledge and experience," they may have never before been encouraged to consider what that means for the school science curriculum. The curriculum itself has its own frame of reference, is, "value-laden and partial," so "schools have the responsibility to help students understand the perspective(s) reflected in and excluded from it." If connections are not made between students' knowledge and experience in the science curriculum, then an opportunity to engage that active learning process is lost. Science teachers can promote engagement by ensuring that "differences among students are

acknowledged and treated as resources for learning" (Villegas & Lucas, 2002, pp. 67–68).

Each of the chapters in this book sheds light on one or more of these features of culturally-responsive science instruction. We will ask you to reflect, after reading each chapter, on what connections you make between that chapter and these features and identify and share what questions and/or further concerns those connections raise. You will also be asked to articulate, after each chapter, what is revealed about the "work" that language is doing, as featured in that respective chapter, in promoting or denying access to science instruction. Finally, you will further be asked to reflect on the implications of your chapter learning for science teaching and to engage in observation, data collection, or other activities to extend your chapter learning beyond university classroom walls. The work you do in association with your chapter learning will enhance your familiarity with three major themes and teaching objectives running through the literature on culturally-responsive science teaching.

The first of these themes suggests that because there is, in fact, a language of science schooling that serves to impact student access to and success in science, the role of the teacher is to help students master this science code. The second is that because there are knowledge and skills related to science that students bring into the classroom from their everyday lives outside of school, the role of the teacher is to draw on and connect to these experiences and related forms of expression. Lastly, the third is that because there are a range of sociopolitical influences that impact students' willingness and ability to adopt science discourse and take up a science learner identity, the role of the teacher is to be cognizant of these influences and, simultaneously, acknowledge and ease student resistance while enhancing access to quality instruction. The authors in this volume come to their research on the work of language in multicultural science classrooms from more than one of these perspectives, therefore, for purposes of clarity and emphasis, we have organized the chapters into the following three thematic clusters: **The Language of Science Schooling, Science Learning Funds of Knowledge,** and **The Development of a Science Learner Identity.**

Chapters within the first thematic cluster, **The Language of Science Schooling,** are those that focus on the intersection of language, literacy, and science. These chapters are particularly interested in documenting students' access to and development of science literacy, or the academic language of science. This development involves more than learning discrete science *words,* but becoming familiar with the lexical, grammatical, and discursive *choices* that produce the kinds of written and oral texts expected at school (Halliday & Martin, 1993; Lemke, 1990; Schleppegrell, 2004; Richardson Bruna, Vann, & Escudero, 2007; Gomez, in Press). In this way, academic language is part of the hidden curriculum that can apportion privilege to those with access to its rules and penalty to those without.

Chapters within this first thematic cluster take as their point of departure the prominence of language use in a classroom community and consider how students and teachers understand and negotiate this often hidden curriculum of science. For example, the chapter by Pappas, Varelas, Ciesla, and Raymond, "Journal and Book Writing in Integrated Science-Literacy Units: Insights from Urban Primary-Grade Classrooms," draws attention to the nature and characteristics of young children's science writing. Then Ku, Yu, and Garcia, in "Developing Expository Writing Skills in Science Through the Integration of Inquiry-Based Science and Literacy Instruction," describe the results of a school reform intervention focused on improving elementary students' expository science-writing skills. Taking us into the high school context, Cole, in "The Writing on the Wall: The Daily Calendar as Science Practice", suggests how even a small instructional routine, like a daily writing segment in science, serves to influence students' access to science literacy development, as well as their perceptions of science and the work of scientists. Sherer, Gomez, Herman, Gomez, White, and Williams in "Literacy Infusion in a High School Environmental Science Curriculum" maintain the secondary focus but direct our attention away from writing and instead toward reading comprehension in science inquiry learning.

While helping students acquire the academic language of science is crucial, so too is valuing and centralizing their *everyday* science talk and related forms of knowing. Highlighting students' "funds of knowledge" (Moll, 1992; Vélez-Ibáñez & Greenberg, 1992; González, Moll & Amanti, 2005) or the skills and other resources students obtain from their home and community contexts, as well as what teachers can do with enhanced awareness of these everyday skills and resources, is the second major theme uniting some of the book's chapters. Many science scholars now agree that, in order for science to be a truly democratic rather than an exclusionary discipline, classrooms must become more respectful of and amenable to students' descriptions, interpretations, and positions about scientific phenomena. Teachers must recognize, value, and affirm these ways of knowing rather than situate science and science talk exclusively within the dominant paradigm.

The second thematic cluster, **Science Learning Funds of Knowledge**, emphasizes research that illuminates how the cultural and linguistic resources children bring to science interact with their classroom learning. Goldberg, Welsh, and Enyedy do this, in "Negotiating Participation in a Bilingual Middle School Science Classroom: An Examination of One Successful Teacher's Language Practices," by providing a case study of a teacher who uses awareness of her students' funds of knowledge to enhance their science learning. Similarly, Solís, Kattan, and Baquedano-López, in "Locating Time in Science Classroom Activity: Adaptation as a Theory of Learning and Change," provide examples, this time of elementary-level science instruction, that in some cases integrates or in others fails to integrate the understandings and experiences of culturally- and linguistically-diverse students. Richardson Bruna, in her work on language play in "'You're Magmatic Now': Language Play & Linguistic Biliteracy in the Science Crossing of Adolescent Mexican Newcomer Youth," argues that high school students' informal literacy practices, conventionally understood as marginal to science instruction, are, in fact, an integral part of their science and language development. The argument that she, and the other chapter authors make, is that, when adequately understood, students' funds of knowledge can be strategically mobilized by teachers to enhance science instruction.

Finally, a third and related perspective suggests that students' content and language learning in science may be influenced by their awareness (conscious or unconscious) of the polarization of everyday science and classroom science understandings and experiences (Calabrese Barton, 2003; Brown, 2004; Warren, Ognowski, & Pothier, 2005). This perspective highlights the important role of students' values, experiences, and beliefs and sees students as active agents in determining if, when, and how they embrace or reject classroom science discourse. Our third thematic cluster, **The Development of a Science Learner Identity**, focuses on practices in classrooms that facilitate or otherwise influence science learner identities.

Within this third thematic cluster, Reveles, in "Academic Identity and Scientific Literacy," examines elementary student assignments to document the ways in which students acquire and demonstrate their science learner identities over the course of a school year. Hansen-Thomas, in "The Math Initiative in a 7th Grade Science Class: How a Daily Routine Results in Academic Participation by ELLs," shares a look at assignment-related small-group dynamics at the middle-grades level and their impact on access and identity. Kuipers, Brendel Viechnicki, Massoud, and Wright, use "Science, Culture and Equity in Curriculum: An Ethnographic Approach to the Study of a Highly-Rated Curriculum Unit," to illustrate the significant role that curriculum itself plays, as well as practices associated with its implementation, in encouraging middle-grade students to try on and take up particular science identities. And, importantly, moving beyond the formal learning environments of schools, Ash, Tellez, and Crain in "The Importance of Objects in Talking Science: The Special Case of English Language Learners," asks us to consider the promising role that informal science settings like aquarium and marine biology discovery centers play in supporting and extending science learning identities out in the community and family contexts.

While the last two chapters of the volume clearly contain elements that fall under the previously-mentioned three themes, we have purposely situated them at the end, in a cluster entitled **Assessing the State of Science Education: Towards Promising Practices for Responsive Instruction** because they are forward-looking and thus well-suited to encourage capacity-building for cultural and linguistic responsiveness in pre- and in-service science teacher education. Stevens, Jeffries, Brisk, and Kaczmarek in "Linguistics and Science Learning for Diverse Populations: An Agenda for Teacher Education," provide a useful overview of the current state of science teacher preparation for cultural and linguistic diversity and point to collaborative work between science teachers, language development specialists, and multicultural educators as an essential first step. Torres and Howe close out the volume with "Experimenting in Teams and Tongues: Team Teaching a Bilingual Science Education Course" by reflecting on their experiences in engaging in just such a capacity-building collaboration. Through sharing the efforts and insights of these and other authors—authors who, in fact, occupy different institutional positions but similarly devote research and teaching energies to understanding and improving science education—we hope this book is a catalyst for more interdisciplinary, collaborative, capacity-building efforts.

The underperformance and under-representation, in science-related courses and professions, of students from culturally- and linguistically non-dominant communities is not, as Gates' remarks help us see, a mere accident or flaw. Such underperformance and under-representation is the system working precisely as designed. We are witnessing the results, in science education, of the systemic marginalization of non-dominant racial/ethnic groups. Our goal is to intervene in and interrupt this marginalization by beginning to think and theorize about science from the lives of the marginalized. This volume then not only contributes to the growing body of work on the role of language in developing science classroom communities of practice, but expands that work by insisting on a multicultural focus. It does so by highlighting the challenges faced specifically by diverse learners and their teachers in joining and nurturing those communities, and showcases exemplary teaching and research initiatives helping to meet those student and teacher needs.

In sum, the evidence from classroom research suggests that students have at their disposal many forms of making meaning in science. It follows, as well, that teachers, with a culturally responsive pedagogy can leverage these forms of meaning-making to connect learners to the ideas and discourse of science. A central claim of this volume is that evidence can legitimately be drawn from students' culture- and language-based science activities to help us not only understand what they already know but, moreover, help us better understand how to teach them well. The research offered here provides strong examples of attending to the work of students' language in multicultural science settings. We invite you to join us in building a world where such attention, for all students, is no longer the exception but the rule.

References

Calabrese Barton, A. (2003). *Teaching science for social justice.* New York: Teachers College Press.

Gomez, K. (in Press). Negotiating discourses: Urban students' use of multiple science discourses during science fair presentations. *Linguistics in Education.*

González, N., Moll, L. C., & Amanti, C. (Eds.). (2005). *Fund of knowledge: Theorizing practices in households, communities, and classrooms.* Mahwah, NJ: Lawrence Erlbaum Associates.

Halliday, M. A. K., & Martin, J. R. (1993). *Writing science: Literacy and discursive power.* Pittsburgh: University of Pittsburgh Press.

Lemke, J. L. (1990). *Talking science: Language, learning, and values.* Westport, CT: Ablex.

Moll, L. D. (1992). Bilingual classroom studies and community analysis: Some recent trends. *Educational Researcher, 21*(2), 20–24.

National Center for Education Statistics. (2006a). *Characteristics of schools, districts, teachings, principals, and school libraries in the United States: 2003–4 schools and staffing survey.* Washington, DC: U.S. Department of Education. Retrieved March 28, 2007, from http://nces.ed.gov/pubsearch/pubsinfo.asp?pubid=2006313.

National Center for Education Statistics. (2006b). *Projections of education statistics to 2015.* Washington, DC: U.S. Department of Education. Retrieved March 28, 2007, from http://nces.ed.gov/pubserarch/pubsinfo.asp?pubid=2006084.

National Science Board. (2006a). *America's pressing challenge—building a stronger foundation: A companion to science and engineering indicators.* Washington, DC: The National

Science Foundation. Retrieved March 28, 2007, from http://ww.nsf.gov.statistics/nsb0602/#challenges.

National Science Board. (2006b). *Science and engineering indicators 2006.* Washington, DC: The National Science Foundation. Retrieved March 28, 2007, from www.nsf.gov/statistics/seind/06.

Rampal, A. (1992). A possible "orality" for science? *Interchange, 23*(3), 227–244.

Richardson Bruna, K., Vann, R., & Perales Escudero, M. (2007). What's language got to do with it?: A case study of academic language instruction in a high school 'English learner science' classroom. *Journal of English for Academic Purposes, 6,* 36–54.

Roth, W.-M. (2005). *Talking science: Language and learning in science classrooms.* Lanham, MD: Rowman & Littlefield.

Schleppegrell, M. J. (2004). *The language of schooling: A functional linguistics perspective.* Mahwah, NJ: Lawrence Erlbaum Associates.

Vélez-Ibáñez, C. G., & Greenberg, J. B. (1992). Formation and transformation of funds of knowledge among U.S. Mexican households. *Anthropology & Education Quarterly, 23,* 313–335.

Villegas, A. M., & Lucas, T. (2002). *Educating culturally responsive teachers: A coherent approach.* Albany: State University of New York Press.

Warren, B., Ogonowski, M., & Pothier, S. (2005). "Everyday" and "scientific": Rethinking dichotomies in modes of thinking in science learning. In R. Nemirovsky, A. Rosebery, J. Solomon, & B. Warren (Eds.), *Everyday matters in science and mathematics: Studies of complex classroom events* (pp. 119–148). Mahwah, NJ: Lawrence Erlbaum Associates.

Part I

The Language of Science Schooling

2 Journal and Book Writing in Integrated Science-Literacy Units
Insights from Urban Primary-Grade Classrooms

Christine C. Pappas, Maria Varelas, Tamara Ciesla, and Eli Tucker-Raymond

Introduction

In this chapter, Pappas and her colleagues draw our attention to the nature and characteristics of young children's writing and drawing as a part of their participation in integrated science-literacy units. The overall goal of this work is to offer evidence for the value of multimodality, or the use of different forms of expressive work as sites for meaning-making, in science. This chapter focuses on two types of texts produced by 1st-, 2nd-, and 3rd-grade children in a multicultural setting: the more informal illustrated written journal entries and information books, as well as their conversations about them, and more formal major written texts.

The Pappas team encourages us to consider the ways that journal writing and illustrated information books serve different purposes in science learning and how science content understanding is expressed in language and images in these writings. Particularly exciting are their analyses of the various sources that students draw from in writing and drawing about their images of science. This analytic approach, grounded in intertextual connections, considers the ways in which writers rely on the substance of texts (including images) that they draw from other textual sources, as well as importing organizational and other stylistic features.

As you read this chapter, think about how in your personal writing you make your own intertextual connections. Can you think of an example in this kind of writing when you've been influenced by other written texts, conversations you've had, or images you've seen? What about your professional writing, particularly the writing you've done related to science? Now, consider how, as a science teacher, you draw on different texts and images to prepare lesson materials for your students. It is important, as the Pappas chapter argues, to provide students with opportunities for intertextual expression and to learn from the intertextual connections they already make.

Students draw on text, talk, and image connections to support their claims and to make meaning in science. How might you open the classroom to science learning opportunities for students to bring in, collaboratively create, and express their understandings through intertextual connections? As Pappas and her colleagues note, an important goal in science is for children to learn scientific content while also learning how to express this content in different ways. This is especially true for students from non-dominant cultural and linguistic backgrounds. They can benefit, then, from strong examples of texts that make connections between ideas using different modes of organization and style. You can help them toward this end by explicitly recognizing the connections that they already make and encouraging them, through consciousness of intertextuality in your own practice, to keep connecting!

There has been a call for connecting science learning with language and literacy learning (Douglas, Klentschy, & Worth, 2006; Saul, 2004), yet very little is known about the nature of young urban children's writing/drawing in the context of science education (Wallace, Hand, & Prain, 2004). This chapter focuses on two types of multimodal texts produced by urban primary-grade children in two integrated science-literacy units (Matter and Forest)—the illustrated written journal entries that the children composed throughout the units, and the illustrated information books (as well as their conversations about them) that they produced as a culminating activity at the end of each unit. These two types of writing—the journals and the books—represent the "genre set" (Bazerman, 2004), or the major written texts created by the children in the two units.

Science is a particular domain of thinking and knowing that involves using a particular type of informational language (Gee, 2004; Halliday & Martin, 1993). Scientists' communication reflects recognizable discourse patterns that linguists called *registers* or *genres* (Bazerman, 2004). Furthermore, scientific informational text includes a variety of visual designs or images (Lemke, 1998). That is, scientific texts are multimodal in nature—thus, when young children read and write scientific texts, they are engaged in using multiple meanings or *semiotic* systems. Language and images afford different ways to express ideas or concepts (Kress, 1997), and in creating their illustrated journals and books in the two units, young children made meanings in both modes of expression. According to Kress and van Leeuwen (1996), "children actively experiment with the representational sources of word and image, and with the ways in which they can be combined. Their drawings are not just illustrations of verbal art, not just 'creative embellishment'; they are part of 'multimodally' conceived text, a semiotic interplay in which each mode, the verbal and visual, is given a defined and equal role to play" (p. 118).

Writing and using visual images, as well as reading, talking, and doing hands-on laboratory work, are integrated activities that professional scientists engage in as part of their practices (Goldman & Bisanz, 2002). They use and construct multimodal texts that are integral to their scientific inquiries. Thus, because writing/drawing is an essential means of *doing* science, young children also need opportunities to employ these modes to inform and express the concepts and ideas as part of learning science (Varelas & Pappas, 2006).

The distinctive linguistic/image features of scientific writing and informational books are best understood by contrasting them from those found in the storybook genre (Pappas, 2006; Pappas, Kiefer, & Levstik, 2006). Table 2.1 lists the major features of each genre.

First, different types of nouns are used in the two genres. Typical stories use particular characters, places, and objects, whereas in informational texts, classes, via *generic nouns*, are employed. Thus, *a chipmunk* found in the beginning of an informational book on chipmunks, is not placing a character on stage, but instead is introducing the class or topic to be covered. Because particular characters, places, and objects are involved in stories, certain reference items (pronouns—e.g., *he, she, you, my, I, they, them, it*—and the definite article (*the* plus a noun construction) are used to refer to them. In contrast, in information books, reference items refer to the classes or generic nouns (e.g., *liquids, earthworms*), with a prominence of plural forms (e.g., *they, them, the solids, the chipmunks*), as well as the use of general pronouns, such as *you*.

Another difference between storybooks and scientific information texts is verb tense. Storybooks mostly use the past tense (except in dialogue), and information books more typically use the present tense. Verbs are also different in other ways. For example, mental process verbs—verbs of cognition (thinking, knowing, and understanding); verbs of affection (liking, fearing, and hating); and verbs of perception (seeing and hearing) (Halliday, 1985)—are prevalent in storybooks because they express characters' thoughts, feelings, intentions, and motivations. Such mental verbs are rarely found in information books. In contrast, information books are filled with *relational verbs*, such as *is, are, has, have, resembles, is called*, that express characteristics or attributes of a class, classify members, or define parts or aspects of them. *Material verbs* of doing occur in both genres. However, in stories, they usually express the various actions of characters, whereas in information texts, they realize the actions related to the topic being addressed—the typical behaviors of animals, the processes of a phenomenon, or the actions that constitute parts of an experiment.

Finally, the lexical wordings in storybooks and information books are different. Those in stories are more "everyday" ones, but information books possess technical terms and vocabulary. And, as Table 2.1 shows, information books have other features that storybooks do not include: lots of visual designs (graphs, charts, tables), minor text (labels, captions, keys), as well as indices and glossaries.

In this chapter, we focus on six students in three urban classrooms (two 1st, two 2nd, and two 3rd graders) as they wrote and drew in their ongoing journals during the two units and created illustrated information books on topics of their choice at the end of the units. First, we briefly describe the major activities in the Matter and Forest units. Second, we explain how these two types of science writing serve different purposes, thereby reflecting different expectations about children's use of language and images. Third, we examine the six children's texts in terms of science content, exploring how these ideas are expressed in language and images. In doing so, we also consider the various sources that may have influenced their writing/drawing, which are called intertextual connections (Jenkins & Earle, 2006; Lemke, 1992). Intertextuality is the juxtaposing of

Table 2.1 Typical linguistic and image features of scientific informational books and storybooks

Storybooks	Information Books
• Particular nouns	• Generic nouns
• Personal pronouns (e.g. "I," "me," "he,")	• General pronouns (e.g. "you," "we") and many plural forms (e.g. "they," "them")
• Past-tense verbs (except in dialogue)	• Present-tense verbs
• Mental verbs Affection ("like," "hate") Cognition ("know," "think") Perception ("hear," "see")	• Rarely found
• *Some* relational verbs	• Lots of relational verbs (e.g. "is," "are," "has," "resembles," "is called")
• Material verbs of character action	• Material verbs of animal behaviors, scientific processes, and experimental actions
• "Everyday" wordings	• Technical terms and vocabulary
• Not found	• Lots of minor text (labels, captions, keys) • Lots of visual designs (graphs, charts, tables) • Indices, glossaries, etc.

texts—how writers rely on other texts as they compose texts. Finally, in the last section, we use the idea of this intertextuality to summarize and highlight instructional implications.

The Integrated Units: Matter and Forest

Both units included a variety of curricular activities and offered students experiences in whole-class and small-group settings—read-alouds of children's literature information books, hands-on explorations, on-going journaling and semantic mapping, literature circles, a home project (the findings of which they shared with the class), and children's composition of their own illustrated information book (the culminating activity of each unit).

Matter Unit

The Matter unit centered on characteristics of matter in different states (solids, liquids, and gases), changes of states of matter, how they take place, and how these changes are related to how rain is produced. Observations and discussions on weather provided one of the contexts for thinking about the latter.

Seven children's literature information books were used for the read-alouds (in order): *What's the Weather Today?* (Fowler, 1991); *When a Storm Comes Up* (Fowler, 1995); *What Do You See in a Cloud?* (Fowler, 1996); *It Could Still Be Water* (Fowler, 1992); *What is the World Made Of? All About Solids, Liquids, and Gases* (Zoehfeld, 1998) (which is read in two parts, with other unit activities intervening); *Air Is All Around You* (Branley, 1986); *Down Comes the Rain* (Branley, 1983).

Children also viewed and discussed a CD-Rom on states of matter that addressed macroscopic properties of each state, along with microscopic properties, namely, the relative position, speed, and bonding of molecules (particles) in each state. Some of the hands-on investigations that children engaged in were: stuffing a napkin or piece of paper towel at the bottom of a cup and submerging the cup straight in one case and slanted in another in a bowl with colored water (the napkin activity); wetting three paper towels and leaving one hanging straight from a table, one laying straight on the table, and one laying on the table crumpled up in a ball (the paper-towel activity); dropping a drop of coloring in cups of water in different temperatures (the colored-water activity); a water bottle taken out of a freezer that "sweats" (the water-bottle activity); ice cubes left in different settings (the ice-cube activity); droplets being formed on a cold cookie sheet that was placed on top of boiling water (cookie-sheet activity). In addition, children engaged in other explorations (e.g., observed and documented the weather for a few days, classified various objects into solids, liquids, and gases), as well as in a drama activity where children acted out molecules in the three states of matter.

Forest Unit

The Forest unit addressed characteristics of plants and animals living (under and above ground) in a temperate forest and relationships between them, related to ideas such as what plants and animals look like, where they live, what they eat, what they are eaten by, and how they protect themselves.

The twelve read-aloud books for the unit (in order) were: *In the Forest* (First Discovery, 2002); *A Forest Community* (Massie, 2000) (with selected chapters read at different times with other unit activities intervening); *Animals Under the Ground* (Fowler, 1997); *Earthworms* (Llewellyn, 2000); *An Earthworm's Life* (Himmelman, 2000); *Seeds* (Sauders-Smith, 1998); *From Seed to Plant* (Gibbons, 1991); *A Log's Life* (Pfeffer, 1997); *Starting Life: Frog* (Llewellyn, 2003); *Look Out for Turtles!* (Berger, 1992); *About Fish: A Guide for Children* (Sill, 2002); *Who Eats What?: Food Chains and Food Webs* (Lauber, 1995).

Children also viewed and discussed several websites on worms, trees, and insects. Some of the major hands-on explorations children engaged in were: exploring whether worms prefer certain conditions (wet vs. dry and light vs. dry); observing and tracking the growth of lima beans; exploring plant growth in different conditions; and examining how various seeds travel by wind, water, and other materials. Children also participated in an activity that explored the idea of "camouflage" and how it is associated with prey–predator relationships, and a drama activity in which they enacted food chains/webs to appreciate the relationships between plants and animals in a forest community. In addition, in small groups, they created animal posters.

In summary, in each unit, children participated in many integrated activities that constituted the curricular context and in which they wrote and drew in their on-going journals and the books they created at the end of each unit.

Journals and Books: Different Purposes and Expectations for Ideas, Words, and Images

The journals and books served different purposes in the two units. Table 2.2 shows these main purposes, and, as a result, the expectations one could have about these two major texts.

The journals were a place for students to respond to a read-aloud book or hands-on-exploration. Children wrote/drew about ideas they found new or interesting in a read-aloud. They wrote/drew about what they were observing and offered their explanations (how ice cubes melted [Matter unit] or how worms behaved [Forest unit]), made their prediction and then recorded the findings of their investigation, or monitored class/small-group explorations (e.g., how water was evaporating in different places in the classroom [Matter unit] or how lima beans were growing [Forest unit]). Thus, the various on-going entries that children created in their journals reflected many different texts or genres. Moreover, although children sometimes shared what they wrote/drew with the class, for the most part, the texts were written mostly for themselves (or the teacher). In some sense, then, the journals could be seen as forms of formative assessment. In contrast, the purpose of the books was different. The books were like science reports (similar to the children's literature information books children had read in the units) on topics of their choice. Children were seen as experts on their topics and the books were written for others to read—other 1st, 2nd, or 3rd graders, who had not studied the two units, as well as their classmates. Students were able to read through their journals, view the semantic maps or charts that may have been created during the unit, as well as the read-aloud

Table 2.2 Different functions of and expectations for journals and books

Journals	Books
• Purpose was to document and respond to various activities during the unit—although students might sometimes share with class, they were written mostly for self (and teacher).	• Purpose was to write books at the end of units on topics of choice—written as experts for others to read.
• Relatively more emergent and tentative ideas of concepts to be included.	• Relatively more developed and stable ideas and concepts to be included.
• Recording of data collected during explorations/experiments and offering possible explanations.	• Presenting findings of explorations/experiments that have been done or offering ideas for future exploration.
• More lifeworld language (or hybrid of lifeworld language and scientific language) to be used.	• More scientific language to be used.
• Child, classmates, and teacher, as well as specific objects in illustrations capture the immediacy of children's unit experiences.	• Generic, not specific, people and objects in illustrations capture the children's synthesis and abstraction of their unit experiences.

and literature circle books, which were usually placed around the room (often on the chalkboard), as they wrote and illustrated their books. However, they did not have access to the books at their desks to copy, refer to, or otherwise study. As a culminating activity in the units, in which the teacher did not explicitly scaffold, the books represented a means of summative assessment.

Because journals and books served different functions in the two units, each was associated with different expectations about the nature of the scientific content, language, and visual images found in them. For example, because the journals were sites to explore, respond, or react to ideas in books or hands-on-explorations, emergent (even incomplete or misunderstood) scientific ideas were likely to be found. However, the expectation for the books was that more stable and more elaborated understandings of concepts would be included. Moreover, in their journals, children were expected to record and interpret data of ongoing explorations, whereas in their books, if children included explorations, they were expected to be ones that had been completed or could be done in the future. In addition, there was a contrast regarding the language to be used in each type of text, with children using their lifeworld wordings (Gee, 2004) or a mixture of both this language and scientific registers ("hybrid" language) in their journals, but mostly employing scientific writing in their books. Similar expectations existed for the inclusion of minor text (Unsworth, 2001)—e.g., labels, captions, and so forth. Finally, visual images were expected to be employed differently in the two types of texts. Pictures of particular classmates, the teacher, or others children knew, as well as specific objects (ice cubes, worms, and so forth) that were used in activities would be common in journals, but mostly generic people and objects would be depicted in books. Using van Leeuwen and Jewitt's (2001) idea that images could be seen as a "record" or a "construct," images were to be produced in journals as *records* of reality, the documentary evidence of the who, where, and what of particular classroom activities, whereas they were to be *constructs* in books—as evidence of how children reconstructed reality according to their understandings developed throughout the units.

It is important to note that although we have described differences regarding the expectations for journals and books, these distinctions are more a matter of degree rather than either/or oppositions. Moreover, even though teachers did share with children the purposes for their writing/drawing in journals and books in informal, general ways, they did not explicitly tell children how to realize these purposes in the detailed ways we have discussed them. That is, the expectations we have indicated here have been inferred from the ways in which journals and books functioned in the units.

Meaning Making in Words and Images in Journals and Books

Both types of texts, journals, and books, offered children opportunities to engage in science, and our aim in this chapter is to illustrate and analyze the ways in which children expressed their meanings in words and images in each of the units. Our approach was to examine first how children wrote and drew about the book topic in each unit, and then look at their journal entries that addressed the same topic and related ideas. Except for a few times when all teachers had students write on certain activities, teachers had flexibility in when and how children used their journals during a unit. As a result, children in the three different classrooms did not always write about the same read-aloud books or explorations. Thus, the availability of journal entries varied in each of the classrooms regarding children's book ideas. Another caveat about journals that needs to be noted is that we did not have any conversations with the children around their entries. Thus, we sometimes had to speculate what children might have meant due to their invented spelling or how they depicted entities in their pictures. In contrast, because we had taped conversations with children as they shared their books with us, we are more confident about our interpretations of the words and images in their books.

Due to length constraints, we offer here typical selections from books and journals to show the range of strategies that the young children employed in writing and drawing these texts. For clarity, we use conventional spelling (not children's invented spelling) and punctuation in the examples we provide, and we put children's writing in quotes and what they said in their book conversations in italics.

Angela

In the Matter unit, 1st grader Angela wrote a book entitled *States of Matter and How the Molecules Move.* Her book consisted of two pages on each of the three states of matter, with one of the two pages having both text and pictures, and the other having only pictures. On the text-picture pages, she wrote about how close molecules are in each state and the fact that they move faster when heated. Her illustrations reflected these two properties. For example, on the first page on liquids (Figure 2.1), she wrote, "The molecules of a liquid are close together and if it is heated the molecules will move faster."

Her picture included two cups depicting molecules, with a red candle next to one of the cups. As she explained, *This one cup that doesn't have a candle, so it's not being heated. And it is kind of moving slowly ... And this cup with a candle by it ... it's kind of moving faster.* Thus, what she wrote was also shown in her illustrations. The other liquid page—the picture-only one—showed clothes hanging (*to get dry*), and next to them, a black cloud, with blue drops of rain falling from it. There were several vertical blue lines above the clothes, which she explained as *water vapor ... it always goes up into the sky.* When asked to tell how the clothes and cloud are connected, she said, *this is the water [from the clothes] goes up into the cloud and then when it gets too heavy, it starts to rain.* Thus, although she did not use the term *evaporate* in her text or talk, she seemed to have a good grasp of the rain cycle.

Several journal entries are related to ideas in her book. One page was a response to the CD-Rom *States of Matter* (New Media, 1997), which she also mentioned in her book conversation. She started this page, using past verb tense, mentioning that her teacher showed it to the class and that "it was fun." The rest of the entry was in present tense. Because the CD-Rom used the term *particles*

Figure 2.1 Angela's *Molecules* book page (Matter unit)

instead of *molecules*, she next explained that "another way to say molecules is particles." Then she wrote how close the particles are in each state—both particles of a liquid and a solid "are close together," but "gas is far apart." She ended the entry with "Everything has particles." She had four quadrants (one of which was blank) marked off in her penciled pictures: the top-left one depicted a "solid," which was a big round object filled with small circles next to each other to depict the closeness of particles; the lower-left quadrant showed a cup, which also included small, close circles; and the final one had a large rectangle, which had five small particle circles that were spaced apart. There was a difference in the degree of differentiations Angela made in her journal and in her book. Although she did not make any distinction in her journal entry about how close molecules/particles are in solids and liquids, she did in her book, where molecules are "close" in liquids and "very close" in solids. In her book, she also included the effect of heat in both the text and illustration.

Two other entries were related to a hands-on exploration on evaporation. Three wet paper towels were in different conditions—one was crunched up, one was lying flat, and one was hanging. In the first journal page, children were asked to predict which towel would dry up first and tell why. Angela wrote: "I think that the one that is hanging will dry first because it is hanged up and that is how most things dry mostly because it is a little windy." In the second entry, they were to report the findings and again explain why, and she noted, "The hanged towel dried first," and gave much of the same reason as before, but this time she dropped the "windy" part and only stated, "because it was hanged up like most people do to get stuff to dry." Her illustrations for these two latter entries were sparse—in each she had a hanging towel, and in the second, she included, below the towel, a small rectangle with small marks in it (perhaps this object was meant to depict the crunched towel that would still be wet). What is salient in these two entries is the relationship with Angela's picture-only book page on liquids where she showed clothes hanging from a clothesline and drying up. Her everyday experiences are reflected in both her book and her journal, but in her book she demonstrated a further understanding of the water cycle. Moreover, her pictures in her journal depicted a particular hanging towel used in the exploration, whereas the clothes in the book were generic ones.

Thus, there were many intertextual connections between Angela's journal ideas and those elaborated ones found in her book, as well as connections to those found in the CD Rom. Both similarities and differences were found in terms of her use of scientific registers. For example, her molecule journal page reflected hybrid language—using first lifeworld language (referring to her teacher and "fun") and then more scientific language. Her "I think," a self-referral pronoun and a mental verb, which is appropriate for what she was asked to write in her journal, was not found in her book, where she used mostly scientific language in the book (e.g., generic nouns and present-tense verbs and the scientific term *molecules* [although she also used this work in her journal as well]).

Angela's Forest book, *Frogs*, consisted of four pages, and related what frogs look like, where they live, what they eat, and what they—as tadpoles and frogs— have as enemies. This was a sparse book, but nevertheless had interesting information in it, especially on what frogs eat. For example, on her first page, she wrote what frogs look like that live in the forest, and then in the last sentence, she wrote that "they eat plants and crickets." Her illustration did not show what frogs eat, but depicted a green frog on the top of one of four trees, which, as she explained, were included in her picture because she had written that *it lives in the forest ... and so that's why the trees are here ... that's why I drew it [the frog] on top of the tree.* On the next page, she wrote: "That means they are omnivorous. Omnivorous are animals that eat meat and plants," and she provided a fairly long rationale for this page. She first argued, *a frog is one of the omnivores and that's ... what eats plants and crickets ... and crickets are made of meat like us.* When asked where she had learned this, she said that her teacher *was teaching us omnivores, carnivores, and herbivores,* and as she reasoned, *I just figured that frogs eat crickets and plants because I knew that from the book ... but it didn't say frogs are omnivores—I just guessed it because they eat both meats and plants.* The fact that frogs eat crickets is information that a class read-aloud book *Starting Life: Frog* states, although the book specifies that it is only tadpoles that "feed on the slimy water plants called algae that grow on underwater stones and leaves" (p. 6). Thus, although scientists would not classify frogs (as adults) as omnivores, Angela demonstrated the kind of thinking that scientists would employ in categorizing animals into classes. Moreover, complex intertextual links were exploited by her in doing so—the book and her recall of her teacher's explanation. Her picture for this omnivore page was a lone, light green frog with darker green spots and four legs with webbed feet. Although her book was short, it reflected the use of scientific language and included some scientific terminology (*tadpoles, omnivore*). She did not include such language in describing frogs ("frogs look like circles and legs"), but her pictures reflected more frog (and tadpole) features, such as the spots to depict frog's speckled skin and their webbed feet (most likely influenced by both text and illustrations from the read-aloud book).

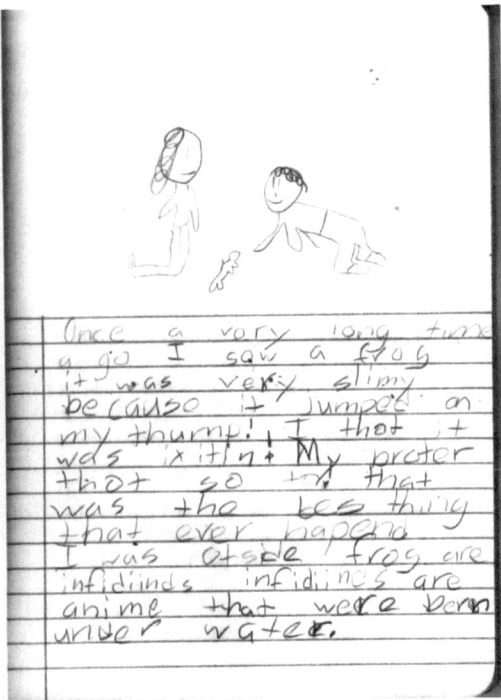

Figure 2.2 Angela's *Meeting-Frog* journal page (Forest unit).

Angela's one journal entry (Figure 2.2) on frogs was a response to the read-aloud book mentioned above.

She wrote mostly about a personal event that she and her brother had with a frog: "Once a very long time ago I saw a frog. It was very slimy because it jumped on my thumb! I thought it was exciting! My brother thought so too. That was the best thing that ever happened. I was outside. Frogs are amphibians. Amphibians are animals that were born under water." Her illustration depicted her brother and herself and the frog they encountered. The personal account is in the past tense, includes particular people and frog, and even a story-like beginning. She then switched to scientific language, using present-verb tense, generic nouns, and the scientific term *amphibians*. However, in general, the genre of her journal—a personal recount of a particular experience with her brother—and her book was quite different. Thus, in many ways, the content, words, visual images in the both types of writing in the Forest unit—and the Matter one as well were in alignment with the expectations listed in Table 2.2.

Miguel

In the Matter unit, 1st grader Miguel, Angela's classmate, wrote a three-page book entitled *Rain*. He wrote about clouds being made of water, that rain occurs when clouds are heavy, that rain is a liquid, and that there are different types of rain (storms and drizzle). His illustrations consisted mostly of rain coming from different colored clouds that seemed to show the different types of rain that was talked about in his text. For example, on the first text page, he wrote, "Clouds are made of water and vapor comes from water. If water comes down fast it is called a storm." Three ideas are expressed in his text—water makes the clouds but also water makes vapor, rain comes from clouds, and water falling fast makes a storm. He explained in his book sharing that *water comes from lakes and it's going up*. Miguel's writing implies a connection between water on the ground, vapor going up, and water in the clouds, but he did not articulate this connection in his book. His illustration included two clouds, with a lot of blue dots underneath them. These clouds were positioned close together, possibly linked to the concept of *fast*, which could have been used as a synonym for "a lot" that he used to define a storm. These clouds were predominantly blue with some gray. Although he was asked why he drew blue and gray clouds, he had trouble articulating an answer. In his last page, however, he seemed to be able to express his thinking about rain, "Rain is a liquid … Rain can be droplets … Clouds are heavy that means it's going to rain."

Miguel also referred to the concept of rain and storms in his journal. For example, in his second weather entry (as a response to the read-aloud book, *When a Storm Comes Up*), he wrote the following, "I learned that hurricanes come from the weather and drizzles are little drips." His illustrations included several clouds—only gray in color here—but other journal entries had clouds of both gray and blue, similar to those in his book. Because the use of only gray in this first picture was unique from all of the other clouds, it could be that in these later entries, he was gaining understandings about how water or rain came out of clouds. Indeed, in his next entry, which was a response to the read-aloud book, *What Do You See in a Cloud?*, this seemed to be the case. He wrote, "Clouds are made of water. Rain comes from clouds. I learned that clouds are made of water." His illustration again included several clouds, but in this entry the clouds were half blue and half gray with slashes of blue rain coming down from them, closer to the clouds that he created in his information book. Although he used the more scientific term, *drizzles*, in his journal, defining it via lifeworld wordings—"drizzles are little drips"—in general he used hybrid language, telling what he had learned from the read-aloud books, in his journal. Miguel used many technical terms and vocabulary in his book than in his journal. In his book he used the following scientific terminology, "*vapor* comes from water," "rain is a *liquid*," and "rain can be *droplets*." He also used a range of relational verbs that define and describe: "clouds *are made of* water," "if water comes down fast it *is called* a storm," and "clouds *are* heavy that means it's going to rain."

Miguel's book for the Forest unit was *Fish!* His book consisted of six pages, three pages consisting of text on the right-side page, with illustrations on the left, and three pages having a combination of text and illustrations on the same page. He wrote about what fish are like, how they move, where they live, the sizes of fish, and what fish eat. He also wrote about the following, "Fishes enemies are people, sharks, puffer fish, other fish, killer whales, seals, polar bears, there are more but I don't want to say them!" When asked why he did not want to say them he replied in his book conversation, *it's too much!*

In Figure 2.3 (below), Miguel wrote about what fish eat and then mentioned enemies again: "Some fish eat plants and other fish. Some fish have a black dot in the back of its body because it is for the enemies to think they will go swimming away."

In his picture, he drew a yellow-bodied fish with black, vertical stripes and two black dots, one each located on the front and back of the fish. Under this fish, is another fish, which had two black dots next to each other (to depict two eyes) in the front and a gaping mouth. Four other fish are found below *eating plants.* Miguel provided a very complicated explanation about the two top fish: *this is the enemy* [pointing to the lower fish] and *he thinks it's* [the top fish with the front and back dots] *going to go away* [making, with his finger, a "going away" gesture toward the top of the page]. He further explained that the dot on the back of the top fish confuses the bottom, enemy fish and *like slips from behind and goes away.* Miguel's ideas about the top fish probably come from the read-aloud book, *About Fish,* which tells, in the Afterword (Pappas, 2006), about the black-striped Foureye Butterflyfish, which has a "black-eyed spot" on the rear, which confuses its predators. Because they are not able to tell which end is the head, they cannot tell if the fish is coming or going. In his first page of the book and on the last, summary page at the end of the book, he referred (in a general way) to this two-spotted fish on this page, by writing, "fish can

Some fish eat plants and other fish. Some fish have a black dot in the back of its body because it is for the enemies to think they will go swimming away.

Figure 2.3 Miguel's *Fish!* book page (Forest unit).

camouflage." Thus, his textual and pictorial account of this type of fish reflected intertextual connections to the read-aloud book. His book was written mostly in the scientific register (generic nouns, present-tense verbs, relational and material verbs, and so forth)—with the exception of "I don't want to say them [fish enemies]" and "I know…," which introduced the summary of things he knew and wrote (and drew) in the book. All fish and other objects in the pictures were generic ones.

Miguel's journal briefly mentioned fish in two separate entries. The first was a very short response to the read aloud book *Starting Life: Frog*, "Today my class learned about fish and frogs." (Although this book is a book about frogs, fish are depicted, too, with some minor text captions about them.) His picture showed two orangish turtles in blue water. The second entry was the first of the three favorite activities he liked about the Forest unit: "My first favorite thing is drawing the poster because I like to look at fish a lot and when Ms. Gill told us the class was going to do a poster I was happy." His picture (Figure 2.4) depicted himself and his classmates around their poster, with one student saying "help, help" in what is called a narrative dialogue bubble (Pappas, 2006).

He might have gotten this idea for the dialogue bubble from some of the books that were used in the Matter unit that employed this type of minor text (e.g., *What Is the World Made Of?* and a literature circle book, *Weather Words and What They Mean* [Gibbons, 1990]). These two entries, especially the latter

Figure 2.4 Miguel's *Poster* journal page (Forest unit).

one, were filled with mostly lifeworld language. Because the second one asked for personal response, Miguel told of specific people and objects (e.g., posters), used the mental verb *like*, as well as the attribute *happy*. Unlike his pictures in his book that utilized only generic fish (and objects, such as plants), his journal pictures also included particular people/objects, and a narrative dialogue bubble. Thus, this journal entry is similar to Angela's recount of her and her brother's personal experience, except that Miguel's centered on what he had done with his classmates.

Maria

In the Matter unit, Maria, a bilingual, ELL, 2nd-grade student, wrote a nine-page book (four pages having only pictures but related to the previous picture-text page) entitled *Weather*. She wrote about different weather conditions (specifically, rain and storm-like weather) and how they might pose safety hazards, as well as the role of meteorologists in informing us about the weather. Often, she employed the general pronoun *you* in her text. For example, on the second text page on tornados (see Figure 2.5), she wrote: "The tornados can pick up houses but be careful it can pick you too! Don't be scared, nothing is gonna happen, okay."

Thus, she first wrote what tornados could do (pick up houses and *you*, who referred to people in general, not a particular person). The possibility of being "picked up" by a tornado is contradicted by her effort to console by saying that

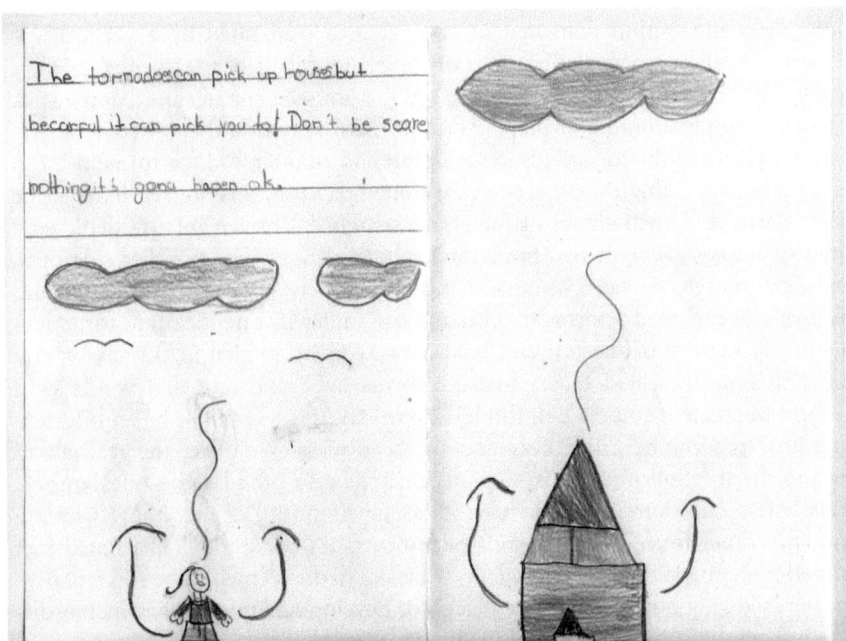

Figure 2.5 Maria's *Weather* book page (Matter unit).

nothing would happen to *you*. When we asked about this contradiction in our book sharing with her, she explained by noting *some people are scared … by the wind*. Under her text on the left-hand page, she drew a picture of two blue clouds in the sky, a girl with an arrow pointing up toward the clouds on either side of the girl, and a squiggly line above the girl. In the right-hand page (with no text), she drew another blue cloud in the sky, a house, also with an arrow pointing up toward the clouds on each side of the house and a squiggly line above the house (which had an open, front door). When asked to explain the clouds, she said, *Because it's cloudy. The tornado comes from the clouds*. She related that the arrows and squiggly lines near the girl and house showed that the girl and house could be picked up by the tornado, and her remark in her text "to not be scared" and that "nothing would happen to you" was because *you can get right out of the house to be safe* (through the open door). Thus, her apparent text contradiction seemed to be resolved. Moreover, although Maria's pictures and words on this page communicated similar ideas, the pictures provided more information, for example, that clouds "cause" tornados, according to Maria. General pronouns were used on other pages as well—on her page on sleet ("be careful because *you*'re going to slip); on the meteorologist page ("the meteorologist warns *us* that it is going to…"); and on her last page on heavy rains ("when it rains so hard that *you* cannot go outside that means *you* have to stay inside").

Two journal pages focused on different types of weather (as responses to the read-aloud books, *What's the Weather Today?* and *When a Storm Comes Up*), which possibly served as places to try out her initial ideas on the topic that she wrote about in her book. On the first entry, she wrote, "I learned that when a lightning storm comes you have to be indoors. And it can't be sunny all of the time." Her illustration consisted of two pictures (separated by a vertical line between them), each depicting each of the sentences of her text: on the left, a house with lightning in the sky; on the right, a house, a sun, a cloud, and a stick figure seemingly jumping or playing on the grass. The second entry covered tornados: "I learned the tornado is made of hot and cold air and the tornado it can pick up houses." Again she had two partitioned pictures, this time each depicting what she said in each clause of the above sentence. The left picture depicted a spiral-like tornado, with two horizontal squiggly lines, one on either side of it, and next to each, the label "cold air" and "hot air," respectively. The right-hand picture also included a tornado, with a house under it. The idea that tornadoes involve a mixture of the hot and cold air was not provided in the read-aloud book (it only indicated that a tornado forms over land and that winds swirl around to create it to be a tall, funnel-shaped column); it might be an intertextual link to something that was part of the discussion during the read-aloud session. At the bottom of this second page, she also posed some questions: "I have a few questions. How do you know all of this? Do you know where a tornado comes from?" Her questions demonstrate that although she stated that tornados are made of hot and cold air, what she wrote seemed to be still tentative for her, showing how journals are places for developing knowledge, which is different from the "final" product of the books.

Maria's Forest book was *Earthworms*. This eight-page book included a Table of Contents, which listed seven sections (*Body, House, Food, Movement, Enemies, Babies,* and *Cool* [this latter one covering what Maria thought was cool or interesting about worms]). Each section consisted of one page each, each having a heading. This book format may have been influenced by one of the read-aloud books on worms, *Earthworms,* which had a similar layout, and by the charts that the class used to document major ideas about the various animals they studied (including the "Interesting" category—"What's really cool about them?"). Maria's book was expressed completely via scientific language, except for one instance of lifeworld language on the body page (Figure 2.6).

After relaying information about the two features of worms and their functions, she used the everyday word, *stuff:* "The worm's body has a mouth and a saddle. The saddle is where worms lay their eggs. Worms have a mouth. They use it for carrying stuff." She did not write about the color of worms or segments on this page, but her illustration complemented her text by reflecting those characteristics in a picture of a worm (light brown, with segments and a dark black section in the middle) with labels that point to the saddle, mouth, eggs, and leaf (to represent her word *stuff*), which were mentioned in her text. When

Figure 2.6 Maria's *Earthworms* book page (Forest unit).

she was asked why the eggs (three oval objects) were positioned next to the worm, she offered an explanation: *Because—I don't know how to say that word—but it's like … rings … come out … It's the eggs, but first it comes out of the rings to make an egg.* The concept and word, *rings,* probably come from *An Earthworm's Life,* another read-aloud book, which uses this term to tell about how "a ring full of eggs comes loose … and becomes an egg case." So, Maria did not recall the word *egg case* (or *cocoon,* which the other read-aloud book uses), but she knew of the process, which she tried to depict in her illustration. On this body page, she did not mention bristles (or draw them), but she did on her movement page, associating a body feature with its function ("Worms move with the bristles or anchor itselfs.")

Maria's journal included five entries related to worms; we cover only the first one here (Figure 2.7), which was a response to the *Earthworms* book. She wrote, "Today I learned that the worms don't have eyes, mouth, ears and bones. The front is more pointy than the back. If you dig a little bit in your garden you will see worms." Similar to her journal drawings described above for the Matter unit, her illustration here was organized into three partitions, separated by vertical lines, with each partition related to each of the three sentences of her text. Thus, it is only in journal entries (except for one page in her *Weather* book) did she create these multi-sectioned pictures that depicted a close correspondence with the wordings in her text—in her books, usually pictures were complementary, depicting different ideas from those noted in the text.

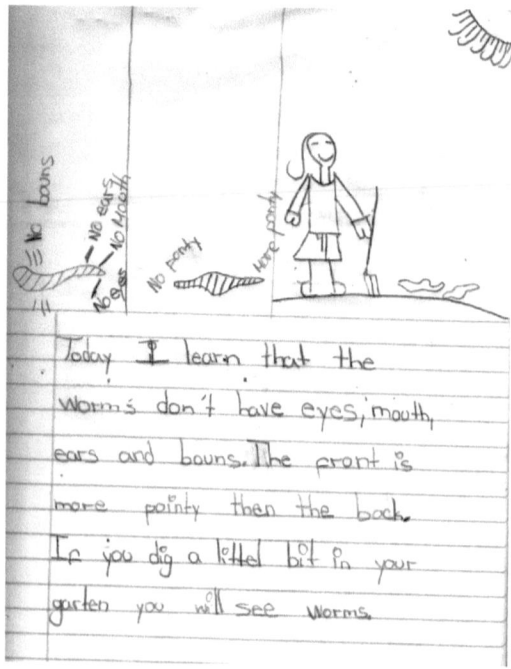

Figure 2.7 Maria's *Worms* journal page (Forest unit).

In the first partition in Figure 2.7, she drew an earthworm with the following labels: "no bones," "no ears," "no mouth," and "no eyes"; in the second partition, she had another worm that had a wide middle and more tapered ends, with the labels "no pointy" and "more pointy" next to each end; and, in the third partition, she depicted a smiling girl with a rake in her hand and two worms on the ground. Unlike Angela's picture of her and her brother and a specific frog (Figure 2.2) and Miguel's illustration of particular classmates and a fish poster (Figure 2.4), here, the girl and worms were generic ones. She used scientific language except for the beginning frame, "Today I learned..." Her "more pointy" was her own version of the read-aloud book's "more pointed" front of a worm, and her use of "negative" labels was interesting—for example, her "no bones" to label a feature that a worm does *not* possess.

Pablo

Pablo was also a 2nd-grade bilingual, ELL student in the same class as Maria. His *Matter* four-page book was simply entitled *Matter*, where he wrote about the three states of matter: solids, liquids, and gases. Unlike 1st grader Angela, who concentrated only on the behavior of molecules in each state, Pablo mentioned molecules, but also focused on the macroscopic properties of the states of matter. For example, on his "Solids" page, he wrote: "(Solids) It never changes unless you do something to it. The molecules doesn't have too many space to move in. It's hard to break it or smush it. A solid has a shape or some don't." On the bottom of the page, he drew a yellow object with small circles, then beneath it, from left to right, a flag, a boy, and a brown dog with small brown objects next to it. According to Pablo, the yellow object was a key and the small circles were molecules that *can't move too much* because it is a solid. He further explained that the flag was included because it *has its own shape*; but that although they are solids, neither *the kid ... doesn't have its own shape*, nor the clay dog, which *if you put in the sun ... all the little particles [get] soft [and] it starts to fall in pieces.* Thus, Pablo had a multifaceted, emergent sense of solids. He knew that the molecules in solids are close to each other and move slowly and these two ideas were linked for him. He also addressed a property of solids, namely, that they do not change when left alone, but he did not explicitly connect this to their shape. He considered the shape of an object, but he concluded that some solids have shape and others do not. The kid and the clay dog are examples of the latter for him. This clay dog example may have been influenced by the read-aloud book *What Is the World Made Of?*, in which the softness of modeling clay is discussed and the idea that you can shape it in different ways (with illustrated examples of an animal and dinosaur), "but if you leave it alone, it will hold whatever it is in" (p. 10). Pablo also considered the hardness of solids, but he also talked about their softness via the clay dog example. Thus, his book page on solids communicated several ideas, some of which were intertextual connections to the read-aloud book, and were expressed in scientific language.

Pablo had two journal entries that had to do with solids and when he wrote about them, he also addressed liquids and gases, which we do not cover here.

The first entry was his response to the above read-aloud book that covered the three states and their properties. For solids, he only wrote: "And the things on earth are solids." Like Maria, his illustration for this page consisted of partitions—three of them, each with "solid," "liquid," and "gas," respectively, noted at the top. Under the solid section, Pablo drew a brick wall (with label "wall") and door (label "door"). Because he wrote so little about solids, his teacher added a remark on the bottom of the journal page ("How do we know it is a solid?"), urging him to think more about it. We think his second journal entry is a response to the sorting activity in which children sorted a range of objects into solids, liquids, or gases. He again had three-partition pictures, with a box of crayons and a book under his solids section. He added a little more about solids here, but again what he wrote was sparse: "Solids are heavy and stay on earth."

Pablo's Forest seven-page book, *Beavers*, was organized similar to Maria's, with a contents page listing the various sections (*Body, Home, Food, Movements,* and *Interesting*), each being covered by one page (except for the last section that

Figure 2.8 Pablo's *Beavers* book page (Forest unit).

consisted of two pages—one a full text-image page, and a second page consisting of only one sentence). His book included scientifically accurate facts about beavers (e.g., that they have brown fur, can use their tail to communicate, hold their breath for fifteen minutes, and so forth), and these ideas were realized by scientific language and visual images. For example, on his Movements page (Figure 2.8), he wrote: "Beavers swim to go places. To swim they need to use their tail and webbed feet."

In this way, Pablo made a connection between body features of beavers and their functions. His picture had a brown beaver, with teeth and a paddle-like tail, a sun, and blue water. On the top, on the right, he had a box in which part of a leg and foot was depicted. He explained, this is so people *could see it closer and bigger ... the feet of the beavers ... and how their feet is webbed.* He also included the caption, "Their feet looks like this." Thus, Pablo used in his book a common visual image utilized by scientists and authors of information books and websites, namely, a magnified inset of a feature of an animal (Unsworth, 2001). This type of image is found in some of the books used in the unit (but not the one [*A Forest Community*] that covered beavers).

The topic of beavers was never addressed in his journal, but when asked where he learned all of this information on beavers he said, *from all the books the teacher was reading and some from just I knew ... [like] they go under water.* Thus, as Pablo notes, his ideas—and the use of the webbed-foot inset image—reflected intertextual connections to books used in the unit or classroom.

Aida

Third-grader Aida's Matter eleven-page book, *Liquids*, focused on one state of matter. As she explained in her book conversation, *This book is only going to talk about one thing.* However, although she does only talk about liquids, she also covers how liquid water can be found or transformed into different states, e.g., the fact that water evaporates, that it can become ice, snow, hail and rain, fog, and so forth. Her book may have been influenced by one of the read-aloud books, *It Could Still Be Water*, which relates the various ways water is used and found in our lives, but also tells how "water isn't always a liquid" (Fowler, 1992, p. 15). However, other books may have also influenced ideas in her book. For example, the idea that "whatever water goes in ... it becomes that form" on one of her pages might have come from *What Is the World Made Of?*, which discusses the properties of liquids. *Down Comes the Rain*, a read-aloud and the home-project book, might have contributed to ideas that she expressed in both words and pictures on her page on evaporation (Figure 2.9).

She wrote: "Water can evaporate from you, plants, animals, ponds, seas, oceans, and more. Everyday water evaporates." Her picture depicted two blue clouds, with a girl, dog, flower, and a blue entity, each of which —except for the girl—had an arrow and label ("dog," "plant," and "pond"). She also included small blue vertical squiggly marks between each cloud to the entities below, and next to these marks on the left-hand of the picture were another arrow and the label "evaporating water." In her conversation, Aida explained that *it's*

Figure 2.9 Aida's *Liquids* book page (Matter unit).

evaporating from the dog and it's evaporating from her, the plant, and the pond, going into the clouds, which she made blue *cause if they were gray, it would have showed it was raining.* The *Down Comes the Rain* book also tells of the sources of evaporation (lakes, rivers, oceans, plants, cows, horses, cats, dogs), as well as "*and* from you and me" (p. 29) that is found in a dialogic bubble near a girl. The book's picture also shows many of the things noted in the text, including dotted-lined arrows emanating from these entities to clouds. Thus, Aida's page was quite similar to that of the book. Yet, it is important to note that although the book depicts a particular girl, referred in the text as *me*, Aida only included the general *you* in her text to stand for the generic girl in her illustration.

Aida had many journal entries about water in various states related to the read-aloud books and various hands-on elaborations she engaged in. For example, she provided a response to the read-aloud book, *What Is the World Made Of?*, describing a property of water as, "If you pour water into something a different size it will be that different thing"; offered a rationale for her sorting of various objects into solids, liquids, and gases, explaining that "I put milk in the liquids because it can spill"; recorded a prediction about what would happen when her teacher puts a cookie pan over boiling water, stating that "the steam is going to go back down and the pan start to sweat"; and so forth. Several journal entries involved evaporation—she wrote and drew about the on-going evaporation exploration, writing about "the air [taking] the water up" from the graduated cylinder, and the three paper towel experiment (where one towel is crunched up, one is lying flat on the table, and one is hanging), explaining that "I learned that if you put a paper towel hanging, it will dry real fast. 24 students guessed the hanging towel." Her illustrations for these evaporation entries included the relevant, particular apparatus, namely, the graduate cylinder (with its markings), and the table having the three towels, plus the prediction and results chart of the class experiment.

In the Forest unit, Aida wrote an eighteen-page book entitled *Earthworms.* She covered many ideas about worms—their characteristics, who ate them and what they ate, how babies are born, where they live, and so forth. She included many scientifically accurate ideas about earthworms, which were expressed in scientific language—e.g., "earthworms have slime on their body called mucus" and "the earthworm has something on its belly called a saddle." She also presented several ideas throughout her book in the form of "Did you know…?" questions, a style that she may have adopted from the book (as an intertextual link) that she read in her literature circle, *Tunneling Earthworms* (Dell'Oro, 2001). For example, on her page on worm bristles, she began with, "Did you know an earthworm has hair on its body called bristles?" She then wrote more about the function of this feature: "The bristles help the worm move comfortably." The literature circle book did not use Aida's "Did you know" question format, but it does include questions, such as "What else do earthworms eat?" (p. 17), which are answered by the subsequent text. Aida's illustrations on the bristle page mentioned above, had a worm approximately the width of the page, with lines to represent segments and a larger, yellow section approximately in the middle of the worm, which she told us was the saddle. On the top of the worm are small vertical lines, which she said were *these little lines going like up and down are the bristles.* Above the worm, she included an arrow with the label "bristles." Although the people (and worms) in her book were generic ones— that is, they are not referred to in her text or her conversation as specific persons or entities—they sometimes made remarks in dialogue bubbles that we have seen other children (e.g., Pablo) employ. For example, on a page where Aida presented what animals eat worms, she has a bird with a worm in its beak, and a boy, saying "poor worm"; on a page where she wrote about fishermen using worms because fish like worms, she has a man in a boat, whose rod has a worm that says "help me." Similar to Pablo's dialogue bubble, Aida's were narrative ones, which are sometimes found in children's literature information books written for children (Pappas, 2006). And, although these narrative bubbles (which are sometimes found in children's literature information books) are not consistent with adult scientific linguistic register, they do support and highlight the scientific meanings that Aida wanted to communicate on these pages, namely, that birds and fish are worms' enemies.

Four journal entries were addressed to worms. Aida's first entry was related to a website that students viewed on the anatomy of worms (Adventures of Herman, Anatomy section, www.urbanext.uiuc.edu/SchoolsOnline/index.html), and she included in her journal entry several ideas from that site—worms' "butt" (which is called *posterior* on the site) and that worms "like the dark better than the light because no one is really out at night" and "live on the top of the ground." Also included was a recount of a personal experience, similar to Angela's account of her and her brother's seeing a frog (Figure 2.2): "I remember when I was gardening and I saw a worm." It is this latter sentence that was illustrated: along with blue clouds, the sun and three fairly large flowers, she depicted herself, holding a brown worm, and her mom. Each person had a narrative dialogue bubble attached to her mouth, which said: "What is this mom!" and "A worm!" Her second entry was a response to the read-aloud book, *Earthworms*. Here, using more scientific language, she included several things from the book that she thought were "interesting"—e.g., "that moles cut worms heads off so they won't escape," "that a worm can have over 2,000 babies in its saddle but only one or two babies survive," and so forth. Her illustration depicted what she wrote about moles—she drew a tunnel, inside which were worms and a mole (with labels for "tunnel," "worms," and "mole"), which was very similar to one found in the *Earthworms* read-aloud book. The second idea that she wrote about, namely, the number of babies made and surviving was partially accurate. The book states that some of the eggs (without specifying the number of eggs) die and that only one or two worms hatch. The book includes the specific approximate number found in a medium-sized garden ("20,000") and the number of types of worms ("1,800"). The number "2,000" seemed salient for Aida as she used it throughout her book for both the number of worms found in a garden and the number of types of worms. It was probably a big number for Aida that she used for both quantities that she knew to be large. However, in her book, she wrote that worms have "over 100 earthworms ... but only 1 or 2 survive." Aida seemed to be making a distinction between on the one hand the large quantities of worms found in a garden and of types of worms, and, on the other hand, the smaller quantity of eggs worms make.

The last two journal entries involved hands-on explorations. The first, two-page one described Aida's observation of worms, using the here-and-now language of "right now the worms are mating. I see the head and the head looks pointy," and so forth. She also wrote about boy and girl worms ("the boy worm keeps on bringing the girl worm back"), which is an emergent idea in that every worm has male and female parts and can mate and reproduce (information found in the *Earthworms* book). Aida's two-page journal entry on her observations of the worms included two pictures: on the left-hand page, a worm showing many body features and labels—"head," "segments," "saddle," "hair," and "butt"; on the right-hand page, the apparatus of the exploration—the pan, the worms, and the napkin that lined up the pan. The last entry addressed two experiments on worms. Per instructions from her teacher, her journal pages were organized with two headings, first a "prediction" and then "actual results" for each of the two experiments ("light vs. dark" then "wet vs. dry"). In each of the four sections, Aida provided a rationale for her prediction or result, all of which used a "because" structure—e.g., for the light vs. dark experiment, she wrote, "I predict that the worms are going to be under the paper because they are sensitive to light."; for the wet vs. dry experiment, she offered, "I think the worms are going to go the wet side of the paper towel ... because it needs to be moist to survive." For the results sections of each exploration, she more or less reiterated the same reasons (since her predictions were confirmed by the results). Her illustrations were rough sketches of the two experiments, with two worms in the light side of the pan but right next to the edge of the dark side, and two worms on the wet side, shaped like a heart. In her writing, she explained that "right now the worms are making a heart for valentines day," a description presumably influenced by the fact that the day she completed this journal entry was one week before this holiday.

In sum, it is clear that the functions of Aida's two types of texts were distinctively different. Regarding the scientific content in her *Liquids* book, what she wrote and illustrated about evaporation was accurate. However, also in her conversation about the book, she also noted that the clouds on her evaporation page *are like sucking up the water*, which was similar language found in her journal where "the air [is taking] the water up" from the cylinder, both of which are consistent with a transportation model for the concept of evaporation (Varelas, Pappas, Barry, & O'Neill, 2001). Thus, although her book showed that she had developed more stable ideas for evaporation (and actually used the term *evaporate*) than her journal, she did not provide any explicit evidence of the transformation process in evaporation (the change from liquid to gas or water vapor [even though she did use this latter term elsewhere in her book]), and gave agency to clouds (in the book) and air (in the journal) in transporting water up from plants, animals, ponds, and cylinders. In her *Earthworms* book, Aida demonstrated that was quite knowledgeable on the topic. Her emergent ideas of boy and girl worms in her journal reflected correct information in her book— she wrote, "Did you know there is no boy or girl earthworm. It is just whatever you call it." Her illustration on this book page had a girl and a boy with frowning faces, each saying in a dialogue bubble "It's a girl" and "It's a boy," respectively, explaining that *they're mad ... cause they're like "no, it's a girl" and "I'm right,"* but *they're actually both right.*

Arturo

Third-grader Arturo's Matter seven-page book was *Storms*, and in his first page, he named the ones he wrote about—"Some storms are called hailstorms, blizzards, tornadoes, snowstorms, and hurricanes." Most of his book told about the characteristics and impact of these storms, similar to the read-aloud book on the topic, *When a Storm Comes Up*. However, Arturo elaborated on information he offered in this book by adding his own. For example, for his page about snowstorms, he sought out information about a particular storm to include: "Snow storm can go up to a door like in 1979. Sometimes snowstorms has 20.7 inches of snow. 60 people got killed in the snowstorms of 1979. It started on Friday night Jan. 12 at 2:00 am." Here, he adds what is called a historical vignette, which can sometimes be found in information books on such phenomena such as storms, earthquakes, erupted volcanoes, and so forth (Pappas, 2006). We are unsure of his source of this information for it is not found in any of the read-aloud or literature circle books in the unit. In his writing, Arturo emphasized the "up" (by underlining) in "*up* to a door," and he tried to depict that fact in his illustration on this page, which was a large rectangle, which he explained as, *it's a picture of a house, but there's snow all the way up to the door.* The rectangle consisted of three horizontal sections: the bottom one was a large blue expanse that represented snow (and a hardly recognizable, penciled-in door underneath), with two smaller sections on top—yellow (the house) and brown (the roof). Arturo's blizzard page also contains his own ideas about storms (Figure 2.10).

The first part was similar to ideas in the read-aloud book, but also Arturo added a reasonable inference about flight cancellation during blizzards: "A blizzard is a lot of snow. Snow can go up to a car. People shovel to move snow out of the road. Blizzards can be very strong. The blizzard can make your flight cancel too." His picture was also interesting in that he showed a hotel that people might stay at if their flight was cancelled: he depicted a dark red car, with blue snow up to it, and a light red hotel with a black sign "hotel." When he was asked why he drew the snow blue, he offered a very reasonable explanation: *Well, because if you*

Figure 2.10 Arturo's *Storms* book page (Matter unit).

drew it white, you're never gonna know what it means or something. If you draw it blue, you'll know it's snow ... I know sometimes when you make a book, snow can be different colors, but I wanted to make it regular colors because it's a nonfiction book and it's talking about real things that used to happen. Here Arturo shows his awareness of the realistic representation of nonfiction books and his intentional use of particular colors. He knows that snow is white, but white on a white book page would not show and the reader would not know what it meant. So, Arturo chose blue for the snow—probably a "regular" color for him as water is often colored blue by many children who incorrectly transfer the blue color they see of ocean, lake, river, pond water (resulting from the reflection of the blue sky) to the entity of the water itself.

Arturo had one journal page on storms, his response to the read-aloud book *When a Storm Comes Up.* Here, he wrote what he learned from the book: "I learned tornados could kill you and they make bad stuff happen. I learned that if it is light raining it is called a drizzle. If there is a flood and it touches your face you could die. If there's a hurricane it could kill you too." The ideas related to his claim that "tornados ... make bad stuff happen" were greatly elaborated upon in his book, where he told of tornados as strong winds "that spin around" and "can suck up cars," and so forth. Both his journal and book included the idea that tornados and hurricanes can kill. His illustration depicted a funnel-shaped tornado, with a small "glum"-faced individual in the top of it, and next to the tornado, what appears to be a vehicle of some kind that may have been broken apart, with the top of another individual, seemingly crying. Thus, his picture, indeed, represents some of the things in which tornados "make bad stuff happen." Moreover, in his journal illustration, he included ideas that he expressed in words in his book page.

Arturo's thirteen-page Forest book was entitled *Chipmunk,* in which he covered the major ideas about them—what they eat, what their babies are like when born, what they look like, where they live and that they hibernate, their enemies, that they are rodents and in the same family as squirrels, and so forth. During his book-sharing conversation, it was clear that Arturo was very deliberate about what he wrote and drew. For example, on the enemy page (Figure 2.11), he wrote: "The chipmunk senses his enemy which is a screech owl. Then he digs under to hide from his enemy."

His picture depicted a striped chipmunk with a fairly prominent nose, in an underground brown burrow (with one part still filled with dirt), with a screech owl (with two large eyes) in a tree with a dark sky and moon as a background. We had initially thought that Arturo was writing about a particular chipmunk on this page (they are generic on all other ones), but when we asked him why he decided to write this page, he said, *I was trying to think that if they [readers] can know what kind of enemies they [chipmunks] have ... if they go outside in the dark.* Thus, we believe that this "the chipmunk" stands for the class, even though he also uses *his* instead of *its* on this page. (On other pages he mostly used the plural noun form, with plural pronouns, and a few times the singular pronoun *it.*) He explained other aspects of his pictures: the black sky is *the night;* the owl has big eyes *so that they can see in the dark;* the chipmunk is *digging ... and this is the dirt*

The chipmunk sençes his enimie which is a screech owl then he digs under to hide from his enimie.

Figure 2.11 Arturo's *Chipmunk* book page (Forest unit).

that's stuck and he's trying to dig into [that is not as yet dug out]; and the square-like black mark next to the owl's head is *where I screwed up.* Arturo seemed to draw on his literature circle book, *Chipmunks* (Whitehouse, 2004) and a chapter on chipmunks in a read-aloud book *A Forest Community* for information for his book. Arturo's last page of his thirteen-page book depicted a desk with a computer, keyboard, and monitor, where he wrote, "If want to learn more, go to www.chipmunks.com." At the time Arturo's text was written, the above website was for "Alvin and the Chipmunks." We do not know if Arturo knew that, but he did say that the website was a "real" one. He could have just found the website to add to his last page, which he also said would be the one that other 3rd graders, who did not study this unit, would think as the best: *it tells them like information where to search a lot about chipmunks.*

Arturo did not really have any journal entries on chipmunks. He mentioned it once in a list of topics (twelve of them, *chipmunks* being the second one, with a star next to it) when he brainstormed topics he had learned and could write about. The other time the word *chipmunks* appeared in his journal was on the page that served as preparation for the drama activity where children enacted a food web toward the end of the unit. He was an earthworm in this activity, which was listed on top of the page. Then, he had two columns entitled "Things I eat" and "Things that eat me." Under the latter column, he wrote "chipmunks" as the top animal on the list (with "moles," "frogs," "spiders," "bats," and "birds"). And, in his book, Arturo included "earthworms" as one of the kinds of food that chipmunks eat. There were no illustrations for these two journal pages.

Implications for Writing and Drawing in the Elementary Science Classroom

According to Wallace and colleagues (2004), children require multiple discourse forms in scientific language use for them to understand science concepts and learn the language of science. The different genre possibilities offered by students' on-going journals during the two integrated units and the books they composed at the end of the units represented such opportunities. We selected examples to illustrate how similar, yet varied the ways in which young children make meaning in writing and drawing in two science domains are. We offer some general summary remarks: First, over the grades, children's texts grew longer and more elaborate. Second, all of their texts reflected developing competence in including scientific ideas and employing relevant language and images to express these ideas. Third, overall, the children's journals and books reflected the expectations noted in Table 2.2. Thus, the two types of texts in each of the two integrated units provided students opportunities to participate in genuine "semiotic apprenticeship" (Hodson, 1998) in learning science.

Throughout the chapter, we have discussed how the journals might have contributed to the unit books and the way both journals and books might have been influenced by other sources, that is, how their writing reflected intertextual connections. Below, we use this concept of "intertextuality" to summarize the main ideas we have brought up and suggest possible instructional implications regarding them. Intertextuality is defined as the ways in which "we build every text upon and out of texts" (Lemke, 1985, p. 275). It is the juxtaposition of texts (Pappas, Varelas, Barry, & Rife, 2003); how texts are related to or "lean on" other texts in social communities (Jenkins & Earle, 2006, p. 47). Thus, because the journals and books were products of the integrated unit activities, and because this chapter specifically focused on how children used these two types of texts, the notion of intertextuality is useful to organize our summary remarks. Two different major kinds of intertextuality are considered—thematic intertextuality and organizational intertextuality (Jenkins & Earle, 2006; Lemke, 1992). An important emphasis in this chapter has also been to shed light on children's use of visual images that reinforce and complement their linguistic messages. Children's pictures are not simply creative embellishments (Kress & van Leeuwen, 1996) or merely "extras" (Fleckenstein, 2002) in children's written texts. Instead, they are just as integral to scientific practice as writing is, and for this reason, we consider both the verbal and visual aspects in these two types of intertextuality.

Thematic Intertextuality

Thematic intertextuality refers to how different texts express similar meanings. That is, it refers to the ways such texts express the same content or is "on the same topic." As you recall, our analytic approach in this chapter was to examine first how children wrote and drew about the topics of their books in each unit and then explore their journal entries on the same topic or related ideas. Thus, our method itself emphasized the co-thematic nature of both types of texts.

Angela's ideas about the "closeness" of molecules expressed in both her journal and her book in the Matter unit are instances of thematic intertextuality. And, according to her (in our conversation with her over her book), so are the ones that were found in the CD Rom shared in the classroom. Another thematic similarity involves the explanation she offered in her journal entries of why she thought the wet hanging paper towel in the hands-on experiment would dry first, namely, "because it was hanged up like most people do to get stuff to dry" and the meanings she depicted in an illustration on her liquids page in her book (Figure 2.1) that showed clothes hanging on a line (*to get dry*). That is, her wordings in her journal and her picture in her book are co-thematic because she is expressing similar content in both modes. Thus, thematic intertextuality can be realized through different modes of communication. Miguels's *Fish!* book reflects a case of both textual and pictorial thematic intertextuality. He used ideas about and illustration of the Foureye Butterflyfish from the read-aloud book, *About Fish*, to tell about and illustrate the black-eye spot on a fish that helps it to confuse predators (see Figure 2.3).

How children's emergent or initial, partially developed ideas in journals get clarified as more stable ideas in their books can also be seen as thematic intertextuality. For example, Aida's initial idea that there are boy and girl worms expressed in her journal entry got changed to her understanding that "there is no boy or girl earthworm" in her book. Another similar, but yet different instance of such intertextuality occurred when Angela constructed a scientifically incorrect idea that adult frogs are "omnivores" in her *Frogs* book by putting together ideas from her teacher regarding what her teacher had told her about *omnivores*, *carnivores*, and *herbivores*, with the fact that tadpoles eat water plants and frogs eat crickets from the *Starting Life: Frog* read-aloud book. Thus, children forge their own meanings by being engaged with content on the same topic from different text sources. Other examples of this type of thematic intertextuality are Arturo's historical vignette in his *Storms* book that related information about a particular snowstorm from 1979 from another text resource and his reference to the chipmunk website in his *Chipmunk* book.

Organizational Intertextuality

Organizational intertextuality relates to how different texts reflect structural compatibility. It involves how the organization of meanings or information is similar in different texts. Maria's and Pablo's use of the similar headings to organize their Forest books (e.g., *Body, House/Home, Food, Movement,* etc.), which were drawn from similar headings found in classroom charts that their 2nd-grade class had created after read-aloud books on different animals in this unit, illustrate such intertextuality. Another instance is Pablo's inclusion of the inset of a beaver's webbed foot (and caption)—Figure 2.8—from books having such visual images (and minor text).

Two examples of organizational intertextuality can be seen in both of Aida's books. Her picture on Figure 2.9 from her *Liquids* book had strong similarities with a page from the read-aloud book *Down Comes the Rain*. Both pictures included a dog, a plant, a body of water (a pond, she called it in her book), a girl, and clouds, as well as dotted (or squiggly) lines to show evaporating water. In her *Earthworms* book, she relied on the use of questions similar to the structure of her literature circle book to organize her text.

Several journal entries were instances of organizational intertextuality because they relate to recounts of personal experiences. For example, Angela expressed a personal event of her and her brother's encountering a frog ("once a very long time ago I saw a frog...") (Figure 2.2), and Aida told about seeing a worm ("I remember when I was gardening and I saw a worm..."). These were both instances of thematic intertextuality, but they also represent cases of organizational intertextuality, for the children relied on narrative or story-like linguistic registers (similar to ones found in Table 2.1). Their pictures of particular people and animals from their lives (Angela, her brother, frog; Aida, her mother, worm) could be seen as visual-image organizational intertextuality.

Finally, in a general sense, children's use of the scientific register (e.g., present-verb tense, generic nouns, general pronouns, technical terms, and so forth) found in journals and especially in their books reflected the generic features of the information books (read-aloud and literature circle books) used in the two units. Also, the use of minor text of labels, captions, and dialogue bubbles in journals and books, which were found in the children's literature books used in the units, could be seen of occasions of organizational intertextuality.

In summary, there are many implications to be drawn from these examples regarding the above types of intertextuality (and other points made in the chapter), which are often interrelated. First, regarding thematic intertextuality, it is important to provide children with experiences and texts on similar meanings, which integrated science-literacy units offer. Journals are useful sites for children to try out their emergent ideas and also develop more elaborated meanings on these concepts. Culminating books offer a view of children's more developed and relatively more stable understandings of unit content. Second, if an important goal in science is for children to learn scientific content *and* to learn how to express these ideas in multiple modes, then they need good generic models—illustrated children's literature information books that can inform and extend their hands-on experiences. And, they need the writing occasions—such as journals and books—to realize this organizational intertextuality in an ongoing way.

These implications also underscore how the different writing/drawing experiences helped produce and sustain "classroom (multi)cultures" (Kamberelis, 2000, p. 278). That is, the journals and books offered "open" approaches for children in these urban classrooms to explore and reinterpret their familiar, life-world ideas in relation to new scientific ideas and ways to express them. The details described in the chapter show that in enacted curriculum in multicultural classrooms, "different students may experience the same curriculum differently" (Hollins, 1996, p. x). Our examples demonstrate how children employed inter-textual connections in various ways—they relied on a range of sources in their meaning making. Moreover, these children's teachers had high expectations for them in these activities. They believe that: children can try out scientific ideas in responding to unit activities in their journals; children are capable of creating an illustrated information book on a topic of their choice; children can be seen as experts about their books, explaining to others how and why they wrote and drew what they did. Thus, the teachers challenged the deficit views that many still have regarding low-SES and ethnic-minority children, and instead practiced what Bartolome (1994) has called a "humanizing" pedagogy that respects and uses the perspectives of students as an integral part of educational practice.

Targeting Enduring Understandings

1   What are the connections you see between the discussion in the Pappas, Varelas, Ciesla, and Tucker-Raymond chapter and the four elements of culturally-responsive instruction, as described by Villegas and Lucas?
2   Reflect on the role of language use in this multicultural classroom where students and teachers engage with both science and literacy?

Deepening the Reflection

1   Besides the ones that we have noted in the chapter, what other intertextual connections can you find in examining the journals and books of the six children?
2   Explore the relationships between the pictures and texts in the six children's journals and books. What ideas seem to be expressed by pictures and words? What meanings are expressed in words or pictures only?

Encouraging Engagement

1   If you had an opportunity to have a conversation or conference with children who might have created illustrated journals or books, what would you address? What questions would you ask to better determine their scientific understandings and how they expressed them through words and/or pictures?
2   When you teach or observe others teach, create or notice various opportunities for children to express on paper their thinking on a science topic. Ponder and discuss with others the following questions: What are children asked to do in each of these tasks? How does each of the tasks facilitate and encourage learning? Which children are more likely to be involved in the task and why? How do children engage with each task? Do they use both

words and pictures, or does one mode dominate? If you can change the tasks after you look what children do with them, how would you change them and why?

The research reported in this chapter has been supported by a National Science Foundation ROLE grant (REC-0411593) to Maria Varelas and Christine C. Pappas. The data presented, statements made, and views expressed in this chapter are solely the responsibilities of the authors.

References

Bartolome, L. I. (1994). Beyond the methods fetish: Toward a humanizing pedagogy. *Harvard Educational Review, 64*, 173–194.

Bazerman, C. (2004). Speech acts, genres, and activity systems: How texts organize activity and people. In C. Bazerman & P. Prior (Eds.), *What writing does and how it does it: An introduction to analyzing texts and textual practices* (pp. 309–339). Mahwah, NJ: Lawrence Erlbaum Associates.

Douglas, R., Klentschy, M. P., & Worth, K. (Eds.). (2006). *Linking science and literacy in the K-8 classroom*. Arlington, VA: National Science Teachers Association.

Fleckenstein, K. S. (2002). Inviting imagery into our classrooms. In K. S. Fleckenstein, L. T. Calendrillo, & D. A. Worley (Eds.), *Language and image in the reading-writing classroom* (pp. 3–26). Mahwah, NJ: Lawrence Erlbaum Associates.

Gee, J. P. (2004). Language in the science classroom: Academic social languages as the heart of school-based literacy. In E. W. Saul (Ed.), *Crossing borders: Literacy and science instruction: Perspectives on theory and practice* (pp. 13–32). Newark, DE: International Reading Association.

Goldman, S. R., & Bisanz, G. L. (2002). Toward a functional analysis of scientific genres: Implications for understanding and learning processes. In J. Otero, J. A. Leon, & A. C. Graesser (Eds.), *The psychology of science text comprehension* (pp. 19–50). Mahwah, NJ: Lawrence Erlbaum Associates.

Halliday, M. A. K. (1985). *An introduction to functional grammar*. London: Edward Arnold.

Halliday, M. A. K., & Martin, J. R. (1993). *Writing science: Literacy and discursive power*. Pittsburgh: University of Pittsburgh Press.

Hodson, D. (1998). *Teaching and learning science: Towards a personalized approach*. Philadelphia: Open University Press.

Hollins, E. R. (1996). Preface. In E. R. Hollins (Ed.), *Transforming curriculum for a culturally diverse society* (pp. ix–xiii). Mahwah, NJ: Lawrence Erlbaum.

Jenkins, C. B., & Earle, A. A. (2006). *Once upon a fact: Helping children write nonfiction*. New York: Teachers College Press.

Kamberelis, G. (2000). Hybrid discourse practices and the production of classroom (multi)cultures. In R. Mahalingam & C. McCarthy (Eds.), *Multicultural curriculum: New directions for social theory, practice, and policy*. New York: Routledge.

Kress, G. (1997). *Before writing: Rethinking the paths to literacy* (pp. 261–285). London: Routledge.

Kress, G., & van Leeuwen, T. (1996). *Reading images: The grammar of visual design*. London: Routledge.

Lemke, L. L. (1985). Ideology, intertextuality, and the notion of register. In J. D. Benson & W. S. Greaves (Eds.), *Systematic theoretical papers from the 9th International Systemic Workshop* (pp. 275–294). Norwood, NJ: Ablex.

Lemke, J. L. (1992). Intertextuality and educational research. *Linguistics and Education, 4,* 257–267.

Lemke, J. L. (1998). Multiplying meaning: Visual and verbal semiotics in scientific text. In J. R. Martin & R. Veel (Eds.), *Reading science: Critical and functional perspectives on discourses of science* (pp. 87–113). London: Routledge.

Pappas, C. C. (2006). The information book genre: Its role in integrated science literacy research and practice. *Reading Research Quarterly, 41,* 226–250.

Pappas, C. C., Kiefer, B. Z., & Levstik, L. S. (2006). *An integrated language perspective in the elementary school: An action approach.* Boston: Pearson.

Pappas, C. C., Varelas, M., Barry, A., & Rife, A. (2003). Dialogic inquiry around information texts: The role of intertextuality in constructing scientific understandings in urban primary classrooms. *Linguistics and Education, 13*(4), 435–482.

Saul, E. W. (Ed.). (2004). *Crossing borders: Literacy and science instruction: Perspectives on theory and practice.* Newark, DE: International Reading Association.

Unsworth, L. (2001). *Teaching multiliteracies across the curriculum: Changing contexts of text and image in classroom practice.* Buckingham, UK: Open University Press.

Van Leeuwen, T., & Jewitt, C. (2001). Introduction. In T. van Leeuwen & C. Jewitt (Eds.), *Handbook of visual analysis* (pp. 1–9). London: Sage.

Varelas, M., & Pappas, C. C., & the ISLE Team. (2006). Young children's own illustrated information books: Making sense in science through words and pictures. In R. Douglas, M. P. Klentschy, & K. Worth (Eds.), *Linking science and literacy in the K-8 classroom* (pp. 95–116). Arlington, VA: National Science Teachers Association.

Varelas, M., Pappas, C., Barry, A., & O'Neill, A. (2001). Examining language to capture scientific understandings: The case of the water cycle. *Science and Children, 38,* 26–29.

Wallace, C. S., Hand, B., & Prain, V. (2004). *Writing and learning in the science classroom.* New York: Springer.

Children's literature, CD Rom, and website

Adventures of Herman. Anatomy section. www.urbanext.uiuc.edu/SchoolsOnline/index. html.

Berger, M. (1992). *Look Out for Turtles!* New York: Harper Collins.

Branley, F. M. (1983). *Down Comes the Rain.* New York: Harper Collins.

Branley, F. M. (1986). *Air Is All Around You.* New York: Harper Collins.

Dell'Oro, S. P. (2001). *Tunneling Earthworms.* Minneapolis, MN: Lerner.

First Discovery Book. (2002). *In the Forest.* New York: Scholastic.

Fowler, A. (1991). *What's the Weather Today?* New York: Children's Press.

Fowler, A. (1992). *It Could Still Be Water.* New York: Children's Press.

Fowler, A. (1995). *When a Storm Comes Up?* New York: Children's Press.

Fowler, A. (1996). *What Do You See in a Cloud?* New York: Children's Press.

Fowler, A. (1997). *Animals Under the Ground.* New York: Children's Press.

Gibbons, B. (1991). *From Seed to Plant.* New York: Holiday House.

Gibbons, G. (1990). *Weather Words and What they Mean.* New York: Holiday House.

Himmelman, J. (2000). *An Earthworm's Life.* New York: Children's Press.

Lauber, P. (1995). *Who Eats What?: Food Chains and Food Webs.* New York: HarperCollins.

Llewellyn, C. (2000). *Earthworms.* New York: Franklin Watts.

Llewellyn, C. (2003). *Starting Life: Frog.* Chanhassen, MN: Northword.

Pfeffer, W. (1997). *A Log's Life.* New York: Simon & Schuster.

Massie, E. (2000). *A Forest Community.* Austin, TX: Streck-Vaughn.

New Media. (1997). *States of Matter, Multimedia CD-Rom for PC and Macintosh.* New York: Facts on File.

Saunders-Smith, G. (1998). *Seeds.* Mankato, MN: Pebble Books.

Sill, C. (2002). *About Fish: A Guide for Children.* Atlanta, GA: Peachtree.

Whitehouse, P. (2004). *Chipmunks.* Chicago: Heinemann Library.

Zoehfeld, K. W. (1998). *What is the World Made Of? All About Solids, Liquids, and Gases.* New York: Harper Collins.

3 Developing Expository Writing Skills in Science Through the Integration of Inquiry-Based Science and Literacy Instruction

Yu-Min Ku, Monica S. Yoo, and Eugene E. Garcia

Introduction

Ku and her colleagues have focused their research and intervention energies towards creating equitable opportunities for science achievement for linguistically-diverse students. In this chapter, Ku and her team argue that expository writing in science can be supported and developed through rich literacy-in-science-infused interventions that aim to build students' understanding of scientific processes and reasoning while providing rich narrative contexts for connecting everyday science experiences to canonical science concepts.

A particularly exciting feature of this chapter is the year-long nature of the intervention and assessment. Their team recognized, in the design of the intervention, that students, especially linguistically-diverse students, benefit from opportunities to experience an intervention over a longer period. Their use of an assessment that measured science knowledge and processes, on the one hand, and expository science writing skills including language use, on the other, offers an example of an authentic periodic assessment that teachers could use to measure change over time in students' science notebook writing.

The insights offered in this chapter help us see the ways that classroom interventions can yield demonstrable change in students' abilities to "show what they know" through writing in science. Currently, policy makers at the national and local levels have begun efforts to design curricular materials and professional development workshops aimed towards improving writing, in general, in K-12 schooling. This is occurring while there is also a growing groundswell of support among classroom science researchers for more reflective and culturally-relevant design of curricular materials, instruction and assessment pedagogies that enhance equitable science and literacy outcomes for all learners. The research featured here by the Ku team offers an important addition to these converging conversations.

In current science education reform documents (AAAS, 1993; NRC, 1996, 2000), inquiry has been highlighted as the most critical component of science learning. As indicated in the National Science Education Standards (NRC, 1996), all students "should have the opportunity to use scientific inquiry and develop the ability to think and act in ways associated with inquiry" (p. 105). While teachers have explored various approaches to implement inquiry-based science instruction, many researchers also have advocated the importance of integrating science with literacy, writing in particular, in classrooms. Although Yore (2000) and Lang & Albertini (2001) have argued that physically engaging in hands-on science does not necessarily promote meaningful learning, the benefits of embedding authentic writing tasks with science inquiry can lead to improvements in students' understanding of scientific concepts (Fellows, 1994; Lang & Albertini, 2001; Rivard, 1994; Rivard & Straw, 2000) and the development of scientific reasoning (Keys, 1999; Yore, 2000). Additionally, Glynn and Muth (1994) have pointed out that "when students write about their observations, manipulations, and findings, they examine what they have done in greater detail, they organize their thoughts better, and they sharpen their interpretations and arguments" (p. 1065). Since incorporating writing into the science curriculum has been shown to positively impact the learning of content and to enhance students' thinking, the question then is: What types of writing activity better support integrated inquiry-based science teaching and learning and help promote the goals of educational equity for all science learners?

Writing as a Learning Tool in Science

Writing tasks in science, according to Rivard (1994), can be generally categorized into two types: expressive and expository. Expressive writing tasks mainly include journal- or narrative-style writing that elicits students' prior experiences and personal responses to particular topics. While much of the writing-to-learn in science research employs such expressive writing tasks, the effectiveness of expressive writing tasks on science learning has not been consistently reported in previous studies (Keys, 1999; Klein, 1999; Rivard, 1994).

Expository writing tasks in science usually require students to write to inform, to explain, to report observations from the experiment, to analyze, and to summarize. Studies that have employed such types of writing tasks have shown to enhance science learning (Rivard, 1994). For example, Laidlaw, Skok, and McLaughlin (1993) deduced from their investigation of 5th- and 6th-grade student science achievement that note-taking improved science outcomes as students learned to take notes about their investigations as scientists do.

Mason and Boscolo (2000) investigated whether the use of writing as a learning strategy improved students' conceptual understanding of the new topic. The study involved thirty-six 4th graders who were divided into two groups: experimental (writing) and control (no writing). While students in both groups conducted experiments on the topic of photosynthesis, students in the writing group were instructed and encouraged to use writing to post their questions, generate hypothesis, reflect upon ideas, and synthesize what they have learned. The results indicated that the writing activities contributed significantly to a better understanding of the topic for the students in the experimental group.

More recently, Aschbacher and Alonzo (2004) found that when 4th- and 5th-grade students were explicitly taught to use a notebook during an experiment in order to record the question for the inquiry, data collection during an investigation, knowledge claims based on the evidence when conducting hands-on science, they exhibited a greater understanding of science concepts than students in the comparison group.

Previously, most studies have investigated whether different writing tasks have assisted students' changes in conceptual understanding; however, our study attempts to document how children's writing skills develop along with their understanding of the science inquiry process.

Goals of the Current Study

The main purpose of this study was to investigate whether students' expository writing skills in science would improve through an instructional intervention aimed at promoting achievement and equity in science and literacy for linguistically and culturally diverse as well as mainstream students. This intervention, in the form of a thematic science curriculum, used household materials for conducting scientific inquiry activities and was a medium for examining language, literacy, and collaborative interactions in the classroom. The research framework recognizes that science learning has its roots in processes both out-of-school and in school. Its foci are on responsive instructional engagement that encourages students to construct and reconstruct meaning and to seek reinterpretations and augmentations to past knowledge regarding literacy and science within compatible and nurturing schooling contexts. Diversity is perceived and acted on as a resource for teaching and learning instead of a problem. A focus on what students bring to the schooling process generates a more asset/resource-oriented approach vs. a deficit/needs assessment approach. Within this knowledge-driven, responsive, and engaging learning environment, skills are tools for acquiring knowledge, not a fundamental target of teaching events.

The instructional intervention focused on two units each for 3rd graders (Measurement and Matter) and for 4th graders (Water Cycle and Weather). The implementation of the science units took place, on average, two to three hours a week for the majority of classrooms. All teachers were provided with complete sets of materials, including teachers' guides, copies of student books, and science supplies.

Before implementing each science unit, two workshops were conducted for the 3rd- and 4th-grade teachers to meet with the research team to discuss the overall goals of the project, review the instructional materials, and receive training on how to implement the units. The workshops were facilitated by two teachers who implemented the units in the previous years. In addition to reviewing the instructional units, the workshops focused on teachers' sharing their instructional strategies and activities to promote science learning with students from diverse backgrounds. Teachers also provided insights on integrating math and literature/literacy into science instruction. This created more space for the discussion on how to integrate the curriculum with students' diverse backgrounds.

The science units were written for promoting the science inquiry process; for example, every lesson embedded a hands-on activity that provided teachers opportunities to assist students to develop their understanding of the five aspects of science inquiry: (a) generate questions, (b) design investigations and plan procedures, (c) carry out the investigations, (d) analyze and draw conclusions, and (e) report findings.

In terms of promoting literacy development for linguistically diverse students, each lesson included short narrative texts to motivate students' engagement at a personal level and expository texts to summarize the main concepts covered in the lesson so that students would gain a more structured understanding. In addition, students were also provided with various writing tasks to reflect on what they learned and observed from the hands-on activities or group discussions.

The researchers involved in the current study designed a writing assessment system that was embedded in the instruction to serve as a learning tool for students and a diagnostic instrument for teachers to evaluate students' progress in achieving scientific literacy. Three distinctive features of the system are noteworthy. First, a lab book was incorporated as a template to guide students' science activity and reasoning. As indicated in Baxter, Bass, and Glaser (2001), "notebook writing in science is an important tool for recording observations, generalizing, hypothesizing, and theorizing—in general for assisting thinking, reasoning, and problem solving during the conduct of science inquiry" (p. 138). The lab book in our assessment system provided explicit instruction on the activities that students would need to do individually or in a group while engaged in experimentation (see further description in the research instrument section). We also hoped to draw upon teachers' use of scaffolding and the structured frames in the lab book to help students develop discourse knowledge about the process of science inquiry.

The second and third features of the assessment system were in the writing prompts that were administered after students completed the experiments. Each prompt was designed to elicit students' understanding of their science experiment and to investigate how students connect their understanding gained from the experiment to the actual science concepts. More specifically, the prompts instructed students to create informational texts that described how they conducted the experiment and then required students to write explanations on how the findings in the experiment played a part in their learning of the targeted science concept.

In order to promote students' development of expository writing skills, the second feature of the writing assessment system directed students to write for a group of scientists. We hoped this feature would increase students' awareness of audience and help them choose a voice/tone and types of vocabulary appropriate to the audience and the purpose of writing.

In response to Prain and Hand's (1996) viewpoints that (a) "writing for learning in science should extend beyond the traditional records of observations and formal reports" and (b) "students should also write to explain, to sort out what they understand,..." (p. 613), another feature of the prompt was the inclusion of a question that required students to provide explanations on how and why the experiment findings were important to their understanding of the target science concept. Writing explanations has been demonstrated as a powerful activity that helps students with "developing targeted science concepts from hands-on activities" (Keys, 1999, p. 127). Because composing scientific explanations involves higher-order thinking and reasoning, we anticipated that writing explanations would be a more difficult task for the students than organizing information to demonstrate their understanding of the science inquiry process.

The primary focus of the following sections will be on the design of an authentic writing assessment system and a discussion of students' development of expository writing skills in science.

Methods

Research Setting and Participants

Participants were fifteen 3rd-grade and thirty 4th-grade students from two schools in a large urban school district characterized by ethnic, economic, and language diversity. The key features of the participating schools are summarized in Table 3.1. Although in these two schools, there were a few classes of 3rd- and 4th-grade students who participated in the instructional intervention, only one class of 3rd-grade students from school 1 and two classes of 4th-grade students from school 2 were involved in the writing assessment.

Research Instrument

The Authentic Science Inquiry/Literacy Assessment System (ASILAS) was generated to gauge students' understanding of the science inquiry process and their ability to write science reports. The design of the ASILAS was based on the framework of the Authentic Literacy Assessment System (ALAS), which was

Table 3.1 Key features of schools

School	Ethnicity (Major Groups) (%)	Free and Reduced Lunch (%)	English Language Learner (ELL) (%)
1	40 Latino 25 African American 23 Chinese	88	41
2	25 White 25 Latino 16 African American 11 Chinese	22	6

collaboratively created by Professor Eugene Garcia (former Dean of Graduate School of Education at University of California, Berkeley), and other Berkeley researchers and teachers in the San Francisco Bay area. Similar to ALAS, which provided opportunities for students from diverse backgrounds to discuss writing prompts within small groups and to use a graphic organizer to plan their responses, ASILAS utilized the idea of a 'lab book' that encouraged students to record the process of their experiment implementation and a discussion of their findings.

Since professional development and the inclusion of teachers was an integral part of the instructional intervention, teachers in the project were included in all facets related to the development and implementation of the ASILAS. From its inception, teachers' ongoing input related to the assessment led to various changes such as the inclusion of a lab book, the addition of a graphic organizer component, and a revision of the prompt, which directed students to write to an audience of scientists. Particularly, the design of lab book which provided explicit instruction on the activities that students would need to do individually or in a group while engaged in experimentation was based upon previous participating teachers' experience and knowledge of how to motivate and support students from diverse backgrounds to engage in a more meaningful and purposeful inquiry activity. That is, by providing a clear layout of when and where to write in the lab book for the question and hypothesis related to the experiment, the materials and procedures involved, and a summary and conclusion based on the results and of the experiment, we hoped to draw upon teachers' use of scaffolding and the structured frames in the lab book to help diverse students develop discourse knowledge about the process of science inquiry.

In order to attain prompts that were age and grade appropriate, teachers were involved in reviewing and discussing the structure of the prompts, the graphic organizers utilized and the most-conducive lessons for writing an expository text. Teachers also provided the research team with logistical information on how long it would take to carry out plans for the experiment and the best viable grouping strategies for optimizing student learning. Additionally, teachers assisted with assuring that grade level standards in science and literacy were being met by verifying that the ASILAS scoring rubric adapted and incorporated aspects of the 3rd- and 4th-grade California State Standards in Language Arts and Science (California State English Language Arts Contents Standards, 2004a, 2004b).

The administration of the ASILAS involved a two-day four-stage sequence that commenced with an experiment from the selected units and lessons. These stages included: (1) Experiment, (2) Group Work, (3) Graphic Organizer, and (4) Independent Writing. Table 3.2 shows the ASILAS administration protocol developed by the research team for teachers participating in the project's writing assessment.

In the first step of the ASILAS, students engaged in activities that provided them with opportunities to think and write like scientists. As they conducted the science experiment, students used a lab book to record their thoughts, answers to

Table 3.2 ASILAS administration protocol

Step 1: Experiment
After introducing the experiment guidelines, student lab book, and main problem, encourage students to think about how they might solve the problem. Explain the scientific model. Ask them to share aloud responding to the following questions:
• What hypothesis can you form?
• What materials will you need?
• What procedures are necessary?

Step 2: Group Work
Have students turn to page 8 in the lab book where the group work activity is found. Have students share their results by answering the following questions:
• What do I want to tell others about the activity?
• What was your partner's hypothesis?
• Are your predictions similar or different from other students in your group? How?

Step 3: Graphic Organizer

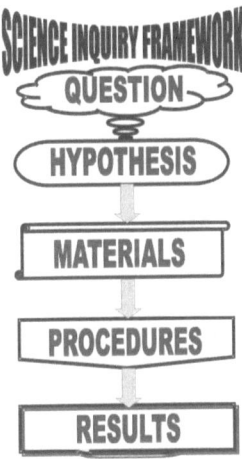

Re-introduce Scientific Inquiry Framework as a guide for student writing.

Step 4: Independent Writing
Students should be given thirty minutes but no more than forty minutes to complete the assessment. Students are not allowed to use dictionaries or thesauruses. Remind students to use the graphic organizer and lab book to assist them with the composition and to proofread their compositions before turning them in to teacher.

---

questions, and hypotheses. Additionally, students were encouraged to freely document their thoughts in relation to conceptual categories within the scientific inquiry framework. The lab book provided space for jotting ideas, generating questions, creating hypotheses, preparing for discussions, and revising former thoughts as students discovered new understandings through the process of experimentation.

In the second stage of the ASILAS, students completed the experiment and proceeded to a discussion about the results. They compared their findings with their classmates and reflected on whether the findings were consistent with their initial hypotheses. This process allowed students to further explore their new-found ideas and modify their initial understandings from the preliminary stages of the science experiment. As students engaged in written and oral group work, they were encouraged to use and practice the language and vocabulary of science, which was embedded in the materials and procedures of the experiment and the scientific inquiry framework.

In the third stage of the ASILAS, which was on the second day of the administration cycle, students were reintroduced to the scientific inquiry framework as a guide and reminder about integrating the language and concepts of scientific inquiry into their writing. The graphic organizer, as an additional tool, functioned as a curricular step promoting the synthesis of ideas from the student lab books, class discussion, and the experiment itself.

In the fourth stage of the ASILAS, a formal writing assessment/prompt was distributed to each student. Students were instructed to use their lab books as aids to independently write in response to the prompt (Table 3.3). After each ASILAS administration, the lab book and writing sample were collected from every participating student.

At each grade level, four prompts (two for every science unit) generated four separate expository writing samples from each participating student. The prompts were consistent with the science unit objectives. For example, in the 3rd-grade prompt below, students were asked to describe to a group of scientists the process of their inquiry on the topic of condensation and then explain the role that temperature plays in the process of condensation. In order to maintain consistency among the prompts and to ensure that the wording of the prompt would not affect the outcomes between units and grade levels, the creation of the prompts followed a generic structure which was then applied to specific steps, materials, and concepts presented in each accompanying lesson.

To measure student improvement and achievement, a scoring rubric, using a 1–6 scoring criteria (6 being the highest), for the writing samples was later employed for evaluation purposes by the research team. The rubrics accounted for science inquiry development (specific knowledge and understanding of science inquiry) and general mechanics and contents of language use including organization, the language of science, and conventions.

Table 3.3 Example of a writing prompt used for ASILAS

Write to a Group of Scientists.
Tell them about how you conducted the experiment on condensation using one cup of cold water and one cup of tap water, including all the steps.

Explain how temperature affects the process of condensation when a gas changes into a liquid.

Use the Science Inquiry Framework to organize your answers and give as much detail as possible.

Procedures and Data Collection

Each participating teacher who administered the ASILAS attended two workshops on the implementation of the instrument. The first workshop addressed the development of the ASILAS instrument, while the second was devoted to understanding the procedures and administration of the ASILAS.

Although every student in the participating classrooms was expected to participate in the ASILAS administrations, which took place a total of four times during the school year, not every student did, mainly for the reasons related to absence or moving schools. Due to student attrition and some teachers' incompletion of the assessments, our final data analysis included fifteen 3rd-grade students and thirty 4th-grade students who obtained ASILAS scores for two science units.

Data Analysis

The research team evaluated each of the student writing samples by scoring it according to four domains (Science Inquiry, Organization, Language of Science, and Conventions) on a rubric. Before scoring the writing samples, five team members were involved in two sessions of ASILAS training on the scoring protocol. The training session included a calibration of the writing rubric and a reliability check. Members scored ASILAS writing samples only after obtaining a 90% inter-rater reliability on ten of the student writing samples. Before rendering scores on the ten calibration samples, team members reviewed the writing rubric, prompts, and the lessons from which the prompts were developed.

Because two ASILAS writing assessments were developed for every unit, each student received a mean score in each of the four writing domains for a particular unit. If a student had only one ASILAS writing sample for either science unit, the individual scores in the four domains would be used instead of the mean scores.

Results

The results from the ASILAS were presented in two sections for each grade level. The first section looks at the ASILAS writing scores in the four domains (Science Inquiry, Organization, Language of Science, and Conventions) for the two science units taught at each grade level. The second section focuses on comparing the percentage of students below and at or above grade level for both units. The scores presented are based on the ASILAS rubric, which is aligned with the state standards for expository writing and science writing. Since the rubric was written to apply to multiple grade level abilities, the benchmark scores are as follows: 1 for 1st grade, 2 for 2nd grade, 3 for 3rd grade, 4 for 4th grade, 5 for 5th grade, and 6 for 6th grade.

Third Grade ASILAS Performance

Figure 3.1 illustrates 3rd-grade students' mean score in each of the four domains for ASILAS in the Measurement and Matter units. The results indicated that students' scores on the four ASILAS domains were better in the second science unit than the first unit, especially in the domain of Science Inquiry and Language of Science. In addition, students' ASILAS scores in the four domains were lower than the benchmark score of 3 in the first unit but the scores related to Science Inquiry, Organization, and Language of Science reached or surpassed the benchmark score during the second unit. Figure 3.2 further demonstrates the change in the percentage of students who obtained or exceeded the grade level benchmark in the Measurement and Matter units.

The Science Inquiry domain evaluated students according to how they (1) stated the question and hypothesis related to the experiment, (2) described the materials and procedures involved, and (3) wrote a summary and conclusion based on the results and concepts of the experiment. The findings indicated that 47% of 3rd-grade students were considered to be writing at or above grade level in the Science Inquiry domain during the Measurement unit. During the second unit on Matter, an additional 40% of students reached grade level expectations, taking the total up to 87% of 3rd-grade students reaching or exceeding the benchmark.

The Organization domain involved assessing paragraph structure and overall cohesiveness of students' written texts, and included examining how students (1) organized ideas over several discernible sections, (2) used a topic and supporting details, (3) wrote paragraphs in correct and logical sequences. The findings showed that 40% of 3rd-grade students were considered writing at or above grade level in the Organization domain during the Measurement unit while an additional 33% of students reached grade level expectations during the Matter unit, with a total of 73% of the students reaching or exceeding the benchmark.

The Language of Science domain reviewed how students (1) used science and inquiry vocabulary, and (2) synthesized multiple pieces of evidence to create explanations. The findings revealed that 33% of students reached the grade level

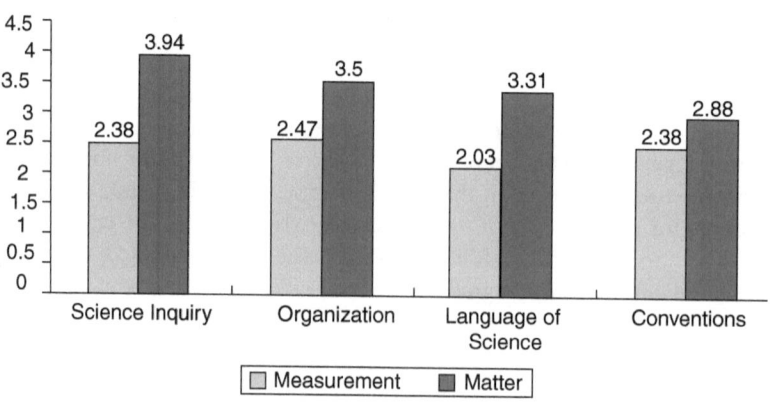

Figure 3.1 3rd-grade ASILAS performance in four domains.

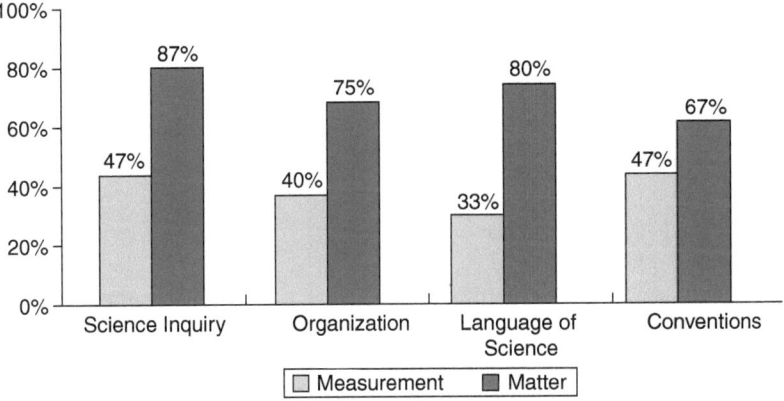

Figure 3.2 Percentage of 3rd-grade students at or above benchmark.

benchmark in the Language of Science domain during the Measurement unit while an additional 47% of students also scored at or above the benchmark during the Matter unit.

The Conventions domain addressed students' correct usage of spelling, punctuation, and grammar. The findings illustrated that during the Measurement unit, 47% of students scored at or above the benchmark, whereas an additional 20% of students were writing at or above grade level standards during the Matter unit. While 3rd-grade students overall achieved the most gains in the Science Inquiry and Language of Science domains, they also improved in the literacy domains measuring Conventions and Organization.

Fourth-grade ASILAS performance

Figure 3.3 shows 4th-grade students' mean score in each of the four domains for the ASILAS during the Water Cycle and Weather units. The same four domains that comprised the 3rd-grade rubric also applied to the 4th-grade ASILAS writing assessment. The results indicated that the mean scores in three of the four ASILAS domains were slightly better in the second science unit than the first science unit. Despite the mean increase, ASILAS scores in the Language of Science and Conventions, however, remained lower than the benchmark score of 4 in both science units.

Figure 3.4 demonstrates the change in the percentage of students who reached or exceeded the grade level benchmark in the Water Cycle and Weather units. In the domain of Science Inquiry, 70% of 4th-grade students were writing at or above grade level during the Water Cycle unit while an additional 3% of students reached or exceeded the benchmark during the Weather unit. Under the Organization domain, the results indicated that 77% of students' writings were considered to be at or above grade level, and the number of students reaching the benchmark during the Weather unit further revealed an additional increase of 6%. In the Language of Science domain, 40% of students reached or exceeded

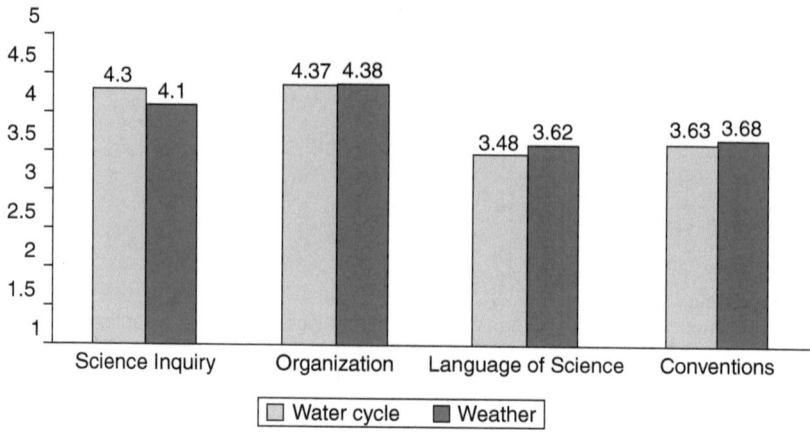

Figure 3.3 4th-grade ASILAS performance in four domains.

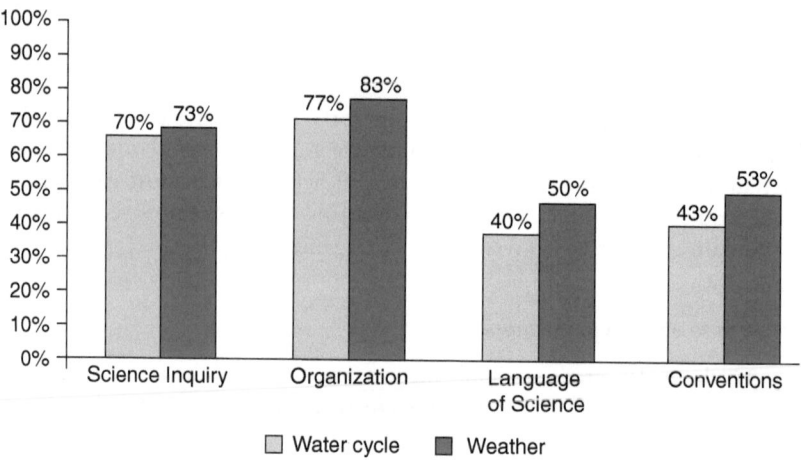

Figure 3.4 Percentage of 4th-grade students at or above benchmark.

the benchmark during the Water Cycle unit, and an additional 10% of students received similar scores during the Weather unit. Within the domain of Conventions, 43% of students scored at or above grade level standards during the Water Cycle unit, and an additional 10% of students met the benchmark during the Weather unit.

Discussion

From our findings, it becomes apparent that students in the 3rd grade achieved gains in all of the four domains. The increase in the percentage of students meeting or surpassing the benchmark in each domain may be due to the practice of various types of writing activities and teachers' scaffolding of science inquiry process as they progress through the ongoing instructional units. Furthermore, it is important to note that in between the first and second units, the curriculum allowed students to participate in a largely hands-on curriculum, to continually engage with scientific methodological procedures, and to have classroom discussion with peers and the teacher.

Overall, the 3rd-grade students in the study achieved the most gains, especially in Science Inquiry and the Language of Science. This pattern of growth could be attributed to the ways in which the teacher embedded science and inquiry vocabulary into the lessons. This particular 3rd-grade teacher has a certificate in BCLAD (Bilingual, Cross-cultural, Language and Academic Development) and has been teaching in bilingual settings for about seven years when participating in the study.

A closer look at 3rd-grade students' writing samples reveals that many students in the 3rd grade first wrote tentatively with a confused use of the terminology related to scientific experimentation and inquiry. For example, many students at the beginning of the year shied away from terms such as "hypothesis" and "procedures"; yet, their subsequent writings revealed more confidence and knowledge about how to correctly use such terms as the year progressed. The increase in comfort levels with specific terms related to scientific discourse could help to explain why students experienced the most growth in these particular domains. Figure 3.5 and Figure 3.6 present a Spanish–English bilingual 3rd-grade student's writing samples from the first and second unit.

Figure 3.5 presents this student's writing sample from the Measurement unit. This ASILAS writing prompt asked students (a) to describe what they did in an experiment where they made their own combination of "doodle juice" from a mixture of different juices and sodas and (b) to explain the importance of using standard units. In this sample, you will notice that the student was unfamiliar with the conventional terminology used in scientific experiments and scientific discourse. The student shied away from using the word hypothesis and instead wrote, "I think" (line 3). Although the student used the word "procedures", she wrote, "We made a procedures" (lines 8–9), revealing that she was not quite sure of what the term meant or how it was used. Likewise, she struggled with transitions. It is possible to read the piece as though she was structuring a narrative, moving from her hypothesis to "then we gather the materials," (line 5) illustrating that this was all part of the experiment's story. Her lexical choice also could signify that she was demonstrating procedural knowledge, which could show that she was moving toward blending the genres of narrative and expository writing, as she developed her skills for writing scientific reports. After explaining the procedures, her conclusion was limited and she summed up her writing by noting: "the activity we did is called doodle juice." In what appeared to be her conclusion, there was no explanation on why standard units were necessary for making doodle juice.

Figure 3.6 presents the same student's writing sample in the second unit on

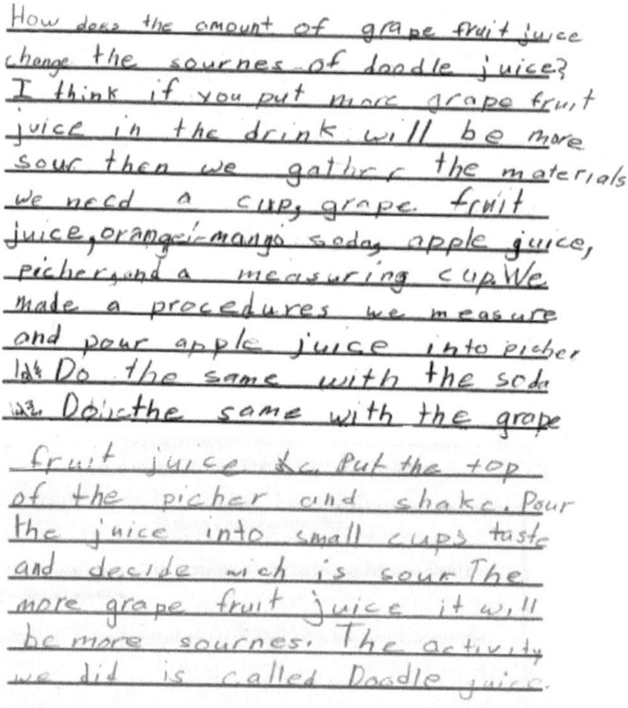

How does the amount of grape fruit juice
change the sourness of doodle juice?
I think if you put more grape fruit
juice in the drink will be more
sour then we gather the materials
we need a cup, grape fruit
juice, orange-mango soda, apple juice,
picher, and a measuring cup We
made a procedures we measure
and pour apple juice into picher
1st Do the same with the soda
2nd Do the same with the grape
fruit juice &c. Put the top
of the picher and shake. Pour
the juice into small cups taste
and decide wich is sour The
more grape fruit juice it will
be more sourness. The activity
we did is called Doodle juice.

Figure 3.5 Writing sample of a 3rd-grade student on the Measurement unit.

Matter. Third-grade students were asked to write to response to the ASILAS prompt shown in Table 3.2. From analyzing this sample, the student has shown an improvement in both science inquiry and the language of science domain. In this second piece of writing, she supplied the word "hypothesis," (line 3) and used it correctly by integrating both the concept of condensation and a prediction about it. This contrasted with her first sample in Figure 3.5, where the student only discussed the notion of what was sour rather than a prediction on how using standard units might affect the making of doodle juice. In the second sample, she exhibited a much clearer understanding of procedures as steps that the students must follow (lines 7–8), rather than as something they make. By the conclusion, this student was able to return to her hypothesis and integrate her finding that yes, condensation only happens with cold water. This student's writing sample for the second unit was written with a sequence that followed the report-like structure of an experiment; and it read much less like a narrative. Although she did not divide the sections into separate paragraphs, her demarcations between parts were much clearer.

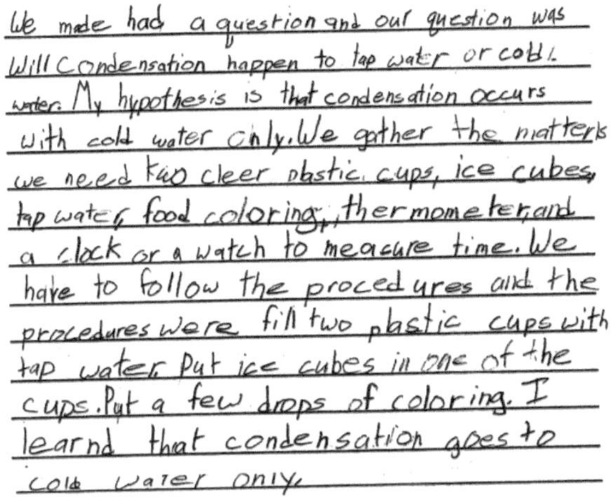

We made had a question and our question was
Will Condensation happen to tap water or cold
water. My hypothesis is that condensation occurs
with cold water only. We gather the matters
we need two cleer plastic cups, ice cubes,
tap water, food coloring, thermometer and
a clock or a watch to meacure time. We
have to follow the procedures and the
procedures were fill two plastic cups with
tap water. Put ice cubes in one of the
cups. Put a few drops of coloring. I
learnd that condensation goes to
cold water only.

Figure 3.6 Writing sample of a 3rd-grade student on the Matter unit.

Comparing the Patterns of Performance Between 3rd- and 4th-Grade Students

Although students in both grade levels improved in the Conventions domain, students, on average, still performed below their respective grade level benchmarks in this area. The reason for this result could be that the research sites involved in the project are schools with diverse populations and high concentrations of English Language Learners (ELLs). Additionally, the writing prompts were primary designed to promote students' ability to conduct science inquiry, thus, in each prompt students received, they were instructed to focus on using the scientific inquiry framework to plan for the writing. Because of the nature of the intervention and the construction of the writing prompts, teachers rarely asked students to pay attention to the conventions of writing when they administered ASILAS. Yet, despite the seemingly low scores in this domain, it is noteworthy that 20% of 3rd graders and 10% of 4th graders who had not reached the Conventions benchmark during the first unit did so during the second unit.

Another worth mentioning finding that emerged from comparing the patterns of performance between 3rd- and 4th-grade students was that the ASILAS scores' gains made by students may have depended on the difficulty of the individual units. For example, the 4th-grade unit on Weather was considered by the teachers and researchers to be conceptually more difficult in comparison to the unit on Water Cycle. While many students had some understanding of the Water Cycle because of their experiences, far fewer students had the same kind of knowledge about the themes (e.g., heat, wind, air pressure, fronts, hurricanes … etc.) covered in the Weather unit, given their relatively young age and their upbringing in an area of the country where weather patterns are less susceptible to seasonal change and extreme climates. For the above reasons, it was likely that only small increases on the ASILAS scores were found from writing sample for these two units.

In searching for the reasons for the disappointing lower scores on the Language of Science for 4th-grade students in both units, students' writing samples were closely analyzed. The findings revealed that students who reached or exceeded the 4th grade ASILAS benchmark in the Language of Science were able to integrate stronger conceptual understandings into their scientific explanations of the experiment. Students who still had difficulty in this area listed procedures and materials, but did not integrate these essentials into an explanation; instead, these students often employed a narrative style in the retelling of how procedures and materials were utilized (see a 4th-grade student's writing samples for the Water Cycle unit and the Weather unit presented in Figure 3.7 and Figure 3.8). Based on further observation, the students who had lower scores in the Language of Science were likely to be the same group of students who also scored around, but not above, the benchmark in the Science Inquiry domain, because they were not able to synthesize the data and create a summary to connect their experimental findings to the targeted science concepts. Since there were greater expectations for 4th-grade students than 3rd-grade students in the area of expository

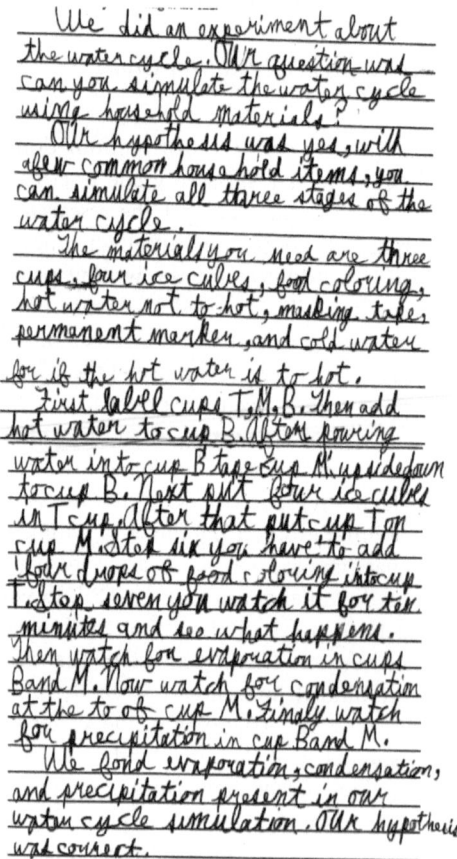

Figure 3.7 Writing sample of a 4th-grade student on the Water Cycle unit.

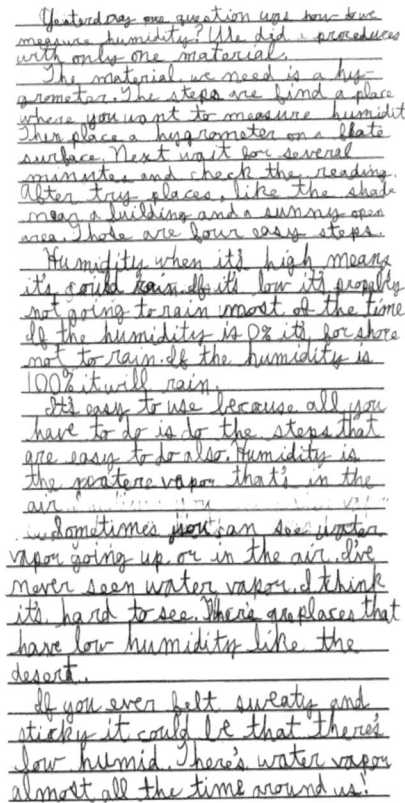

Figure 3.8 Writing sample of a 4th-grade student on the Weather unit.

writing, an analysis of students' samples in this domain spoke to the challenge that students still face difficulties when moving from primarily writing narratives to writing expository essays and reports.

Conclusions and Implications for Teaching and Learning

Despite the limitations mentioned in the discussion of the 4th-grade results, by the second unit, more students in both grade levels reached or exceeded the benchmark writing scores in all four domains. Based on the comparison between the two grade levels and on the substantial gains made by 3rd-grade students on every ASILAS writing domain, it is evident that 3rd-grade students are not too young to successfully produce scientific reports if the curriculum is tied to meaningful activities and writing-to-learn opportunities. Before students are asked to make a leap on grade level writing standards, more opportunities to practice writing scientific reports should be encouraged in order to assist students in the transition between writing narrative and expository texts.

The ability to use scientific inquiry to explain and reason is very important in students' science learning. In this study, it is noticeable that students, even at the 3rd-grade level, could master the skills of "writing to inform" others if provided with practices of science inquiry; however, students' ability to write explanations still remained underdeveloped. This finding supports other researchers' propositions that students need to be explicitly taught and provided with detailed guidelines on how to write explanations for others.

In summary, the current study investigated students' development of expository writing skills in science "naturally" while participating in an instructional intervention. Although the results indicated substantial gains in all four domains for 3rd graders and small to minimum gains for 4th graders, we propose that teachers should be invited and trained to use the rubric to score their students' writing samples so that the results of each ASILAS writing could be utilized to inform the teaching of science writing. In other words, ASILAS could serve as an assessment/evaluation tool for teachers who intend to promote students' expository writing skills during science experimentation. Additionally, based on the results of 4th-grade students, future research should consider including instructional topics that are related to students' preexisting knowledge and experiences and from the perspectives of multiple cultures. In essence, learning is enhanced when it occurs in contexts that are culturally, linguistically, and cognitively meaningful and relevant to the students.

This work is supported by the National Science Foundation, US Department of Education, and National Institute of Health (Grant No. REC-0089231). Any opinions, findings, conclusions, or recommendations expressed in this publication are those of the authors and do not necessarily reflect the position, policy, or endorsement of the funding agencies.

Targeting Enduring Understandings

1   What are the connections you see between the discussion in Ku, Yoo, and Garcia's chapter and the four elements of culturally-responsive instruction, as described by Villegas and Lucas?
2   How would you describe the "work" that language is doing in this multicultural context?

Deepening the Reflection

1   In writing about science, how is expository writing different from expressive writing? Why is it important to provide all students with opportunities to engage in expository writing about science? What difficulties might students face when they are asked to write in an expository style about scientific concepts and procedures?
2   How have you seen teachers support and scaffold all students, especially students from diverse backgrounds, in writing and thinking about science? What techniques are you learning in your program of preparation?
3   How is an assessment/evaluation system like the ASILAS similar to or differ-

ent from other types of assessment/evaluation typically used in the teaching of science? How could the implementation of an ongoing assessment/evaluation system like the ASILAS help improve the instruction of science and writing and promote equity?

Encouraging Engagement

1   Go into a multicultural science classroom and observe to what extent the teaching of writing is integrated into science instruction. Are all students' needs being met? What recommendations would you make to the teacher and, further, what kinds of statements would you offer in support of your recommendations?

2   Design a writing prompt that takes into account an existing science curriculum framework and an authentic science writing task. Now create a lesson plan around that framework-based science writing task that incorporates activities that will help all students be successful when writing in and about science.

References

American Association for the Advancement of Science (AAAS). (1993). *Benchmarks for science literacy.* New York: Oxford University Press.

Aschbacher, P. R., & Alonzo, A. C. (2004, April). *Using science notebooks to assess students' conceptual understanding.* Paper presented at the annual meeting of the American Educational Research Association, San Diego, CA.

Baxter, G. P., Bass, K. M., & Glaser, R. (2001). Notebook writing in three fifth-grade science classrooms. *The Elementary School Journal, 102,* 123–140.

California State English Language Arts Content Standards, Grade Three, Writing Applications. (2004a). Retrieved from www.cde.ca.gov/be/st/ss/enggrade3.asp.

California State English Language Arts Content Standards, Grade Four, Writing Applications. (2004b). Retrieved from www.cde.ca.gov/be/st/ss/enggrade4.asp.

Fellows, N. J. (1994). A window into thinking: Using student writing to understand conceptual change in science learning. *Journal of Research in Science Teaching, 31,* 985–1001.

Glynn, S. M., & Muth, K. D. (1994). Reading and writing to learn science: Achieving scientific literacy. *Journal of Research in Science Teaching, 31,* 1057–1073.

Keys, C. W. (1999). Revitalizing instruction in scientific genres: Connecting knowledge production with writing to learn in science. *Science Education, 83,* 115–130.

Klein, P. D. (1999). Reopening inquiry into cognitive processes in writing-to-learn. *Educational Psychology Review, 11,* 203–270.

Laidlaw, E. N., Skok, R. L., & McLaughlin, T. F. (1993). The effects of notetaking and self-questioning on quiz performance. *Science Education, 77,* 75–83.

Lang, H. G., & Albertini, J. A. (2001). Construction of meaning in the authentic science writing of deaf students. *Journal of Deaf Studies and Deaf Education, 6,* 258–284.

Mason, L., & Boscolo, P. (2000). Writing and conceptual change. What changes? *Instructional Science, 28,* 199–226.

National Research Council (NRC). (1996). *National science education standards: Observe, interact, change, learn.* Washington, DC: National Academy Press.

National Research Council (NRC). (2000). *Inquiry and the national science education standards: A guide for teaching and learning.* Washington, DC: National Academy Press.

Prain, V., & Hand, B. (1996). Writing for learning in secondary science: Rethinking practices. *Teaching and Teacher Education, 12,* 609–626.

Rivard, L. P. (1994). A review of writing to learn in science: Implications for practice and research. *Journal of Research in Science Teaching, 31,* 969–983.

Rivard, L. P., & Straw, S. B. (2000). The effect of talk and writing on learning science: An exploratory study. *Science Education, 84,* 566–593.

Yore, L. D. (2000). Enhancing science literacy for all students with embedded reading and writing-to-learn activities. *Journal of Deaf Studies and Deaf Education, 5,* 105–122.

4   The Writing on the Wall
The Daily Calendar as Science Practice

KimMarie Cole

Introduction

In this chapter, Cole documents the (de)evolution of a text practice as it is implemented over the course of a school year in a multicultural science classroom. Cole claims that text-based literacy routines, such as the practice of the Daily Calendar, serve to subtly, though powerfully, shape the ways that students think about science. Mr. Walsh's linguistic choices in the Daily Calendar practice, as Cole demonstrates, implicitly position students in particular ways with respect to their science learning and, ultimately, position the teacher and classroom with respect to the debate over traditional vs. reform-based science instruction.

As you read this chapter, think about the daily instructional routines of classroom activity you have experienced as a learner or as a teacher. What is their explicit instructional objective? What might be a more hidden, or implicit, instructional outcome? How does the supposed impact of that routine situate the practice with respect to traditional vs. reform-based notions of science learning as either an act of transmission or of knowledge construction? Also, as you are reading, think about the science teacher and how he might have benefited from more information about the connection between science learning and language practices. Classroom science teachers do, indeed, need mentoring and professional-development support on an ongoing basis to authentically integrate literacy in general, and writing, in particular into science teaching on a daily basis. If science teachers, like Mr. Walsh, are to become agents of change, they will need detailed and descriptive classroom examples of how to support science learning in culturally-responsive and connected ways for all learners. How then would you, if you were now this teacher's mentor, support him, given what you've learned, in recognizing how to best use the Daily Calendar writing to enhance students' science learning?

When it comes to describing what science is, there is no shortage of notions. Many authors, including those in this volume, have argued that reformed understandings of science and science instruction in schools (Lemke, 1990; Lee & Fradd, 1998; Lynch, 2000; AAAS, 2000) are essential in ensuring equitable science education for all students. The analytical perspective I bring to this chapter, Critical Discourse Analysis (CDA), requires that the researcher engage both with the local level of science understandings, in this case the texts students write in their classroom, as well as larger meanings about those texts that circulate within societal understandings of science. While I will say more about these methodological issues later, I would like to begin by outlining some of these societal and cultural understandings of science that circulate and influence how we understand what science is.

Most lay people hold similar and related ideas of the nature of science. These notions are rooted in our histories as students in science classrooms and are reinforced in our everyday lives whenever we see, for example, a reporter interview a scientist or engineer about Space Shuttle safety, hear radio reports of research that appears in prestigious medical journals about health, or flip through the pages of our favorite magazines that now contain big ads for medications with full pages of very fine print that details the benefits and dangers of the latest prescription drugs. This "normal" view of science is characterized by Mikulecky (2001) "as drawing on a model of Cartesian Reductionism and the machine metaphor" (p. 342). From this perspective, science is the effort to identify the distinct parts of the "machine" or to reduce a problem to its component parts and eliminate all doubt. O'Connor (1999) also describes science in this way, adding that this model of science

> privilege[s] the role of scientifically trained experts in organising the pursuit for knowledge, constructing the valid categories of knowledge, arbitrating about Truth and opinion, and furnishing reliable information inputs for policy…. The scientists are the sifters, discoverers and 'integrators' par excellence; they push back the frontiers of knowledge and, progressively, tame the uncertainties.
>
> (p. 673)

Within this typical framework, science classrooms are places for two important things to happen. First, and perhaps foremost, it is necessary to train students, as members of the general public, to accept these assurances and take the "known" at face value because it has been offered by experts who, through their efforts, have made sense of major mysteries. A second goal would be to find the few, capable students who could be initiated into the ranks of scientists and who could become these expert managers of uncertainty and complexity.

Lemke (1990), Gee (1996), and Kelly, Crawford, and Green (2001) all argue that "science" does not exist as this kind of monolithic presence but is instead a constellation of ideas about science that are created and recreated in classrooms through the ways that students and teachers read, write, and talk about science. In their book that examines how science explanations are built in classrooms,

Ogborn, Kress, Martins, and McGillicuddy (1996) observe that much of the work done in science classes is that of building the objects of study. Through language use, teachers and students take commonsense understandings and convert them into abstractions that can serve as the tools of theory building. Lemke (1990) offers a nine-point list of language features that characterizes language use in science. He suggests that science tends to be seen as universal, not because it is universal but because of the way language forms are used to present it. He cites that there is frequent use of technical terms and statements of fact, relatively little informal speech or writing, and few metaphors or sense of agency (p. 133). So, it is through the language used that science becomes what it is.

Many who argue for reforms of science practice in schools not only want to change the way that science is taught, but further wish to alter these commonsense notions that the public accepts about science. Their view is that changing the way that students experience science in the classroom will provoke long-range changes in public perceptions. Lynch (2000) argues that these "normal" ideas of science actually bear little resemblance to the work of scientists, which is grounded in deeply held personal beliefs and experiences and draws extensively on informal understandings to arrive at the ultimate abstractions that are the product of the scientific process. Using these kinds of findings and revelations, those active in science reform call for practices in schools which train students in the ways of forming questions and for talking through processes for arriving at conclusions and presenting these results to others, rather than simply accepting and presenting facts on demand (Lemke, 1990; Wells, 1993; AAAS, 2000; Lynch, 2000; Vosniadou, Ioannides, Dimitrakopoulou, & Papademetriou, 2001). These reforms would not only change the ways that students and teachers practice and understand science but would, presumably, provide a greater number of students access to the material and apprenticeships to scientific disciplines. Through these kinds of efforts, teachers take the lead in changing not only their own practices but sow the seeds of change for larger shifts in public understanding. Teachers who adopt this reform perspective believe classrooms have the potential to instigate social transformation.

These distinctions in the ways that we think about science teaching and learning are central to this chapter. In what follows, I will describe how one teacher and his students started their school year using a reform model for a short time, during an everyday literacy practice that became a focus of their science practice. After a few weeks, they shifted to another way of talking and thinking about science, retreating to a more traditional perspective in their writing about science. Through this work they did together, I show how the teacher reinforced the models of teaching and learning described above, but how he also positioned the students as certain kinds of learners with far-reaching consequences for the ways they practice science and think of themselves as potential scientists. In order to connect the dots, as it were, between literacy events and this idea of "positioning students," I will draw on a theoretical tool that describes how and why studying literacy practices like the Daily Calendar texts can indicate what kind of content is presented to the students and what it has to do with the kinds of scientists that students can imagine themselves becoming.

CDA and Modes of Address

Language is a tool that people use to structure, shape and act on their environments. This agency, or the ability to affect the course of one's life exists in dynamic tension with the histories that come before us. That is, at the same time that we use language to act on the world within our interactions, historical meanings of words, expressions, and ways of communicating also give shape to those interactions. In this way, we can think of language as a tool to be used to mediate relationships between people and ideas. This mediating force is often called power—not the brute force of weapons or violence—but a kind of power that often goes unnoticed in our day-to-day interactions with others. It is crucial to understand that words, phrases, and expressions have histories, and both shape and are shaped by our experiences with them.

CDA is a strand of research that takes power seriously, thinking of it always as a participant, although sometimes a silent one in most interactions. Both Pennycook (1994) and Gee (1996) draw on the work of Foucault to talk about the ways that power circulates in Discourses that resonate in societies and cultures and which help us interpret words and actions we encounter in the living of our lives. Discourses are these constellations of meanings or understandings that position us and allow us possible roles in the interactions. In the case of science curricula, there are at least the two competing Discourses circulating—the "normal" discourse of science and the "reform" discourse. They offer students, scientists and the public very different ways of being a part of the conversation about science and its practice. Luke (2002) points to an important methodological consideration in CDA for an "orchestrated and recursive analytic movement between text and context" (p. 100). That means that CDA must analyze a text and then resituate that analysis in a broader context in order to make complete sense of it. In this study, this means looking closely at a literacy event in a science classroom and showing how the text functions, then taking those ideas back out to our understandings of science more generally.

Ellsworth (1997) introduces a very useful tool for doing this kind of analysis in classroom settings. She brings the idea of "modes of address" from film and media studies to the field of education. Mode of address, as Ellsworth describes it, is a tool for examining "something that is 'in' a film's text, which then somehow acts on its imagined or real viewers, or both" (p. 23). She goes on to say that its effects are neither linear nor causal, but that by looking at how films are written, directed and put together, it is possible to see what a society believes to be "normal" or "expected" and how a film's features position its viewers in terms of those expectations of normalcy.

As Ellsworth describes her use of this theoretical tool, she refers to the ways that people can study the elements of film construction to see what is being created as a text and how it is meant to be interpreted. In the discipline of film, these elements include staging, lighting, camera angles, and shots, as well as editing techniques. They also include things like the script language and the director's choices about movement and the ways that people move through and around the sets. When she refers to the use of this concept in educational settings, she talks specifically about material aspects of classrooms like the architecture and structures of the physical place, as well as elements of the lesson design and delivery.

In this chapter, modes of address are used to examine how the shift in the tasks of the daily literacy event signal a shift in what might be considered "normal" for the class. There is not one single way of thinking about science in this classroom. Rather, there is one way that is presented initially that gives way to another perspective after a few weeks. This analysis shows how our idea of "normal" has to be similarly flexible and context-specific. In the course of the daily practice we can find the common "Discourses" of science at work, shaping and being shaped by the texts the teacher provides for the students to write. Through their writing, the students actually build understandings of science and their possible roles within those Discourses. These actions tell us what kind of science students are being socialized into through their use of everyday language. In this chapter, I document this process as it is played out in a literacy event that this class called "the Daily Calendar."

Research Methodology and Data Set

The data[1] for this chapter are a smaller slice of the larger Project on Academic Language Socialization (PALS) research project that took place at Jefferson High School[2] over a four-year time period, from the fall of 1996 until the spring of 2000. The study took place in the diverse, Midwestern urban context of Center City and its high school. This longitudinal ethnographic project focused on the ways that English Language Learners (ELLs) were socialized to academic language use in their mainstream content area classes. Our research team traveled to the school twice each week during the school year. Two members of the research team were in the classroom for each observation and video-taped the classrooms using a two camera set-up. One camera focused on and followed the teacher around the room. The other faced the students and remained stationary during each class period. One member of the field team took running field notes, which included anything that was written on the board, as well as the main activities that happened during the class period. The team compared impressions and wrote a summary and discussion of each class period following the observation. The two video tapes were combined using a process called Picture-in-Picture (PiP) into a single video, and detailed transcripts of the verbal interaction were generated from the video (see Zuengler, Ford, & Fassnacht, 1999 for a discussion of this process). The data for the analysis presented in this chapter were the field notes which included the Daily Calendar texts.

In the field notes, field team members wrote down the Daily Calendar texts as they were written. Whenever possible, the video also zoomed in on the board to capture the teacher's handwriting, the location of the text on the board, and any additional marks or notes that were written alongside them. In the analysis that follows, I consider different linguistic elements that are present in these texts and describe the ways that these language forms typically function as the students complete the work assigned to them. Specific points of analysis include the sentence form the Daily Calendar utterances take, i.e., whether they are questions or commands. I also look quite closely at two of the elements Lemke (1990) holds up as critical in the creation of scientific languages: abstract noun forms and the presence of agency in the work. This close linguistic analysis allows us access to the modes of address that Ellsworth (1997) highlights as critical to understanding classroom curricula.

The Classroom Teacher—Mr. Walsh

From the ethnographic interviews and ongoing observation, the research team learned that Mr. Walsh began his teaching career in 1997, when he was hired as a science teacher at Jefferson High School. He had completed his degree just months before at a regional college outside Center City and looked forward to the opportunity to teach in the urban school. He found out he would be teaching at Jefferson High less than a week before the semester began and so spent much of his time in the early weeks of the school year trying to organize himself and his materials, building his curriculum and familiarizing himself with the school and students. On the first day of class, he put pieces of paper on all the student desks with different quotes about wisdom, research, and knowledge from different societies and cultures around the world. Originally, he had hoped to be able to develop his curriculum around critical thinking skills and multicultural education, but he soon felt that he had all he could do to prepare for his classes each day. Clearly, these actions suggested that Mr. Walsh was prepared to be an agent of change and a teacher who adopted the reform view of science in his classroom.

Eager to become part of a network of professional teachers, Mr. Walsh took advantage of the district's mentoring program, which assigned an experienced teacher to work with new teachers in the schools. Each week Mr. Walsh and his mentor met and talked about teaching strategies and possible materials to use for each unit. On occasion the mentor visited the class to observe his teaching. Mr. Walsh expressed appreciation for his mentor's work, finding materials and making room decorations to accompany the teaching units he prepared.

The origins of the Daily Calendar, the literacy event I study in this chapter, grew out of a recommendation from his mentor. Mr. Walsh noticed that he needed to provide a focus for the students at the beginning of class while he took care of the administrative details that are often a part of the high school setting. His mentor suggested that he give the students something to do that would help them settle in and focus on their science work during these early moments of the class. This would provide him with the quiet time he needed to complete the roll-taking he was required to do at the beginning of every class hour.

The Students

This roll-taking activity was a challenge for Mr. Walsh initially. During the video-taping of early class sessions, his frustration was obvious. The research team learned that part of the problem for Mr. Walsh stemmed from the fact that there were thirty-eight students on the official roster of his Integrated Science class. He used a computer program to make a seating chart of the room and required that students remain in their assigned seats. There was a great deal of student mobility in the school. Of the students on the roster, some joined once the term was in progress. Others were there at the beginning of the year but stopped coming for a period of time, only to reappear later in the year. On any given day there were regularly between seventeen and twenty-five students in the class, out of the total of thirty-eight he had to account for.

Like Mr. Walsh, these students were novices to Jefferson High School and to their assigned learning community within the school—the Allied Health Family. Many of them, however, had attended school in the district and were from the neighborhood. The African-American students in the class, however, all lived in another part of the city and arrived at Jefferson by bus each day.

This class, like Jefferson High in general, had a high degree of ethnic and linguistic diversity. Using the school demographic categories, this group of students included six Whites, nine African Americans, eighteen Hispanics, and three Palestinian students. In end of semester questionnaires collected by the research team, many of the students in the class also described themselves as bilingual, identifying their primary languages as English, Spanish, Arabic and "street talk" or "ghetto." When asked to describe what they meant by street talk, the students all described features referred to in scholarly literature as African-American Vernacular English (Lippi-Green, 1997). English was the language of instruction of the class, and there was no evidence that Mr. Walsh spoke any other language. Often during class periods, I observed students speaking to one another in other languages. The students ranged in age from fifteen to eighteen, and some of them had been out of school for either brief or extended periods of time. Of the students, twenty-eight were female; eleven were male.

This diverse group of students and Mr. Walsh came together each day for this 5th period science class. Through the wide halls, they moved among their peers during the five-minute passing period, and entered Mr. Walsh's classroom. New to the school and to the expectations of this teacher and of their new "disciplinary" schedules, together this group formed a unit. The teacher was responsible for organizing a curriculum and for training the students to learn science. Ultimately, the Daily Calendar became a significant tool for both of those tasks.

The Daily Calendar

Elsewhere, I illustrate in great detail how the Daily Calendar came to have the meaning it does in this class and how it functions as a physical and conversational reference point for the teacher and students for almost every activity they do in class (Cole, 2002). Here I describe the origins of the Daily Calendar and how it came to function for the class. The Daily Calendar is a literacy event that crucially shaped the teaching and learning of science in this Integrated Science Classroom. It figured prominently in the ways that the students thought about and talked about what they did in class. For example, on an end of year questionnaire, one student, Lashanda, answered that a typical day in the Integrated

Science Class was to "Copy calendar. If talking, get name written on board. Follow calendar. At the end of class, he [Mr. Walsh] picks one person to check calendar." From the beginning moments when students copy it from the board until the end of class when it is checked, the Daily Calendar is recognized, mentioned, and responded to as important by the teacher and students. Like "story time" or "show and tell" in earlier grades, and lab work or tests for other classes, the Daily Calendar serves as a point of focus and becomes one of the primary ways for members of the class to structure the lesson and their participation in it. As Lashanda's comment shows, the Daily Calendar goes beyond a single event like story time though; it actually serves to frame the entire class hour and becomes a resource to which the students refer. Although it is not the only science discourse that the students and teacher participate in together, the role the Daily Calendar plays in organizing the whole class session provides an excellent way to investigate which "Discourses" of science were present in this classroom.

In this way, the words on the chalkboard each day figure prominently in what the students do as they write the Daily Calendar as part of the science class. The Daily Calendar texts provide a record of what materials the students cover over the course of the year and how they are asked to engage with those materials. Since the Daily Calendar time was used as a way to "jump start" class while the teacher was otherwise occupied taking roll, the texts of the calendar were the students' introduction to each day's material, giving important information about how they were to organize themselves and their thinking about science. The texts of the Daily Calendar in this science class helped to create the constructs, materials, and ways of working that would become "science" for this particular "discourse village" (Mercer, 1995).

Within this classroom community, the Daily Calendar introduced the students to the material they would encounter during the class period. It was usually made up of a short list of two or three items, which students were instructed to copy into their notebooks. Then, depending on the content of the Daily Calendar, the students began answering questions, defining terms, or gathering their thoughts about the topic of the day.

Composing or Copying—The Role of Questions in the Daily Calendar

Once the existence and relevance of the Daily Calendar was established within the classroom community, the texts themselves become the focus of study. The first type of analysis considers what the students are asked to do, and therefore, what kind of literacy and what kind of thinking the Daily Calendar texts ask the students to engage in. What students were asked to do shifted markedly between the first and second grading periods of the year. The type of text changed as the Daily Calendar became a routine in the integrated science class. In linguistic terms, the sentence types are the place to begin this discussion.

Each day the students received verbal instructions that usually included copying the Daily Calendar text itself and getting started on the assignment, or answering the question on the board. In this section, I provide comparison of the Daily Calendar texts from the first two marking periods[3] of the class year. I chose these marking periods because they powerfully illustrate some differences in the conceptions of science that were presented to the students.

In the early days, each Daily Calendar included several points that the students should copy in their notebooks, and each additionally posed a question that the students are asked to respond to. Table 4.1 includes the texts from the Daily Calendars during this first marking period of the class and shows the kinds of questions that the students were assigned. These questions are indicators about the type of science that is being brought into the classroom at this point. Notice that there are a number of open-ended questions, all of which allow the students to expand on their own ideas, relate previous experiences, and tie the material of the classroom into their existing understanding of the concepts they are encountering.

For all of these Daily Calendar texts, the students were first asked to respond to some kind of prompt and write their own texts after they had copied down the whole Daily Calendar. In the case of the journal prompts for September 29, the students had completed a lab, in a previous class period, about the tools of science and ways of measuring, and they had also drawn a concept map of the lab, where they were asked to link kinds of measurements to the tools that would give them that information. This, and other similar journal assignments, provides students with an opportunity to apply their understanding from the lab and concept mapping activity and create a new representation of their knowledge. Similarly, on October 1 students were asked to reflect on previous class activities. The journal invoked material covered and raised a previously shared topic that could be used in the class by those participants who attended class the day it was raised.

The third calendar, on October 6, indicated that students should "write what fall means to you." On this day, there are two questions on the board, but Mr. Walsh emphasizes that the students should write about the first question, and not the second. The activity did not draw on previous class material as the other two did, but instead worked as an advanced organizer of sorts, or a schema-building activity, for a leaf collection and analysis project the students would

Table 4.1 Daily calendar texts for marking period 1, 1997

Date	Item 1	Item 2	Item 3
9/29/97	Why is it important to measure accurately in science?	B	—
10/01/97	Begin journals in notebooks.	Hurricanes!	—
10/06/97	What does the season of fall remind you of?	Why do we have different seasons?	Leaves are falling project.
10/08/97	What is the difference between simple and compound leaves?	Draw and label an example of each.	—

conduct in the days ahead. In this writing assignment, students had an opportunity to mention their feelings or observations about a season. For some of them, this question may seem strange in a science class, since the question ostensibly had nothing to do with the science of seasons. The fourth calendar of this marking period, on October 8, worked similar to the one of September 29. The students were posed a question asking them to draw on material covered previously in class. In this case, they needed to refer to the packets of information they gathered about leaves, and either search for or recall the difference between these two types of leaves. Unlike the earlier journal entries which allowed students to generate their own ideas, this journal required the students to provide specific content information in the journal they write.

On the face of these Daily Calendar texts, it appears that the science that is being introduced in the classroom is of the reform variety (Lemke, 1990; Lynch, 2000) in that students are being asked to locate important issues and connect them to their own experiences. Additionally, they provide a teacher, who reads these texts, with an indication of the students' concept development, which is an effective way to know how to scaffold material in science classrooms (Vosniadou et al., 2001). The writing tasks that lead off the Daily Calendars in at the beginning of Mr. Walsh's class are elements that Carrasquillo and Rodriguez (1996) have suggested are helpful for language minority students in mainstream classrooms in the way these activities allow students to activate prior knowledge or ask about relevant vocabulary items, as well as to begin making links, in the target language, between important concepts. Lynch (2000) in her review of science reforms agrees, not only that these tasks are good for language minority students, but that they are good for all students in the classroom.

Lemke (1990) and Hanrahan (2006) have argued that students are alienated from science when there is a rupture between common sense and personal experience on the one hand and scientific explanation on the other. They would likely agree that finding ways either to lessen this gap or to build bridges over it would be a useful strategy in teaching science. Whether the students actually do the reflecting, expressing, recalling, or comparing asked of them is not certain, but from a modes of address perspective, the design of the questions positions the students as people who can do those activities and importantly, also frames science as a discipline that involves this kind of thinking.

Through their writing during this time period, students are asked to use their literacy as a means of developing ideas and relating concepts. Writing is a form of composing, of creating, and of representing knowledge. The role for writing is a crucial one to keep in mind during the discussion that follows.

At the end of the first academic marking period, which took place after the class on October 8 illustrated above, the Daily Calendar texts change in a striking way. Table 4.2 presents the texts of the Daily Calendar for the next six-week marking period. I present this comparison of the two marking periods to show a specific difference in the types of texts the students encountered and to discuss the difference in the modes of address (Ellsworth, 1997) these texts suggest. Because the mode of address that is established in the second marking period is carried throughout the rest of the year, I simply present the smaller sample here.

Table 4.2 Daily calendar texts for marking period 2, 1997

Date	Item 1	Item 2	Item 3
10/13/97	Check the seating chart	Final project day, projects due 10/14/97	—
10/15/97	Falling leaves quiz game	—	—
10/20/97	Earthquake video	B	—
10/22/97	Define terms in your notes	Finish earthquake video	Introduction to earthquake search lab
10/27/97	Review Earthquake search lab	Vee diagram (what is the focus question of lab?) →	Plotting earthquakes and volcanoes
10/29/97	Continue plotting volcanoes	B	—
11/03/97	In your notebook write the meaning of these words: magma, lava	Plot earthquakes and volcanoes	Observation, analysis, conclusion questions on J50
11/05/97	Read the lab intro and explain the theory of hot spots in your notes	Hot spots demonstration	Begin lab
11/12/97	Examine the picture on the overhead and write observations in your notes	Phases of the moon demonstration	—
11/17/97	Pick up a copy of the data and the survey	Begin survey immediately	Introduction to the TI-83[a]
11/19/97	Get copy of decision form and begin reading the directions	Individual Score Total Group Score	Data table

Note
a The TI-83, Texas Instruments model 83, is a graphing calculator that students were shown how to use by a district-wide science and math resource person.

What should be apparent from even a cursory glance at these items is the change in the language and directions of the Daily Calendar texts. Whereas the early texts all begin with questions, none of the Daily Calendars start that way in the second marking period. There is one lone question among these texts, and that one is marked parenthetically as part of another item, October 27. For some reason, then, the way that the Daily Calendar is structured and how it organizes the construction of science has shifted. And that shift means not only a difference in the form of the Daily Calendar but also in what students do with it and how they interact with it as students of science.

The texts of the Daily Calendar indicate the direction of that shift. While the early Daily Calendars contained questions and a few phrases, there were almost no imperatives, or command forms, contained within them. This second set of Daily Calendars is full of administrative comments. Both items from the October 13 Daily Calendar convey instructions, real and implied, to the students. They must make sure they are in the correct seats for the first item, and that they have finished or nearly finished the leaf collection project they had begun several class periods before in order to submit it by the deadline. Similarly, the Daily

Calendar entry for November 17 instructs students to "pick up a copy of survey," "begin survey," and have an "introduction to TI-83 [a graphing calculator]." It is quite clear what tasks the students will complete and in what order.

In this way, the early Daily Calendar and its journals that allow students to compose and construct their understanding of the material and their own actions in the class give way to a Daily Calendar which structures the class hour, telling the students what they will be doing during their time in class. No longer are they writing their own texts in response to prompts; rather, they are copying down the agenda for the day. The Daily Calendar on the board and in the students' notebooks becomes primarily a record of activities and a schedule. The literacy demands of this task are quite different from those of composing a response from personal experience. While copying words from the board does require decoding of the items, it does not require an advanced level of literacy. On the surface, these Daily Calendars have a primary mode of address which positions science and the students quite differently than the previous texts. Instead of presenting science as a series of reflective questions and the students as people who respond to those questions, these entries are directed at recorders of ordered information. The students' notebooks are sites of copied words and phrases, a different, perhaps more mechanical, literacy than that of earlier journal entries. Any student in the class who could see the board, or who could look at another student's Daily Calendar could complete the task. The underlying sense of what it means to do science then is moving from one that is personal to one that places a series of pre-established procedures to be copied, not composed or constructed in interaction with others.

Verbs in the Daily Calendar: Telling Students How to Think

The above discussion focused on some of the differences in the sentence types used in the Daily Calendar and how they worked to position the students as a particular kind of young scientist. Now, looking even more closely at the grammatical forms that appeared in the Daily Calendars of the second marking period provides insight into the modes of address at work in the room. These verb forms allow us an idea of the agent positions that the Daily Calendars create for students. One way to consider these verb forms is to look at the National Science Standards (NCSAS, 1996) glossary of key terms of science. It separates words into two sections: the themes of science, all of which are nouns, and the "terms unique to science," a list of words, primarily verbs that have field specific definitions. One could expect that this list would figure prominently in the work that is done in science classrooms since these verbs are, in some respects, the stuff that science is made of.

Starting with a more general overview of the phrases of the second marking period, of the twenty-five different phrases or items the students write in the Daily Calendar texts, thirteen of them contain imperative verbs, some with multiple examples, for a total of sixteen command forms. The students are thus verbally directed to write the Calendar, and what they write are tasks and commands, things for them to do. They are on the receiving end of these actions rather than on the constructing end. They are recipients of others' decisions about who they will be as they develop scientific expertise and what they will do as they develop this competence.

Some of these instructions are related to the administrative aspects of the class and help regulate where the students will be and what they will be doing. They are to "check the seating chart," "pick up a copy..." etc. Several more directives emphasize the process of the activities that the students are doing. They see on the board and write down the information: *finish, continue,* and *begin* (three instances). In these cases, the attention is placed on where students should be in any given task. Rather than indicating the type of reading the students should do, or how they should review a lab, or what they should be attending to as they watch the video or plot the points, the emphasis is on the act of doing itself, whether they are beginning or ending a task. In these cases, the verb forms serve the function of supporting the curricular coherence that the Daily Calendars begin to establish and reinforce in the second marking period. This kind of process language was absent in the first marking period but is quite prominent in this second marking period.

For example, on October 29 students are instructed to *continue plotting volcanoes,* there is a frame of reference (continue) in place that this activity is one that had been started and that should be familiar to them. In the next class period, November 3, while the students are indeed continuing the plotting (and perhaps by this time plodding) work, the Daily Calendar does not include any information about the process. Instead their text reads *plot earthquakes and volcanoes.* Certainly the link to the previous activity is present in the repeated words *plot* and *volcanoes,* but there is no evidence about where they are in the process, which might be crucial given the large numbers of absences and the amount of mobility within this student population.

This becomes a potentially significant omission because when they have information about the process, students are also able to place themselves and their progress vis à vis these instructions. These uses of imperatives correspond with Mercer's (1995) notion of "educational" Discourse. For him, this kind of talk in classrooms serves to organize students and is a necessary part of the socialization to school language and knowing how to carry on with a group of people in the institutional setting of the classroom (p. 80). These uses of organizational language are not necessarily particular to science although some patterns of verb usage might be.

Other imperatives that appear in the Daily Calendar texts function somewhat differently. They tell the students what science tasks they should engage in during the class hour, how they should approach them and what kind of science students they should be. Referring back to Table 4.2, the imperative verbs that frame the science tasks for the students as they write their Daily Calendar are: *define, review, write, plot, read, explain, examine.* Of these words, none figures on the National Standards list as a term uniquely used in science. Yet, it may be useful to examine the ways that these words position students' writing tasks.

On November 5 and 12 for example, the students are greeted with matched pairs of imperatives by the Daily Calendar: *read* and *explain,* and *examine* and *write*, respectively. These two-part instructions first point the students' attention to a particular place, the lab introduction for the reading task, and the overhead transparency for the observation. These first items focus their attention on an external source of knowledge to form their understanding. Importantly, the second part of the command tells the students what kind of work they should do after they read and examine. They need to "explain the theory" and "write observations." During the process of copying down the Daily Calendar, the students are being prepared for the activities to come and are also previewing the kind of thinking they will do later during the class. These two tasks, explaining and writing, challenge the students to go beyond their first understanding of the materials and do something with it. They both ask the students to take information and transform it, in this case in their notes.

These entries from 11/5 and 11/12 asked students to transform material in some ways. Three other Daily Calendar texts during this time period ask the students to simply record factual data coming from other sources. The students encounter the verbs "define," on October 22 and "write the meaning" and "plot" on November 3. In these cases, the students are asked to take information from other texts, namely their text book or a work sheet and record it in another form. For the plotting task, the students must interpret data charts and put a mark, which corresponds to the longitude and latitude information they have, on a map. All the information they need is given to them, so their effort is one of classification rather than creation.

Similarly, the definition tasks of "define" and "write the meaning" ask the students to find words in a glossary and then put them in their notes. They are not asked to interpret what they read, rephrase the words, find synonyms or use the terms in any way. While it may be Mr. Walsh's hope that the students understand the words and are able to use them, it is unlikely that the act of copying the definitions from the glossary to their notebooks will give the students enough opportunity to practice these new terms or develop their own understandings of the terms. Instead, it is a time when the students learn that the meanings of words in science class are found in a dictionary, an outside source. Like the use of the video or the readings and observations described above, asking students to refer to published definitions reinforces the need for precise information and a reliance on someone else's understanding of material.

As this discussion shows, there are a number of imperative commands in these Daily Calendar texts. Certainly, command forms tell people what to do, and perhaps to some degree how to do it. My examination of the verbs of the Daily Calendar texts suggests that the students are being asked to act and think in ways that echo Mercer's (1995) separation of classroom discourse into "educational" and "educated" (p. 80). For him, there is a dichotomy between language used for the sake of organizing people and language used to teach content and its constructs. He acknowledges the somewhat false dichotomy between the categories but demonstrates how teachers and students can participate in ways that organize them in these two distinct ways. Sometimes the primary focus of the exchanges is to move people and materials around in the room, what he calls the "educational" forms of talk. At other times, the "educated" discourse overtly models the cognitive work of processing content information to make it accessible to students or to help them move from one type of cognition to another. The verbs in the Daily Calendars in Mr. Walsh's class tend to emphasize the "educational discourse." They impart information about what the students should do, or how they should be organized. And for the most part, these actions do not require complex cognitive work, with the few exceptions I highlighted earlier. Instead, they tend to focus more on the process of doing work than on the work itself. A simple check of whether the student has completed the task would serve as an evaluation. There would be no need to check the quality or quantity of the work. Thus, there are very few examples of verbs that emphasize the specific science content or "educated" language.

Ultimately, the students are not asked to do many things that would count as "science" outside their own classroom community nor do these verbs reflect a scientific register. Even in terms of the cognitive tasks the students engage in, the emphasis of the verbs here is on the lower order rather than higher order thinking and knowledge construction. That is, the verbs position students to record and describe more often than evaluate, analyze, or synthesize, which are considered more cognitively complex tasks (NCSAS, 1996).

Nouns in the Daily Calendar: Creating Density and Abstraction

The verbs in the Daily Calendar texts do not tend to emphasize a scientific register in the actions, or thinking, they require of the students. Their agency as young scientists is limited. This finding is not entirely surprising given arguments made by Lemke (1990) and Gee (1996) that one of the reasons science is so inaccessible to many is that it relies on little-understood facts about abstract concepts and processes that many people do not often consider. According to these authors, if actual science practice is in the details, it will likely be found in the nouns of the Daily Calendar texts rather than the verbs. So, while there are reasons to be concerned about the verbs that appear in the Daily Calendar texts of the second marking period, before jumping to conclusions about the modes of address, it is important to take a closer look at the companion to the verb forms of the Daily Calendars—we have to look at the nouns. Bolinger (1980) shows that nominalization often reduces a sense of agency as well and renders the resulting noun forms as opaque. That is, by turning a verb into a noun, the reader's common sense understanding has to undergo a shift to something that is less concrete. For example the noun form distribution takes a concrete action "to distribute" and locates the force of the word in the process rather than the action of its content. In this way, agency can be displaced as well because with the verb forms, it is necessary to include who distributes what to whom. In the noun form distribution, that agency may not be needed to convey the ideas of process.

Therefore, the noun phrases in the Daily Calendar are important because they figure prominently in the Daily Calendar texts, and they seem to work differently than the imperative verbs in the information they convey and how they position the students vis à vis that information. A total of eleven of the twenty-five Daily

Calendar items during the second mark period are noun phrases, or about half of the items.

In some cases, the nouns are quite informative and transparent, such as the aforementioned example on October 13, which announced "final project day," followed by the deadline for the work. On one occasion, October 15, the "falling leave quiz game," which looks like a simple object actually includes specific information about the actions the students will take during the class period. In this case, they will participate, perhaps actively, in a game that will establish eventual winners. From the words that make up this phrase, we also know that this "game" is about the leaf project they have done for several class days and that in some ways, the game will either serve as a quiz or perhaps as a review for a quiz they will do in the near future. The students learn that in some way, this game-like activity will also be an assessment tool, presumably to evaluate their knowledge or understanding of the material they have studied. These four little words then provide a tremendous amount of information about what will be happening in class.

Another noun phrase entry in the Daily Calendar, however, provides much less certain information about what the entity is, let alone what the students will do during this time in class. On November 19, the third item in the calendar says "Data Table." The students copy this term from the board to their notes, but they may not have prior knowledge for this to be a meaningful term. This term does not necessarily come with the same sense of activity that accompanies the word "game" for students in a beginning science course. They may not yet have developed the scientific connotation of "data" as synonymous with scores or results in this context nor a definition of "table" as a means of presenting such information in an organized fashion. This entry is quite "scientific" or mathematic, but it stands alone with no contextual information to ensure its interpretation. Unlike the "Falling leaves quiz game" which draws on familiar concepts, so students can likely interpret what they will be doing, this example of "data table" is precisely what Bolinger (1980) suggests makes scientific language so difficult to interpret. Also unlike my earlier example of "distribution" where a reader could work back to the concrete action "to distribute," this "data table" does not provide a novice in the field easy or direct access to the actions or materials they will need to interpret this item. Students will likely resort to copying the term and waiting for a demonstration or explanation in order to understand, not only what the item is but what they are meant to do with it. If the content is loaded in the nouns, and the nouns are not accessible, the content is likely not accessible either.

In most of the other cases of noun phrase items in the Daily Calendar texts, their degree of explicitness falls somewhere between these types of two examples. The students are presented on October 15 with "Earthquake video," on October 27 with "Vee diagram"[4] and on November 3 (a day Mr. Walsh is absent) with "observation, analysis, conclusion questions." For the students, each of these three nouns suggests some information about what they would do: in the case of the video, they would sit and watch; for them "Vee diagram" meant that they would draw and complete this graphic organizer; and in the third case, they would perform a common student activity by responding to text questions.

The November 3 example, "observation, analysis, conclusion questions," is particularly interesting because there is no command about how the students will answer the questions. It is not clear if they should answer them in their notebooks for themselves, like a journal entry, or if the answers will be submitted to the teacher. The noun "question" is the head, or most important word of this Daily Calendar item rather than the actions of observing, analyzing or concluding, all significant items in the list of verbs from the National Science Standards glossary (NCSAS, 1996). Instead, the weight of the task is placed on the noun, the questions in the textbooks. The important cognitive tasks that would train or guide students to recognize their actions as a particular way of doing science are embedded in nouns, which are acting as adjectives. The really important stuff is reduced in this phrase to a description of the questions. This example truly supports the finding that science in this classroom relies on nouns to deliver content more than verbs.

Two other noun phrases appear several times in the Daily Calendar texts—introduction and demonstration. On October 22, the students receive an introduction to a laboratory activity they will spend several days completing, and on November 17, the Daily Calendar text informs them that there will be an introduction, this time abbreviated to "intro" to something called a TI-83. The demonstrations students observed figured into the Daily Calendars on November 5 and 12. In both cases, new concepts are introduced on these days, and valuable information about these concepts is presented to the students through the model of a demonstration. In these cases, the teacher uses props to illustrate the concepts before giving students a lab task to complete. As with other words and phrases, the concepts of introduction and demonstration must be locally brought to life in order to have meaning within a given classroom context. In this classroom, when students encounter the words "demonstration" or "introduction" in the Daily Calendar text, initially it is just another word to copy. With repeated use and the subsequent activities that follow during a class period, the students are socialized to both the definition of these words and what they mean in terms of their action and participation. In the same way that the phrase *Daily Calendar* indexes an important practice for the members of this classroom community, some of the words that appear in the Daily Calendar texts also index sets of actions. Both "demonstration" and "introduction" in this class refer to an activity where the students will watch someone else perform a science activity. They will subsequently try to reproduce the findings in a lab setting. The questions they are given as well as the results they should obtain are known at the beginning in the science they practice in this classroom.

Using this same division between "educational" and "educated" discourse (Mercer, 1995) to discuss the way that science is made "normal" in this science class, the nouns tend to be where the "educated" discourse is put into the curriculum. In these noun phrases much of the subject matter is conveyed. A partial explanation for why the content may appear in the noun forms may be found in Ogborn and colleagues (1996). Their study of science classrooms concluded that much of what happens there involves explanations. For those explanations to be effective the teacher and students must have access to the entities of explanations, the conceptual building blocks that help students make connections (p. 52). While they do not suggest that nouns are the only building blocks, they do say:

> Scientific texts are well known for their high concentration of events and processes presented as if they were things.... Their presence is not due to the barbarous linguistic habits of scientists. They exist in texts and talk as entities because they exist in the thinking of scientists as entities. They are, as we said before *things* with which to think.
>
> (p. 51, emphasis added)

So, an important aspect of work in science class may well be the development of these "things." In high school science, the entities that get built and established may differ conceptually from those of practicing scientists, but the process of building and recognizing entities may be an early apprenticeship to science fields. This need to have the "things" available can also account for the definition tasks that appear in the Daily Calendar. Defining is perhaps one of the first steps toward creating the "things" that ultimately lead to building explanations. When Mr. Walsh writes "analysis, evaluation, and conclusion questions," or "falling leaves quiz game," this use of noun phrases may be a subtle, if unconscious choice. Through these phrases the students are introduced to some of the linguistic and cognitive practices of scientists. The Vee diagram is a tool for explaining the scientific method, and its very name on the board creates an object that reflects a set of actions and ways of thinking the students must engage in to successfully complete the task.

While the work of Ogborn and colleagues (1996) suggests that the heavy emphasis on noun forms is "natural" in the practice of science, allowing us to understand that its use in a classroom might well be socializing students to the actual discourses of science, Lemke (1990) offers a different perspective on that same phenomenon. He argues that this use of nouns, specifically abstract nouns has the effect of presenting science as an objective fact rather than a "human social activity, an effort to make sense of the world," which, he argues, is how actual scientists talk about their work (p. 131). In this way, science texts, and we can include the Daily Calendar as one kind of science text, construct and uphold the "mystique of science" (p. 129) as inaccessible except to a few talented individuals who can interpret the abstractions and find the actions within them. He claims that students find it difficult to engage with material or to find meaning in it unless they are presented explicit guidance in interpreting these abstract forms. Like Lemke, Lynch (2000) expresses concern with a science curriculum that relies heavily on abstractions to deliver content. Her investigation of equity in science reform confirms reports that the linguistic and conceptual abstraction prevalent in science texts has tended to favor students from upper and middle class, Anglo backgrounds, leaving language and ethnic minority students to confront a subject matter that has little to do with them, their history or their knowledge unless special efforts are made to unpack the abstractions and highlight how they function. Given that most of the students in Mr. Walsh's class fit the demographic profile Lynch describes, the Daily Calendar texts may indeed reflect and maintain a perspective on science that constructs students as "outside."

Discussion and Further Implications

At the beginning of this chapter, I argued that it was relevant to examine the Daily Calendar texts as texts to see how science is being presented to the students and what their mode of address is (Ellsworth, 1997). The results of this examination do not lend themselves to a single, straightforward interpretation. In sum, the early texts of the Daily Calendar during the first marking period, with their emphasis on journals, seem prepared to socialize students to a discipline of science where they should draw on previous material, make connections between previous material and current activities, actively express themselves in writing and evaluating their life experiences and observations, highlighting a science of social activity and inclusivity. The words of those Daily Calendar texts prompt reflection and action on the part of the students and ask them to fill their Daily Calendar time generating and using knowledge and concepts.

The emphasis shifted for the second marking period and the rest of the year. Students were no longer creators of knowledge but rather copiers of information. The source of their information went from being their own experiences to relying on objective external experts along the lines of Lemke's (1990) observations about how science becomes established as a truth rather than a practice. Through their writing of the Daily Calendars, the students in Mr. Walsh's class are being socialized to these notions of science quite subtly. Rather than simply saying that the field of science is inaccessible, this analysis shows that there are ways that "inaccessible" language is used to construct this fairly simple, yet important classroom activity, the Daily Calendar. As I described above, the combination of verbs with their emphasis on process and controlling behavior and nouns that create abstract entities heavily loaded with content lead to a time of copying and following rather than constructing. The science focus is on actions and definitions generated by others and brought into the classroom to be transferred to notebooks. While the students may have a better sense of the organization or relationship of the materials, those decisions are made by someone else and presented to them as final rather than in process.

This shift in the Daily Calendar arrived at an administrative boundary of the school year as the students and teacher completed the first academic marking period. In their local context, the changes also happened at the same time that Mr. Walsh introduced a different purpose for the Daily Calendar. He began using it to help signal the transition from the administrative portion and record-keeping portion of the class to the content-oriented time. For him, the Daily Calendar allowed him a physical place to point at the board when he wanted to shift the focus on class. It became a contextualization cue (Gumperz, 1982), or a place for all participants to focus their attention for a brief moment. In this way, the Daily Calendar itself had multiple functions to fill, and the transition of the texts may have tied to this shift. At this point, Mr. Walsh also began asking students to check the Calendars at the end of class rather than doing that work himself. For those reasons, perhaps it was more expedient to have Daily Calendar texts that could be checked based on completeness rather than content.

It could be that the shift also offers students more guidance in terms of the focus of what they will be doing with the increase in the use of "educational" discourse. Another possibility is that Mr. Walsh realized that his students needed more scaffolding than he had anticipated initially. It may be that he felt the need to provide more guidance and structure through the Daily Calendar than he had in the early days. This idea would resonate with Delpit's (1995) description of the difference between two reading programs, one that was considered progressive, the other behaviorist. In the behaviorist program, there was a tremendous amount of explicit instruction that helped students with little background information build the schema needed to learn to read. In the progressive model, a fair amount of background knowledge was presumed. Perhaps Mr. Walsh's readings of the initial calendars gave him this feedback and led him to surmise that students were not yet ready to compose their own versions of science.

Another possibility is that Mr. Walsh was fighting the dominant "Discourse" during the early days, and that the struggle was difficult to maintain. There are significant obstacles for teachers who would become agents for change. There are institutional and bureaucratic pressures that Mr. Walsh faced, such as the need to take roll and have the computer print out available each class period. There were significant time pressures and lack of support for ongoing collaboration. Although Mr. Walsh had a mentor, she was working with a number of new teachers throughout the district and could not be available to consult with him on a regular basis. The early Daily Calendar texts, with their questions and opportunities for connections came during the time that she was working most closely with Mr. Walsh.

You'll notice that the above comments are all speculative. These are my researcher's supposings, thought through and written down long after I left Mr. Walsh's classroom. The speculations are grounded in what I observed and recorded, but the reader does not see Mr. Walsh's own response to these ideas. A researcher with the luxury of time, hindsight, and access to theoretical tools is sometimes able to see patterns that are not obvious in the day-to-day interactions of the classroom. And this common research practice of gathering data and making sense of it after the fact offers another glimpse at ways that transformative science practices could emerge and be maintained in classrooms like Mr. Walsh's, where the teacher allows researchers access to the unfolding activities. The outcome of this project speaks volumes for the rich potential of research projects that are designed for teachers and researchers to investigate together the literacy practices of the classroom and the meanings that they may convey to students.

It is not possible to know the precise reason for the change, nor is it productive to point fingers at the teacher, the students, the administrators or the researchers for a failing. In the perfect world of 20/20 hindsight, Mr. Walsh could be encouraged and supported to continue the questions, the writing, and the providing of opportunities for the students to connect the content to their lives even as he offered them a greater structure within the Daily Calendar practice. With more extensive mentoring, active engagement with researchers who were working with him or even a greater preparation in the role of writing across the curriculum, perhaps those steps would have been possible.

This research illustrates powerfully how content in classrooms does not exist in isolation or as an objective set of information. Rather, participants in classrooms construct a set of social practices, which make their interactions meaningful in a local context and use those social practices as well as their interactions with one another to develop understanding about the subject matter and give it meaning locally. Then, those local meanings have to be considered again in the light of the dominant discourses or typical practices of the field. In this case, the Daily Calendar in Mr. Walsh's classroom was talked and written into being through a series of interactions. The actual texts themselves were the focus of this chapter and were used as a means to illustrate the ways that language is used in these texts contributes to both the social practice itself and to the ways that the students are socialized to the subject matter. Importantly, this socialization is not just to what the students understand science to be, but also how they are presented with ways of thinking about and responding to the material. In this case, the content of the texts moves to greater levels of abstraction, and the students are positioned through a mode of address that limits their opportunities for active learning and interaction with the material.

While Ogborn and colleagues (1996) argue that science requires a certain number of building blocks be put in place for students to be able to understand explanations, it is not enough for students to copy and have their notes checked off in a grade book simply as a completed task. Those limited literacy events construct science as learning someone else's facts and doing just enough to write them in a notebook. The transformative power of the content that science reformers demand for culturally- and linguistically-diverse students like those in Mr. Walsh's classroom requires them to identify and engage with the material and a personal level (Lee & Fradd, 1998). In his study of younger children, Corden (2000) concurs, arguing that students must engage with and transform texts to learn from them. Without concerted and ongoing efforts, the mystique of science that Lemke (1990) describes will likely continue without these students actively involved in its transformation.

Targeting Enduring Understandings

1   What are the connections you see between the discussion in Cole's chapter and the four elements of culturally-responsive instruction, as described by Villegas and Lucas?

2   How would you describe the "work" that language is doing in this multi-cultural science context?

Deepening the Reflection

3   Table 4.3 includes three Daily Calendars used later on in the school year in Mr. Walsh's classroom. After attempting your own analysis of the modes of

address of these Daily Calendar texts using the approach outlined in Cole's chapter, would you say that a further shift occurred over the remainder of the year? Are the patterns that Cole describes still in place? Why or why not?

Table 4.3 Three daily calendars

02/18/98	(1) Asthma video	(2) Finish homework— Seed Vee diagram 100 pts —due Friday	(3) Notebook check 1/28/98 —minimum ten stars	
02/23/98	(1) Read tobacco, smoke, and asthma	*Write one paragraph with opening, body, and conclusions; summary	*Include at least four ways to quit smoking	(2) Question and Review article–60 pts. Due at end of class
02/25/98	(1) Continue working on diagrams of the respiratory system —title, label parts, neat— demonstrate the exchange of gasses from lungs to blood—color it			

4   Ellsworth (1997) claims that part of the power of modes of address as a theoretical tool is that we can revision a text for a different audience. With this perspective in mind, decide on a way you would like to position a group of students in a Daily Calendar activity and re-write the contents of the three Daily Calendar texts above to target your imagined audience. How do you use: Questions? Commands? Verbs? Nouns? Be aware of your conscious language choices as you design your texts. Discuss both the process and the intended outcomes.

5   In this chapter, information about the experiences of Mr. Walsh's students with the Daily Calendar is largely absent from the discussion. Cole focused her discussion on the texts of the Calendars and from that made theoretical claims about what was the likely experience of students in Mr. Walsh's class. What more would you have liked to have known about the students' experience? What kind of data would fill in the gaps for you? Why do you suppose Cole made the choices that she did?

Encouraging Engagement

1   Cole's chapter begins with some strong claims about a particular view of science. According to her discussion, science is often understood as a domain of experts who manage uncertainty and disseminate "truth" to the rest of us. Whether you agree with this perspective or not, spend some time

looking at different forms of popular media (web sites, television, magazines, newspapers, etc.) and research how often and in what ways science is presented in these media. How it is used rhetorically? For what purposes is it invoked? Finally, if you know someone you consider a "scientist," ask to what extent the claims about science Cole presents at the beginning of this chapter resonate with their own views about science.

2   Ellsworth's (1997) concept of modes of address gives us an analytical tool to evaluate how aspects of any educational setting are designed for an audience. Her model includes both the conceptual and the physical elements. Identify one or two different science classrooms that you can examine in depth. Draw pictures of the layout and placement of things like desks or chairs, media equipment, books, resources, etc. Think about and discuss what those design choices tell you about the presumed audience. Then extrapolate beyond this first level of discussion to examine what conceptions of science teaching and learning are further embodied in these physical spaces.

Notes

1   Data for this chapter come from the project, "The Socialization of Diverse Learners into Subject Matter Discourse," Jane Zuengler and Cecilia Ford, Principal Investigators. The project is part of the Center on English Learning and Achievement (CELA), which is supported by the U.S. Department of Education's Office of Educational Research and Improvement (OERI Award#R305A60005). However, the views expressed herein are those of the author and do not necessarily represent the views of the Principal Investigators, CELA, or the U.S. Department of Education.

2   The name of the high school and all names referring to the research site are pseudonyms.

3   At Jefferson High School, the school year is divided into six-week marking (or grading) periods. These administrative designations are assigned by the school district. At the end of each marking period, teachers figure grades and the school issues report cards at this time.

4   In Mr. Walsh's science class, the Vee diagram was used as a way to account visually and spatially for the content traditionally covered in a discussion of the scientific method. The students received a handout that contained a large letter V. At the beginning of the year, before he began the Daily Calendar, Mr. Walsh spent most of one class period and part of another teaching the students how to use this diagram. Along the left edge of the V they were to write the questions they wanted to answer along with the materials and predictions they had before they conducted a lab. Afterwards, they would write their observations on the right edge of the V along with their conclusions. Although I can not assume that all students understood the demonstration equally nor that they could reproduce the Vee diagram as requested, within this classroom community, this form of representing an experiment or lab was well established and somewhat familiar to its members.

References

American Association for the Advancement of Science (AAAS). (2000). *Designs for science literacy [project 61]*. New York: Oxford University Press.

Bolinger, D. L. M. (1980). *Language, the loaded weapon: The use and abuse of language today*. London and New York: Longman.

Carrasquillo, A. L., & Rodriguez, V. (1996). *Language minority students in the mainstream classroom.* Clevedon, UK: Multilingual Matters.

Cole, K. (2002). *Negotiating the "daily calendar": A language socialization perspective of diverse learners in a mainstream science class.* Unpublished PhD dissertation, University of Wisconsin.

Corden, R. (2000). *Literacy and learning through talk.* Philadelphia, PA: Open University Press.

Delpit, L. D. (1995). *Other people's children: Cultural conflict in the classroom.* New York: W. W. Norton.

Ellsworth, E. (1997). *Teaching positions: Difference, pedagogy and the power of address.* New York: Teachers College Press.

Gee, J. P. (1996). *Social linguistics and literacies: Ideologies in discourses.* London: Taylor & Francis.

Gumperz, J. (1982). *Discourse strategies.* (Studies in Interactional Sociolinguistics 1). Cambridge, UK: Cambridge University Press.

Hanrahan, M. U. (2006). Highlighting hybridity: A critical discourse analysis of teacher talk in science classrooms. *Science Education, 90,* 8–43.

Kelly, G., Crawford, T., & Green, J. (2001). Common task and uncommon knowledge: Dissenting voices in the discursive construction of physics across small laboratory groups. *Linguistics and Education, 12,* 135–174.

Lee, O., & Fradd, S. H. (1998). Science for all, including students from non-English language backgrounds. *Educational Researcher, 27*(4), 12–21.

Lemke, J. (1990). *Talking science: Language, learning and values.* Norwood, NJ: Ablex.

Lippi-Green, R. (1997). *English with an accent: Language, ideology, and discrimination in the United States.* New York: Routledge.

Luke, A. (2002). Beyond science and ideology critique: Developments in critical discourse analysis. *Annual Review of Applied Linguistics, 22,* 96–110.

Lynch, S. J. (2000). *Equity and science education reform.* Mahwah, NJ: Lawrence Erlbaum Associates.

Mercer, N. (1995). *The guided construction of knowledge: Talk amongst teachers and learners.* Clevedon, UK: Multilingual Matters.

Mikulecky, D. C. (2001). The emergence of complexity: Science coming of age or science growing old? *Computers and Chemistry, 25,* 341–348.

National Committee for the Standards and Assessment of Science (NCSAS). (1996). *National science education standards.* Washington, DC: National Academy Press.

O'Connor, M. (1999). Dialogue and debate in a post-normal practice of science: A reflexion. *Futures, 31,* 671–687.

Ogborn, J., Kress, G., Martins, I., & Mcgillicuddy, K. (1996). *Explaining science in the classroom.* Buckingham, UK: Open University Press.

Pennycook, A. (1994). Incommensurable discourses? *Applied Linguistics, 15,* 115–138.

Vosniadou, S., Ioannides, C., Dimitrakopoulou, A., & Papademetriou, E. (2001). Designing learning environments to promote conceptual change in science. *Learning and Instruction, 11,* 381–419.

Wells, G. (1993). Working with a teacher in the zone of proximal development: Action research on the learning and teaching of science. *Journal of the Society for Accelerative Learning and Teaching, 18,* 127–222.

Zuengler, J., Ford, C., & Fassnacht, C. (1999). *Analyst eyes and camera eyes: Theoretical and technological consideration in "seeing" the details of classroom interaction.* Albany, NY: National Research Center on English Learning and Achievement.

5 Literacy Infusion in a High School Environmental Science Curriculum

Jennifer Zoltners Sherer, Kimberley Gomez, Phillip Herman, Louis Gomez, Jolene White, and Adam Williams

Introduction

At this point in your reading of this volume, stop and consider the different kinds of skills you are using in drawing meaning from text. You are certainly decoding the letters on the page and recognizing that, in now familiar groupings, they produce segments of sounds that, forming words, represent particular concepts. You are also navigating particular kinds of organizations of those words in larger, sentence-based, units of meaning. Now, aside from the mechanics of the reading act itself, what else have you been doing? You are likely, drawing on your previous experience with academic texts, using some important text-attack strategies. You are looking for the main idea of the chapter in the introductory and concluding paragraphs and, in between, using headings and subheadings to help you create a mental outline of the development of the authors' arguments. And when the text is not prose, but pictures and presentation of text in alternative forms, such as in tables or student work, you are looking for the relationship between those standard and alternative, supplementary forms of text. Out of all this work, you create your own understanding of what the author is saying. Where and how did you learn to do all this reading work?

The skills you have been drawing on, probably completely unconsciously, in reading this book are those that challenge many culturally- and linguistically-non-dominant students who have limited opportunities to learn, practice, and routinely apply those skills in their home and school contexts. The work of Sherer and her team recognizes the need for explicit instruction in such strategies as a crucial element of equitable science instruction for all students. In this chapter, they describe their design and infusion of reading-to-learn tools into an inquiry-based environmental science curriculum and document its effects on teacher practice, and student behavior and student achievement.

As you read this chapter, think about the literacy strategies that Sherer and her colleagues have incorporated into science instruction. Do they look familiar from your own practice as a learner or as a teacher? What others have you or your students tried? And, importantly, why is it that some kids seem to be able to read the science textbook and others can't? Is it just the English Language Learners (ELLs) who struggle? Who else is being left behind? And, finally, how ready are you to think about yourself as a science *and* literacy teacher? What opportunities do you think this new identity provides? What additional challenges and responsibilities does it bring?

Project-based, inquiry approaches to learning science allow students to develop scientific habits of mind through longer and deeper engagement with scientific ideas. When participating in inquiry science investigations, students learn to develop questions about interesting scientific phenomena, synthesize multiple sources of information, analyze data, and communicate results to their teachers and peers. Through participation in inquiry science activities, students have opportunities to do more of the "work" of scientists instead of simply reading about scientific discoveries in traditional science textbooks (Loucks-Horsley & Olson, 2000). Inquiry science materials are increasingly being adopted by local school districts for use in K-12 classrooms. Yet, many students come to science classrooms ill-prepared to engage in a critical element of scientific inquiry: reading science texts. For this reason, attention to developing reading comprehension skills is an important aspect of equitable science instruction for all learners.

Inquiry science activities require students to read and analyze a variety of texts (e.g., NASA public reports about global warming), apply this information to investigations, and document their learning. However, these texts, often written by science writers for highly literate audiences, are frequently inaccessible to students with poor reading skills. Many American students struggle to read at grade level. The problem is even more severe when students have a history of being educationally underserved. Further, schools that seek to serve these young people often face a myriad of difficult instructional challenges such as large populations of ELLs, high student mobility, few literacy resources (staff, texts, quality teacher professional development), and high teacher turnover. Understanding science text well enough to engage in inquiry science is a huge challenge for these urban populations, particularly at the high school level as the science gets more challenging and students' reading levels fail to keep up. The U.S. Department of Education estimates that one-third of students currently enter 9th grade with reading skills that are two or more years below grade level (2006).

Table 5.1 highlights the "mismatch" between what many 9th graders can read independently and what they are regularly expected to read in inquiry-focused classrooms. The text on the left is typical of what 9th graders can read independently if they are two years below national norms in reading. This is the average reading ability of the students we work with, based on their scores on a standardized reading test (Degrees of Reading Power—DRP). The text on the right is taken from the inquiry curriculum that the students in our program are using in their 9th-grade science classes.

The text that poorly-prepared 9th-grade students *can* read independently has short sentences with very basic vocabulary words such as bird, feathers, and wings. The text this same student is *expected* to read has longer, more complex

Table 5.1 Comparison of what students can read independently and what they are expected to read in 9th-grade science classrooms

Text a 9th-Grade Student Reading at a 7th-Grade Level can Read Independently	Text That our 9th Graders are Expected to Read Pulled from the Inquiry Curriculum They are Using
"A bird's wings are well shaped for flight. The wing is curved. It cuts the air. This helps lift the bird. The feathers are light. But they are strong. They help make birds the best fliers. A bird can move them in many directions. Birds move their wings forward and down. Then they move them up and back. This is how they fly."	"Beginning about 75 years ago, hundreds of small dams began to be 'decommissioned.' In the last decade, 177 dams were removed nationwide, with 26 of these in 1999 alone. Salmon conservation was not the sole reason for decommissioning these dams. Many were in poor condition, dilapidated from lack of maintenance and they posed a flood risk for areas downstream."

sentences with more advanced vocabulary words such as decommissioned, dilapidated, and maintenance.

This "mismatch" between the complexity of texts that students can typically read independently and the complexity of texts they are regularly expected to read in inquiry settings is a major challenge for teachers and students that raises several important questions. If students cannot understand what they read in science class, how can they learn the science? How can science teachers, who often lack preservice or inservice training in supporting literacy in the content areas, better support students' reading in science? How can teachers modify inquiry materials and activity to serve the needs of struggling readers? How do teachers retain a focus on science concepts and scientific processes while helping students learn to more deeply engage with text? Ultimately, we need to know what kinds of support can really make a difference for students and teachers who are committed to learning from science texts.

Our research focuses on just these questions. Specifically, we have set out to better understand the interplay between reading and science in high school classrooms. All students need to develop competency in reading content-area texts to participate in ambitious learning. This is a skill that predicts success in school, as well as success in our twenty-first-century society. Traditionally, high school science teachers have been trained to "do science." But if students cannot comprehend text and thereby engage with scientific reasoning, just doing "hands-on" science will lead to an impoverished experience of science learning for too many students. In addition, students who can only "do science" but do not understand how to make sense of science writing, be it in newspaper accounts, science magazines, or other public periodicals, will be sorely lacking in the knowledge economy. Teachers must be supported in bringing text back into the "doing" of science. The goal of the program of research we describe here is to build a set of supports that help students and teachers deeply integrate text into science instruction. We work within a constructivist framework of teaching and learning. Our approach is to use the design of reading supports to uncover students' and teachers' needs for, and perspectives on, effective reading tools. From the perspective of the learner our approach uses reading supports as a means through which students can monitor and reflect on their learning. From the perspective of the teacher, our approach provides a mechanism for helping students take a metaperspective on a primary vehicle of science instruction: the texts. We also use the tools as a way to study the relationship between reading achievement and science achievement.

We begin this chapter by elaborating on the connection between reading and science learning. Next, we present a description of our program that includes a careful analysis of the texts in the curriculum as well as the development, adaptation, and support for reading-to-learn tools that are a critical focus of our work. We then describe our pilot year implementation in an urban Chicago high school and discuss some preliminary findings. Finally, we discuss how the first year implementation results influenced our thinking and plans going forward.

Purpose of our Design Work

To paraphrase Block and Pressley (2002), we teach students to read because we want them to gain knowledge through texts. That is, we want students to comprehend what they read. To successfully use text for learning, students need to have a toolkit of resources available to them whenever they read. Such a toolkit should benefit adolescents by making reading strategies explicit (Gomez & Gomez, 2006). We have designed a series of tools that aim to assist students and teachers in constructing meaning from science texts and integrating their constructed meanings into science instruction. Students learn to identify, analyze, and synthesize the structure and content of science texts using the tools while always monitoring their comprehension. Following the work of Yore (2000) and others, we call these "reading-to-learn" tools.

We designed a program of supports around an inquiry-based environmental science curriculum that was intended to help struggling readers develop these reading-to-learn skills. We brought together three critical ingredients: a text-rich curriculum, three categories of reading-to-learn tools, and a collection of individuals with diverse expertise (including science teachers and education researchers). We first identified and analyzed the textual elements of the curriculum that might likely be problematic for struggling readers. We then modified the curriculum to highlight and integrate the tools and show teachers where and how use of the tools could best support their science learning goals. We introduced, supported, observed, and studied the impact of using the modified curriculum in twelve classrooms (three teachers) during the 2005–2006 school year.

Elements of our Design Work

In this section we describe the three main elements of our design work. First, we describe the elements of the inquiry-based text we adopted, highlighting how text support would enhance the inquiry process. We then describe the three reading-to-learn tools that we chose. Finally, we describe our collaborative redesign approach that integrates textual support with inquiry learning and which led to the development of the modified curriculum.

The 'Investigations in Environmental Science' Curriculum

Investigations in Environmental Science (IES), formerly piloted as "Learning about the Environment" (LaTE),[1] is a high school inquiry-based environmental science curriculum intended for 9th graders. It makes use of geographic visualization and data analysis tools. The curriculum invites students to take on the role of environmental scientists; they are introduced to several real world problems that must be addressed through an environmental science decision-making model. For example, Unit 1 challenges students to learn about various environmental concepts related to land, such as ecosystems, leading up to a project where students propose a location for a new school in Florida. Their proposal must take into account the amount of space the school will need, as well as the importance of protecting land and native species that live in and around the potential locations for the new school.

Students build understandings around issues of demand for space and other environmental and policy demands through classroom discussion. Many of our students live in neighborhoods that are changing due to gentrification. In some ways similar to the Florida school example, students experience changes in availability of space and place in their own communities. These connections are important for students to consider as they learn about other biological entities (in this case gopher tortoises) that are competing for space with humans. The curriculum offers students and their teachers to consider how they might become agents of change, not only with respect to the local content of the unit, but with respect to the communities in which they live and learn.

The IES curriculum involves three distinct investigations: land use, energy use (generation of electricity in four different regions of the U.S.), and water use. Students learn about the importance of intelligent resource use in sustaining the environment. Students participate in activities that aim to develop their higher-order thinking skills: they analyze data and text-based information and learn how to use scientific evidence to support environmental decision-making; they read a variety of materials for information that will inform their decisions, use computer tools for visualization and analysis of geographic data, and learn to synthesize information from several sources accumulated over the course of the unit; they make claims and are expected to provide evidence for their claims; finally, students learn to communicate this information in useful ways to their peers. At the end of each unit, students must make recommendations for the sustainable use of resources.

The IES activities address authentic environmental issues and examine concerns that currently face the human population and that will face all students into their adult lives. The opportunity to read about and analyze data and text from a variety of sources is potentially of great benefit to students. However, students must have good reading-to-learn strategies in order to take full advantage of the units' activities (i.e., decision-making backed with evidence, data analysis and synthesis, and communication) and to ensure their development of the conceptual underpinnings of environmental science.

Reading-to-Learn Tools

Reading-to-learn tools are explicit strategies and processes that students can use to understand text. Our efforts center on three of these tools: annotation, double-entry journals, and summarization. Each tool supports knowledge transformation in that they each structure ways for readers to actively rework text information to improve their understanding. The tools support individual reflection and reorganization of ideas, giving learners a much more interconnected understanding of a domain of inquiry (Scardamalia & Bereiter, 1986). We chose these tools for our program because each is linked to one or more reading-to-learn skills. They are ideal for peer editing and peer assessment and can also provide teachers with concrete and quick feedback about how carefully students have read the science text and understand the science issues addressed in the text.

Annotation

Annotated texts are readings that have been subjected to content analysis. They serve as a set of texts whose structure and content is well-understood and connected to learners' current reading levels. Expository text often contains explicit or implicit hypotheses, claims, evidence, inferences, predictions, and evaluations—all of which are critical science skills that the Illinois Literacy Standards identify as necessary for 9th-grade science learning. These skills are also evident in the National Science Education Standards. However, this textual information is not always explicit or easy for students to find in the text. Good readers know what to look for and prioritize this kind of information. It is critical that teachers explicitly teach these skills to struggling readers rather than assume that their students know them and can apply them.

Text annotation is a strategy to make the author's message more explicit to the reader. Students must be explicitly taught how to identify important information and disregard irrelevant information. Annotation is a strategy that makes textual elements explicit. Teachers can use this annotation process to scaffold readings for their students, building to the place where students are independently annotating text. Students typically annotate (by marking on the text) one or more of the following items:

- difficult vocabulary words;
- main arguments;
- evidence/supporting ideas;
- transitional words and other signposts;
- difficult sentence construction;
- inferences;
- conclusions.

Annotation supports active reading and allows students to focus on the critical concepts or "intended learning" of the text. The teacher must first identify what aspects of the text are critical for the students to understand. She then crafts the annotation exercise to support that focus. It may be that students annotate a few critical paragraphs, or, if the text is an introductory reading, students may only focus on identifying key vocabulary. Students are usually asked to use a standard format for annotating, for example, circle headings, underline main arguments, double underline evidence/supporting ideas, triangle difficult vocabulary words, and write the letters "SC" over sentences with difficult sentence construction. Figure 5.1 shows what such an annotation looks like.

Typically, the first step of annotation is to quickly skim the text to gain a big picture view of the reading. As students skim the text they identify the title, headings, and subheadings. Next, they annotate key paragraphs for key elements. Students then review their annotations with the teacher or as a peer editing activity. Students can also use the annotated text as a study guide for future activities, tests, etc. The act of annotating difficult to understand text allows students to

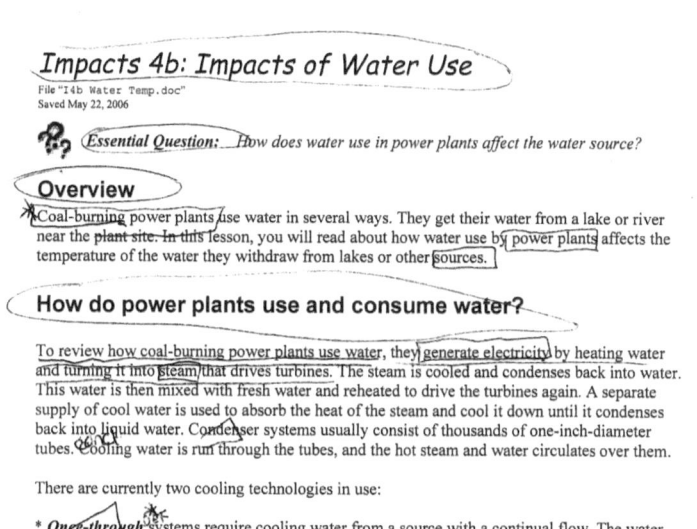

Figure 5.1 Example of student annotation of text.

slow down and pay attention to the structure and important details of a science text, as well as query words and sentences that are difficult for them to comprehend.

Double-Entry Reading Journals

A double-entry journal is a reader-response journal that provides a structure for students to monitor and document their understanding of science texts. Completing a double-entry reading journal provides students with the opportunity to actively read and reflect on what they have read. The variety of double-entry reading journal structures allows teachers to focus student reading on an important idea or skill unique to that particular text (vocabulary, main ideas with supporting ideas, etc.). Here again, the science is explicitly linked to the reading task. For instance, vocabulary journals are commonly used for introductory texts, while argument/evidence journals are used for articles in which evidence is given to support scientific arguments.

Table 5.2 provides two examples of double-entry reading journals for one article about a population of non-native reindeer being introduced onto an island by biologists. In the first example, students are asked to write the main argument and the supporting evidence of an article. The student in this example identified what he believed was the main argument from the article and listed two supporting ideas.

Student work on the double-entry journal gives the teacher critical information that can inform her instructional practice. The journal provides teachers

Table 5.2 Sample double-entry reflection journals

Argument/Evidence Double Entry Journal
Directions: Please write the main arguments you find while you're reading on the LEFT side of the page and write evidence to support those arguments on the RIGHT side of the page.

Main Arguments From the Article	Evidence From the Article to Support Each Main Argument
(1) St. Matthew Island could easily support 1,600–2,300 reindeer.	(1) The biologists believed that the carrying capacity 1,600–2,300 reindeer. Biologists still believe that if they bring the reindeer they would thrive. Their estimated capacity includes the factors of enough food, shelter, areas of giving birth and kind predators.
(2) The population of reindeer got out of control fast. So fast that it almost caused them their extinction.	(2) By 1963 the population grew to 6,000. Between the years of 1963 and 1966 the population kept on growing causing the exhaustion of food resources on the island. This caused starvation and only forty-two survived.

and students with an assessment of what the student understands and perhaps what must be examined in more depth. This student demonstrates his ability to identify main ideas, and he has clearly connected supporting ideas to each main idea. The teacher can then assess both the quality of the main and supporting ideas. By glancing at all the double entry journals in the class, the teacher could quickly determine when the class was ready to move on to substantive class discussions about the main ideas of the text.

In the vocabulary double-entry journal (Table 5.3), students were directed to identify all the vocabulary words in the text that were difficult or unknown to them on the left side. On the right side, students were directed to write what they thought the words might mean based on prior knowledge and contextual clues.

In this example, the student's definitions suggest he needs further support if he is to develop a firm grasp of the concepts in the article and in the curriculum. Words like abundance and availability appear in several other texts in the unit, so if the teacher does not clear up his misconceptions about these words, he will likely struggle throughout the unit. For instance, his misunderstanding of the word abundance will likely impact his ability to understand future lessons on population issues. Like the annotation task, double-entry reading journals allow the teacher to focus on identifying critical learning goals she wants to prioritize. The construction of the double entry journal encourages students to slow down and reflect on what they have read.

Summary

Summarization is a critical scientific inquiry skill. It requires the reader to effectively digest new information and communicate it in writing in a way that makes sense to him, as well as an external audience. Summarization is a particularly difficult task when students are reading texts far above their reading level. In summarizing, students must comprehend the text, identify main ideas, differentiate secondary ideas, and condense the information in a succinct and logical way. Though summarization is an important component that supports students' reading comprehension as well as writing skills, it is time-consuming for teachers to grade frequently written summaries. In our reading-to-learn approach in

Table 5.3 Vocabulary double-entry journal

Directions: While you're reading, please write the vocabulary word or words you find hard to understand on the LEFT side of the page. Then write what you think the word means on the RIGHT side of the page.

Hard to Understand Vocabulary Words From the Article	*What I Think Each Word Means (Student's Own Definition)*
Thrive	Will be able to live
Degrading	Producing, increase
Abundance	Decrease of…
Availability	Availability, can do something

science, we give students opportunities to summarize science text in two ways: first, through teacher-guided summarization skill development, and second, through the use of Summary Street,[2] a web-based program that gives students immediate, machine-generated feedback on their summary writing. Summary Street allows students multiple opportunities to revise their summaries until they reach a specified standard. Summary Street supports student summarization by giving feedback on content, spelling, redundancies, irrelevancies, and plagiarism. The program allows students to get instant and private feedback on their work. Because the tool provides high levels of student engagement, the teacher's class time is freed to have one-on-one conversations with students about their summaries and their understanding of the text (Kintsch, Steinhart, Matthews, Lamb, & LSA Research Group, 2000).

Science teachers benefit from explicit guidance about how to help students write good summaries (for one example, see Table 5.4). The guidelines in Table 5.4 give students an opportunity to craft a summary with support from both their teacher and peers. Teachers can repeat this whole-class guided activity as often as is needed to help students grow confident in capturing and reporting the gist of a science text.

These three reading-to-learn tools can be used individually or in concert. For example, a teacher may begin by asking students to annotate the first reading in a new unit, and the next day ask students to complete a vocabulary double-entry journal about the text. Similarly, a teacher may ask the students to complete an argument and evidence double-entry journal about a particularly important article, assign students to write a draft summary for homework, and then ask students to revise their summary using Summary Street the next day. In this way the students build on their work; if they identify critical main arguments and evidence from the text, they then have an outline of sorts with which to write their summary. The combination of reading-to-learn activities allows teachers to quickly understand what students have learned about the content and what might need to be reviewed before delving into more complex ideas in the unit.

The goals for the tools are three-fold: to get students to read more carefully, to teach students explicit skills for how to read science text for information, and to help students be better prepared to reflect on and think critically about what they read during class discussions. Using the reading-to-learn tools as a way to access science texts, students are better able to participate in the work and talk of science during classroom discussions and inquiry activities. Since many urban students struggle with reading, developing a better understanding of the text allows them to better engage in the inquiry element of science class.

In the next sections, we describe how we collaborated with teachers to modify the curriculum to support use of the tools. We also describe the ongoing professional development that is needed to successfully implement the work in science classrooms.

Table 5.4 Summary writing steps

Activity	Classroom Formation	Tools Needed
Brainstorm with class elements of a good summary	Whole class	Chalk board
Show students the list of elements Summary Street thinks a good summary has	Whole class	Summary Street's list of elements of a good summary
Synthesize these elements with what they have done in the past for other teachers	Whole class	Information from other teachers
Pass out article	Whole class	Paper copy of article for each student
Do some pre-reading: read titles, sub-titles, make predictions about what the article will be about	Whole class	
Read the first paragraph together	Whole class	
Have each child, independently, identify the main idea in the first paragraph	Individual work-share with class	
Discuss as a class	Whole class	
Continue this exercise with next several paragraphs	Whole class/ individual work	
Have students work independently to read the article and identify main points	Individual work	Pencil for marking up text
Have them craft a draft of a summary of the article	Individual work	Paper and pencils
Peer edit the drafts	Partner work	Peer edit sheet
Enter summary into tool (teacher will choose if they can get feedback as they go, or if they have to type the whole thing in and then get feedback)	Individual work	Summary Street tool

Integrating Literacy Into Instruction

Integrating literacy strategies into science instruction puts additional demands on teachers. Teachers must continue to focus on their science learning goals while more actively supporting students as they read difficult text. In order to do this well, a teacher must know the curriculum and the science concepts underlying the unit, understand the nuances of the literacy tools and their use, and strategically integrate the tools into curricular planning. Because integrating literacy strategies with science teaching is difficult, we brought together diverse expertise in the initial planning stages of this work to work on modifying the IES curriculum to better support students and teachers. In this next section we describe the collaborative design structure we used, which we labeled a *work circle*. We then describe the resulting supports that were built to support reading in inquiry implementations.

Work Circle

A work circle brings together people with diverse expertise to design materials and address challenging issues like reading in science. Our work circle consisted of five individuals: three high school science teachers and two researchers. The teachers taught at three different urban high schools, had experience teaching inquiry science curricula, and incorporated explicit literacy supports in their pedagogy. For example, Ms. Brown[3] encouraged students to document the words they did not know. Mr. Ellway worked closely with the special-education teacher and the ELL teacher in his school to support students' science text reading. Ms. Davis, with her honors students, sought to provide challenging texts for students to read. Additionally, Ms. Brown and Ms. Davis had taught the IES curriculum for several years, and Ms. Brown and Mr. Ellway had worked with our research group the previous year piloting Summary Street (the web-based summarization tool described earlier). The teachers were deeply committed to the belief that all students, regardless of background and prior experiences, could actively and skillfully learn science. The teachers often drew on students' prior knowledge to make text to self connections during class readings and discussions.

The two researchers (Gomez & Sherer) brought experience in curricular design, literacy, classroom teaching, cognition, and literacy infusion in science to the group. We met monthly in full-day meetings for six months, crafting the program and discussing challenges and hopes.

The work circle identified critical curriculum components and modified the curriculum to integrate literacy support at these critical points. The group also identified explicit strategies and developed support to help the pilot teachers implement the reading-to-learn tools.

Identifying Critical Texts

The work circle began by identifying the most important readings throughout the IES curriculum. Important texts were identified as those in which critical content is presented that is necessary to develop a conceptual understanding of the unit. Next, we identified the key science learning goals in each text. Finally, we linked the texts to the appropriate reading-to-learn tool that best connected reading activity to learning goals.

Table 5.5 illustrates the organizational guide we constructed based on this work to integrate the text, science learning goals, and choice of tool. The guide includes five elements. First, we identify the texts and length of texts. We list the main concept or goal of the reading. We present those main concepts (or goals) in the form of a question, and provide brief answers to those questions. In column four we identify the recommended literacy support for that particular

Table 5.5 Excerpt from teacher chart: How tools are integrated into the unit

Lesson	Length	Main Concept Question and Answer	Literacy Support Tool	Other Suggestions
New School 1a: Overcrowded?	Three paragraphs	**What is the environmental problem identified in this letter?** Overpopulation and the resulting competition for resources	DEJ (Double-Entry Journal)— What is the problem?/What does it have to do with the environment? DEJ—vocabulary	The DEJ template for this lesson is in the teacher binder
Populations 4a: Reindeer on St. Matthew Island	Six paragraphs with map, picture, graph	**What are the factors that limit the population?** Availability of food and water, shelter, availability of a place to give birth, predators	**Annotation** DEJ: Main idea/ supporting ideas Summary Street	Comparing human carrying capacity to this graph. Where are we on that graph?
Resources 1d: Arable Land	Three paragraphs, one map	**How much of our land is economically productive?** Only a small percentage of our land can be used to grow crops	DEJ: Main arguments/ supporting evidence	Help kids with vocabulary and concepts

text (if a tool has been suggested) and add any other suggestions to support the teacher in the final column of the tool.

For example, the teacher can see—in using this chart—that the first reading, "Overcrowded?", is a three paragraph letter that identifies an environmental problem. The information in the chart can help her plan the timing of her lesson (it is possible to read and do the double-entry journal in one class period), what to focus her students' attention on as they read the text (the environmental problem), and where she can find the template to copy for her students. If she does recognize the environmental problem in the reading, she can see that it is listed in the chart as well. The design of this guide assumes that the teacher will have read the text before teaching the lesson, but this chart can serve as a quick reminder of the main elements with which she can focus her teaching. This serves as a practical tool to help teachers design lessons. It also serves as a framework for thinking about how to integrate literacy support into science learning.

Annotation Templates

We created an annotation template, or marked-up version, of each important IES text. Through this identification of critical elements (essential vocabulary, main ideas, supporting ideas, etc.) within each important text, we help the teachers "dissect" each reading. While some articles clearly have main ideas and supporting ideas, other texts are not as clearly structured and readers may differ in their choice of main and supporting ideas. Therefore, there are not always "right" answers when it comes to annotated text (although there are always answers that are not right). In light of this, we designed these annotation templates to act as a guide (not answer key) for the teachers. We recommend that before annotating a text with students, teachers annotate the text on their own. This process enables the teachers to clearly understand the text themselves and anticipate areas where their students will struggle.

On-Going Professional Development for Teachers

Finally, we built on-going professional development sessions that were highly responsive to the needs of our pilot teachers. Good professional development is rooted in the content to be taught (Cohen & Hill, 2001), occurs close to the classroom in providing learning that teachers can immediately apply to their teaching practice (Garet, Porter, Desimone, Birman, & Yoon, 2001), and is ongoing.

Our research group, the pilot teachers, and Ms. Brown met once a month. The overarching goal of our professional development was to support teachers in their science teaching practice, with a focus on literacy, through consistent work sessions by providing a balance of support, information, and opportunities to plan together. We designed the professional development sessions around the following four sub-goals:

1   Introduce teachers to the literacy supports in the modified IES curriculum.
2   Explore main concepts of environmental science relevant to implementing the IES curriculum.
3   Explore assessment and how it supports their work as science teachers, both with the literacy support tools and the final projects.
    a   Co-construct assessment tools (rubrics, checklists, etc.) for the science notebooks, double-entry journals, analysis questions, annotation, and the final project.
    b   Discuss how to use Summary Street as an assessment tool.
4   Provide time to plan together.

The teachers were learning a new curriculum, as well as the new literacy tools, and this professional development supported their ongoing efforts to integrate the literacy tools with their teaching practice. Changing teaching practice and teacher beliefs is difficult work. In order to support these changes in our pilot teachers, we built opportunities for teachers to feel dissonance around their beliefs about students' capacities to work with content-area text, as well as their beliefs about their own learning and teaching practice. Additionally, these professional development sessions provided an opportunity for our design team to understand strengths and weaknesses of our design.

We now turn to a discussion of our pilot year implementation.

Pilot Year Implementation

In this section we discuss the context of our work in more detail, report on the first year implementation and data collection efforts, and end with highlights from our initial findings.

Context

Our pilot efforts took place in twelve 9th-grade science classrooms at a large public high school in Chicago, which we will call "Classic." Classic serves 2,100 students. Approximately 90% of the student body is considered low-income based on eligibility for free or reduced-lunch. The students are 68% Hispanic, 28% African-American, and 2% White. Nine percent are Limited English Proficient.

For the 2005–2006 school year only 21% of the students met or exceeded standards in reading based on the Prairie State Achievement Exam. Only 10% of students in science met or exceeded standards on the same test. Clearly, the majority of our Classic students struggle with reading.

Pilot Year

Earlier we noted that we piloted the work in twelve classrooms at Classic. This set of classrooms represented a bit less than half of all incoming 9th graders in the school. There were approximately 329 students in our program. For comparison purposes, we collected reading data on the other 9th graders in the school. In the pilot year, we were particularly interested in three broad categories of impacts: How did student behavior change around texts?, How did teacher practice change?, and What was the impact of this work on students' reading and science achievement?

To learn about classroom practice, we observed classroom implementations throughout the pilot year with a focus on classroom practice, student achievement and science learning. Two observers were located at the school full-time and worked cooperatively with the teachers to support the use of tools, the overall implementation of the program, and to systematically observe classroom practice in part through use of an electronic laptop observation protocol. The electronic observation tool was essentially an electronic checklist of teacher and student practices around text and inquiry activity. To understand student achievement, and particularly reading comprehension, we administered the Degrees of Reading Power (DRP) standardized test of reading both in October and May to all 9th graders. This allowed us to measure changes in reading comprehension over one school year. To understand science learning, as well as to understand how students and teachers used the reading-to-learn tools, we collected a wide array of student work to observe growth over time. This work consisted of written responses to questions about text and science learning (labeled analysis questions), textbooks with annotations, double-entry journals, summaries, Summary Street results, science lab reports, and project work. We administered a "tool use assessment" to capture student work with the reading-to-learn tools at two points in time: January and May. In the two-day tool use assessment students read an unfamilar text, annotated the text, completed a double-entry journal identifying main ideas and supporting ideas, wrote a summary of the article, answered a content analysis questionnaire, and filled out a survey. We also interviewed our focal students immediately after they finished each tool use assessment. Finally, we interviewed teachers three times throughout the year to understand their beliefs about science and literacy and to learn how their practice changed over time. We also used the interviews as an opportunity to learn how our design could be adjusted to better support teachers and students.

Initial Findings

We now turn to findings about our three foci: teacher practice, student behavior, and student achievement.

Teacher Practice

When teachers slow down and emphasize reading, they run the risk of covering less of the curriculum. Coverage is always a concern, particularly in an era of standards-based assessment and high-stakes testing. Teachers in underserved schools are particularly vulnerable, as their schools often face reconstitution or closure, under No Child Left Behind Act and other accountability protocols.

We argue that teachers' anxiety about coverage is real but that they, and their administrators, need to be convinced that mere coverage, without students being able to read the texts that are covered, will not likely increase science learning.

Integrating literacy support tools into a science curriculum will slow down "coverage," particularly in the beginning of the year as the students need time to learn to use the tools. If a teacher reads an article with her class and asks the students to annotate that text, this will take longer than if they just "read" the text together. The difference, of course, is the degree to which students understand what they have read, with and without supports. In our pilot year, we were concerned with how much the use of the literacy tools slowed down the pace of science instruction, and what the tradeoffs for understanding were between formally covering more science content and increased comprehension of important texts. While we have no control classrooms per se, and this is the first year these teachers taught IES, we do have evidence that the use of the literacy tools as a part of the science inquiry instruction did slow the pace down. One example is the time they spent on a critical article. Rather than reading the article in one class period, students spent three days on one article, reading and annotating the article, building a double entry journal, and finally writing a summary. Teachers used Summary Street three times with their students, spending two or three class periods on each article.

While this intensive time was more than they would have spent on text in the past, we hypothesize the tradeoff to be that students had a deeper conceptual understanding of the science that they could apply to analysis questions, project work, and classroom discussions. In order to initially test this hypothesis, we sought to determine if teachers and students found the tools to be useful. (We explore achievement results later in the chapter.)

Survey data results indicate that students found the tools to be useful. Table 5.6 shows a sub-set of the results of a survey given to students in May of the pilot year, as the fifth and final task of tool use assessment. The survey asked students to circle one answer for each tool in response to the question: "Of the reading activities you did this year in science class, how useful were each of the following reading supports in helping you better understand your science readings this year?" The numbers in each cell indicate the number of students who circled that response. A vast majority of the students found all of the reading-to-learn tools to be useful or very useful. The highest number of students surveyed found annotation to be useful (85%).

We also analyzed our interview data to determine what students and teachers thought of the tools. The patterns are relatively consistent: they find the tools to be time consuming in the beginning but useful in the long run. One student shares, "At first it [annotation] was kind of hard to understand, but as we kept doing it, it actually helped me a lot." Her teacher would agree. Ms. P admits, "It's hard to find a balance between reading and science," (KP, November 2005) but later in the year she states:

Increasing (the reading and writing) is a huge advantage for the students because those are basic skills that no matter what field they get into, they've got to learn to read and they've got to learn to write correctly, which is a big problem (with our students). They're also going to understand the material, and they are more independent in their understanding. It forces them to actually read.

(KP, April 2006)

Table 5.6 Student survey data: How useful is each of the following?

	Not at all Useful	Not so Useful	Useful	Very Useful
16 Annotating	2	13	65	20
17 Doing a double-entry journal	4	18	53	26
18 Writing a summary	3	15	56	27
19 Using Summary Street	4	16	46	35
20 Answering analysis questions	1	15	58	26

While learning to use the tools slowed down content coverage early in the year, our results show that teachers and students believe that investment is worthwhile in the long term. Overall we find that teachers believe that the initial investment in time using the literacy tools leads to more rigor and less time spent re-teaching.

Literacy Tool Use: Complement or Distraction?

As science teachers learn to support literacy in their teaching practice, we were concerned that the reading-to-learn tools would distract from the science learning; that teachers and students would perceive literacy tasks to be separate from science tasks. We heard evidence of this disconnect in teacher talk: "today is a science day" or "today is a literacy day." This was also reflected in some of the student talk too, "We aren't in English class—why do we have to write in science class?!" We realized that this was, in part, our fault, as we did not make explicit enough that science and literacy are intricately connected. We want teachers to understand that the literacy work is always *in support of* the science learning. To help teachers make this connection, we had tied each reading task to the science content. The list of main concepts (see Table 5.4) acts as a key for the pilot teachers to use to build a concept map for each chapter, as well as for the entire unit. This concept map ties the day-to-day work into the "bigger picture" of the chapter and the unit. But this tool was not enough. We needed to make explicit, in every activity, that the literacy work was in the service of the science learning.

Although the teachers initially saw the literacy tasks as add-on activities, as our professional development support progressed we began to find evidence that teachers understood the ways in which science and literacy are connected. Over time, teachers became more adept at integrating the tasks into their science teaching practice. One teacher articulates this ongoing struggle: "I do see (literacy and science) as separate. The literacy has to be there though, in order to better understand the science. I tried to use them together in the classroom, so that it didn't seem like separate literacy days, and separate science days," (KR, June 2006).

When we interviewed students in the pilot year, we found that some of them were able to see the connection between the literacy tools and their science learning. Here is an example of a student who is able to articulate that connection:

> It [the double-entry journal] helps you in reading, (be)cause in reading you do a lot of that, too. So like answer questions and stuff, but it helps you in science (be)cause it's about the article, and the article's about science ... (The three tools) help you not only in science but in reading, too. In articles people have trouble reading and defining vocabulary, and that really helps you.
>
> (Female student, February 2006)

Change in Student Behavior Around Text

Early results suggest that the literacy support tools are helping students, especially struggling readers, become more conscious of what they are reading. For example, during a visit to a pilot classroom we saw students use Summary Street while constantly referring to their annotations of an IES article. When we asked students about this activity, several said that underlining (marking up) the main argument and supporting ideas helped them plan their summaries better. On another occasion we saw students using their double-entry journals to hand-write summaries. Again, when we asked the students they indicated that the double-entry journals helped them see and remember the main ideas. Additionally, their teachers believe that students are beginning to read science text more purposefully, mentioning that some students use their literacy supports as study guides for tests.

Student Gains in Reading and Science Achievement

We are interested in how our work impacts student achievement in science and reading. We hypothesized that students who use the reading-to-learn tools, and consequently read more carefully, will improve their reading skills and learn more science. Preliminary results give some support to this hypothesis.

Students in the pilot classrooms varied significantly in their growth in reading comprehension from October to May. This led us to explore whether such variation could be related, in part, to depth of implementation of our program. We classified our three teachers along various dimensions that we believe indicate the depth of implementation of the program. Specifically, we gathered evidence about frequency and consistency of tool use throughout the year, attendance and participation in the professional development settings, the degree to which (as determined through observation and teacher interviews) the teacher takes "ownership" of literacy as a problem of practice that she wants and needs to address in her science teaching. We classified the teachers along these dimensions and labeled the teachers "most consistent", "somewhat consistent," and "least consistent" in terms of the criteria specified above. Table 5.7 below presents reading results based on the DRP exam. Students of the "most consistent" teacher (Ms. Perry) showed the greatest gains in reading. Her students increased the most, and that increase in reading (DRP scores) was significantly higher than the students in the teacher we labeled "least consistent" (Ms. Jones). Ms. Perry's students increased from being at the 19th percentile in reading nationally in October to being at the 28th percentile in May. That is a substantial increase, even though the students obviously have more ground to make up to reach even average performance based on national norms. These results suggest that students did benefit from their participation in the reading pilot program. Though this finding might be attributed to a wide variety of factors, we are encouraged by the growth in reading comprehension of those students.

Overall, we know that high school students do struggle with reading inquiry science text. We also know that consistent use of reading-to-learn tools in classroom practice helps students navigate difficult text, requires that they read more carefully and reflectively, and builds on their independent learning skills. We know that the tools must be carefully tied to the intended science learning and not haphazardly or inconsistently applied. On the other hand, we learned that supporting students to use the tools takes time and careful scaffolding that puts

Table 5.7 Growth in reading achievement based on degrees of reading power (DRP) scores from October (Pre) to May (Post)

	Ms. Perry (Most Use of Intervention) N = 43		Ms. Kelly (Mid-Range Use of Intervention) N = 37		Ms. Jones (Inconsistent Use of Intervention) N = 30	
	Pre	*Post*	*Pre*	*Post*	*Pre*	*Post*
Total items correct	32.44	38.77	41.78	46.97	31.53	34.47
DRP score, $P = 0.90$ (Independent reading level)	41.47	46.44	48.81	53.35	40.37	42.90
Percentile rank (based on nationally normed sample)	19.49	27.93	33.11	42.84	18.73	22.07

added demands—both cognitive and temporal—on teachers. If teachers are well supported and they are invested in helping their students deeply understand what they are reading, then the hard work is worth it.

Implications and Future Work

Lessons from the pilot work will guide our future plans. Teachers struggled to learn to teach a new inquiry curriculum and learn to infuse literacy in their pedagogy simultaneously. In an ideal world, we would have only piloted the literacy supports with teachers who have taught the curriculum already for several years. However, it is neither realistic nor, ultimately, as useful to only implement these kinds of programs in "stable" settings in which curricula are not changing. Curricula are always changing; teachers are always facing new challenges. Literacy support cannot wait.

As we continue our work, we hope to refine our teacher training and support materials as we move into the domains of biology and chemistry. Finally, we hope to find additional ways to support teachers so that they implement the tools more effectively and are better able to use the resulting student work to inform their on-going instructional decisions. Toward this end, we have begun analysis of the benefits of using the literacy support tools for ELL students in the 9th and 10th grades at Classic.

We are working to better understand how use of the tools can be related to changes in science achievement. We are improving the quality and number of science assessments during the current school year. We have also recruited a comparison Chicago high school that is implementing the IES curriculum without our literacy supports. We hope to better describe how science achievement and reading achievement varies for students who participate and who do not participate in our program.

Teachers can infuse various literacy supports, including those we described in this article, into their daily classroom practice. Teachers often worry that there is barely enough time for science learning without the addition of reading activities. Our collaborative work circle group and research team has taken this challenge to heart. We believe that when students learn to read more effectively, they will learn science more deeply. Thus, when students are more effective readers, less time needs to be spent on reiterating concepts that students should have internalized but did not, reducing both teacher and student frustration. Developing all students' reading-to-learn skills in content areas such as science is a "win" for teachers and students alike.

Targeting Enduring Understandings

1   What are the connections you see between the discussion in the Sherer team's chapter and the four elements of culturally-responsive instruction, as described by Villegas and Lucas?
2   How would you describe the "work" that language is doing in this multi-cultural science context?

Deepening the Reflection

1   Describe your experience as a science learner with regard to reading in science. Do you remember teachers supporting your reading in science? How do you support your own reading in science now? Think about what skills you actively engage in as a "good reader of science."
2   To what extent do you think developing reading comprehension skills is part of your science teaching job description? Why or why not?
3   How might you identify the "struggling readers" among your science students? How will you support those students? Brainstorm ways that each reading-to-learn tool, as described in the Sherer's team chapter, can support the science learning and reading skill development of those students.

Encouraging Engagement

1   Select one of the reading-to-learn supports and implement its use in a science classroom. Questions to consider include:
   a   How will you select the article? Since selection of the article is critical, what makes it an "important article"?
   b   After reading the article with the reading-to-learn supports, what do you want your students to have gained? What is the link between this learning and the literacy tool you selected?
   c   How will you know that your students gained the learning you intended them to? For example, if there is classroom discussion after the reading, what indicators of conceptual understanding and reflection will you look for in their classroom talk?
   d   Return to class prepared to describe the tool you used, the decisions you made in using that tool with a particular text, and how you structured your lesson(s). Also describe how you assessed student understanding of the text, based on students' work with the tool, and what you will teach next in response to that assessment.

2   Interview a local science teacher about the extent to which they include the teaching of reading comprehension as one of their science instruction goals. What do you learn from his/her discussion about his/her perspective on the role of reading in science? Now imagine that you, like the Sherer team, aim to provide professional development support related to reading in science. What kind of support does this teacher need? How would you deliver that support to best meet his/her needs?

Acknowledgments

The authors would like to thank the National Science Foundation-Research on Learning and Education program (under grant #REC-0440338) for its support of the project "Understanding the Connection Between Science Achievement and Reading Achievement." All findings, opinions, and interpretations are solely the authors' and not necessarily those of the National Science Foundation. We would also like to thank the principal, head of the science department, and the teachers with whom we worked closely. Without opening their doors and making their practice public, we would not be able to make progress on the critical issue of how reading and science achievement are connected.

Notes

1 www.worldwatcher.northwestern.edu/late/LATEpublicpage/Index.html.
2 http://.lsa.colorado.edu/summarystreet/ also http://lsa.colorado.edu/summarize.
3 All teacher names are pseudonyms.

References

Block, C., & Pressley, M. (2002). *Comprehension instruction: Research-based best practices.* New York: Guilford.

Cohen, D. K., & Hill, H. (2001). *Learning policy: When state education reform works.* New Haven, CT: Yale University Press.

Garet, M.S., Porter, A.C., Desimone, L., Birman, B.F., & Yoon, K.S. (2001). What makes professional development effective? Results from a national sample of teachers. *American Educational Research Journal, 38,* 915–945.

Gomez, L., & Gomez, K. (2006). Preparing adolescents to read-to-learn in the 21st century. *Minority Student Achievement Network Newsletter.*

Kintsch, E., Steinhart, D., Matthews, C., Lamb, R., & LSA Research Group (2000). Developing summarization skills through the use of LSA-based feedback. In J. Psotka (Ed.), Special Issue of *Interactive Learning Environments, 8*(2), 87–109.

Loucks-Horsley, S., & Olson, S. (Eds.). (2000). *Inquiry and the national science education standards: A guide for teaching and learning.* Washington, DC: National Research Council.

U.S. Department of Education. (2006). Every young reader a strong reader. High School Leadership Summit Issue Paper. Retrieved October 30, 2006, from www.ed.gov/about/offices/list/ovae/pi/hsinit/papers/reader.pdf.

Yore, L. D. (2000). Enhancing science literacy for all students with embedded reading instruction and writing-to-learn activities. *Journal of Deaf Studies and Deaf Education, 5,* 105–122.

Part II

Science Learning Funds of Knowledge

6 Negotiating Participation in a Bilingual Middle School Science Classroom

An Examination of One Successful Teacher's Language Practices

Jennifer S. Goldberg, Kate Muir Welsh, and Noel Enyedy

Introduction

Imagine that you have been hired as a bilingual science teacher in a school where the majority of students speak Spanish and are learning English as a second or additional language. Because you and your students have access to Spanish and English as languages of instructional conversation, you will need to consider if and how you choose to regulate these two languages in your classroom. Do you insist that your students only use English because that is the school's "official" language of instruction and you believe that English proficiency is important for their futures in and out of school, or do you allow them to use Spanish because it is the language through which they can most easily communicate their science-learning ideas and insights and you believe permitting their Spanish use fosters a more inclusive classroom community? The answer you come to regarding these questions will no doubt be informed by your school's language policies, your personal beliefs, and your professional understanding about the goals of science instruction.

In this chapter, Goldberg and her colleagues share with us the case of one teacher, Ms. Cook, who teaches in such a bilingual setting. While Spanish was not recognized as the "official" language of instruction in the school, because Ms. Cook believes that there are many ways to learn and because she understands that science-learning is her central objective as a science teacher, she allows Spanish to be used as a legitimate means of meaning-making among her students. Goldberg and her colleagues illustrate how Ms. Cook's enacted teaching philosophy about Spanish use in her classroom facilitated two important science goals, that of building a co-inquiring community of learners and of connecting science and the "real" world. In this way, Spanish works as a significant science-learning resource and helps Ms. Cook achieve successful outcomes with her students.

As you read about the case of Ms. Cook and her students, notice in what ways her teaching exhibits aspects of culturally-responsive pedagogy. Also consider how her story challenges or reinforces your own experiences, views, and practices related to language and science teaching and learning. Lastly, returning to the hypothetical scenario we posed at the start, how does hearing about Ms. Cook's success help you re-imagine your own successful science practice with bilingual students? If you, unlike Ms. Cook, are not bilingual, what steps might you take to nevertheless build on students' home languages as an important resource? As the population of the U.S. becomes increasingly linguistically-diverse, you are very likely, sooner or later, to find yourself in a situation in which your need to consider these issues will move beyond the hypothetical into the very, very real.

I teach because it [science] is exciting, meaningful, thought provoking, natural. My students have a real interest in science and it is my obligation to bring science into the classroom. Science is so exhausting—the planning, shopping, preparing, doing, questioning, listening, cleaning up! If the results weren't so absolutely powerful, I probably would spend my energy elsewhere.

(Dinner #1 email, 3/4/02)

Ms. Cook,[1] the teacher quoted above, struggles to find time for science and lacks confidence in her science background, yet she has a firm belief in the power of science as a way to understand the world. She teaches bilingual (Spanish and English) 7th-grade students who are primarily English Language Learners (ELLs). She believes that science classrooms can be environments "of acceptance and learning together, and from each other, where we can bring our identities into the classroom, reshape our knowledge and go on from there" (MSE Definition, 5/02). Because of her belief in the power of science, Ms. Cook is willing to spend her energy teaching in a way that is, as she describes it, "meaningful" and "thought provoking" despite the challenges that teaching in this way can present.

Ms. Cook's views of science are constructivist in nature, focusing on inquiry and discovery, rather than strict adherence to laboratory protocols. Furthermore, Ms. Cook focuses on different ways of learning so that all can participate in classroom activities.

I don't want to give them answers. And I don't really care if they can answer a multiple choice question right. I want them to have some deep understanding and I want them to be excited.... I want them to see that there's all different ways we can learn.

(Conversation 1, 11/4/01)

Through Ms. Cook's constructivist approach to teaching and learning science, she emphasizes multiple approaches to "doing science." Rather than discovering science in one manner, students from various backgrounds, educationally and culturally, may inquire in a number of equally valid ways.

In this chapter we draw on our earlier research with Ms. Cook in order to identify how her classroom practice was informed by her constructivist and affirmation-oriented beliefs, particularly with respect to language use in instruction. We have chosen to analyze and highlight Ms. Cook's classroom practices because, in our previous work, Ms. Cook's students out-performed others exposed to the same curriculum at the same school site. Intrigued by the differences we saw, our larger purpose is to investigate elements in the classroom community that led to greater student "success." Here we are specifically focused on detailing the discursive environment of Ms. Cook's classroom to see how language was used to negotiate participation and shape the types of learning opportunities available to students. Instead of focusing on the scientific content of the teacher's talk (i.e., what was covered and how accurate it was), we are interested in how the teacher's talk positioned her and her students and how these positionings created or hindered opportunities for student learning; that is, we are interested in how the language established roles, created expectations for behavior, and "placed" people in recognizable categories within this framework. Additionally, we coordinate our analyses of this negotiation with Ms. Cook's own words about her goals and beliefs to make clear the impact of teacher identity on classroom practice. Our analysis is grounded in a sociocultural framework with a focus on learning science as inquiry within a classroom community-based social context.

To situate our study and analysis, we start by reviewing some theoretical constructs related to classroom communities and classroom discourse. We then analyze Ms. Cook's classroom community and discourse in terms of her role in the class as a co-inquirer, ways she connected science to outside the classroom, and the implicit and explicit ways she valued Spanish as a resource to learn about and talk science.

Understanding Participation in a Social Context

In the last two decades, the education community has increasingly paid attention to the social context of learning. In particular, many scholars have begun to investigate how an individual's learning and development is tied to social participation and interaction (Greeno, Benke, Engle, Lachapelle, & Wiebe, 1998). The ways teachers organize activity and discourse in their classrooms, and, particularly, interactions between the teacher and the students, has profound effects on how students come to understand science (Lemke, 1990). For example, Fairbrother, Hacking, and Cowan (1997) found that an explicit orientation toward getting the right answer in lab experiments led teachers to oversimplify the content and students to doctor their results. A conceptual understanding of science, on the other hand, involves more than just the acquisition of a stable body of scientific facts. It involves an understanding of the process of doing science and the adoption of a set of beliefs and assumptions about the nature of scientific knowledge (Duschl, 1990).

One place where a teacher's beliefs about science can be seen to organize activity and discourse is in lesson planning. Although published lesson plans may provide teachers with directions and suggestions, teachers and students both contribute to the construction and negotiation of these lessons in ways that are not constrained or predicted by the curriculum. As teachers create lesson plans, they often begin with the national or state standards, in addition to a personally developed set of instructional goals and objectives. These goals are then embodied in some sort of plan for activity. However, as every teacher experiences, a plan does not script out every instructional interaction. The enacted plan is typically responsive to what the students bring to the situation and to what they say and do.

Beyond planning *what* will be done in class, teachers consider *how* they will interact with their students and *how* students will interact with each other. Teachers constantly make conscious and unconscious decisions that create a social "space" for learning. The organization of this space lays out the boundaries for what will be discussed, the different roles that students will take on, and what counts as science talk and what doesn't. Thus, the organization of this social space creates a landscape that shapes how any particular science activity will unfold, much like a physical landscape may create micro-climates that shape what plants and animals live there.

Inquiry teaching, which values the development of conceptual understanding over explicit instruction, can be particularly challenging because it demands that the teacher and students constantly negotiate and strike a balance between the sometimes conflicting goals of: (a) having the students pursue their ideas using their own, often invented, strategies; and (b) having the teacher help the students learn the concepts and skills that the teacher, the standards, and the curriculum, as developed by more competent members of society and the discipline, want the students to learn (Hammer, 1997). This tension can be productively managed and inquiry experiences can be woven into the regular fabric of classroom activity (Wells, 1999). However, there is also the potential for teachers to fall back on traditional patterns of school discourse that transform inquiry-oriented activities into more explicit, didactic, procedural, "cookbook" science activities (Enyedy & Goldberg, 2004). In a classroom, particularly a bilingual setting, a teacher can cultivate a community of learners and use language to encourage full participation during science lessons.

Creating Communities

One productive way to think about how social structures shape interaction within a classroom (and ultimately shape what and how well students learn) is to think of a classroom as a community. Communities can be defined in many ways, but one compulsory characteristic is the unity of its members. For example, political parties are a type of community unified by some abstraction or set of beliefs such as "fiscal conservatism" or "social liberalism"; even though the members of these communities may never even meet one another, they are unified by shared values that have social significance. Communities can also be defined geographically, such as a neighborhood community. In this case, people who live near each other are unified by their concern about issues facing their neighborhood. In classrooms, many researchers and teachers also talk about communities and adopt a Community of Practice (COP) approach. From this perspective, the unifying factor of the community is the participation in and identification with a "practice"—a set of activities, shared tools, roles, etc. that help the group accomplish something (Lave & Wenger, 1991).

The "practice" of many science classrooms is the mastery and memorization of a body of scientific facts or developing skills in following procedures and protocols of experiments. These two practices can certainly organize and unify a community. However, for a large number of students this type of practice often unifies them in opposition to the discipline, rather than serving to enhance their identification with it. Similarly, in mathematics, Boaler & Greeno (2000) found that classrooms organized around procedures and calculations alienated students. Even students that could perform the procedures did not want to identify with the discipline of mathematics as presented in this manner. Similar arguments have been made in science (Brown, 2004).

A beneficial approach for science classrooms that can unify the community is knowledge production (Wells, 1999). The practice of knowledge production means that students themselves are the ones producing scientific knowledge— through experimentation, observation, presentations, debates, etc. Further they are the audience for this knowledge and the arbiters of what counts and does not count. This does not mean they have to be doing cutting-edge science. What counts is if the knowledge produced is news to them. If science teachers wish to promote a conceptual understanding of the material and an accurate view of the nature of science as a discipline, as well as attract students to the discipline, then engaging students in authentic inquiry projects where they are invested in producing rather than merely learning knowledge is an effective method (Bransford & Donovan, 2005). From a COP viewpoint, then, learning is seen as part and parcel of a person becoming part of a community as they participate in socially-organized activities or practices (Linehan & McCarthy, 2001). Because of this, the details of that social organization matter. Students learn how to act and speak during classroom interactions depending on the particular rules and roles established within classroom communities. Depending on how the "practice" of science in a classroom is understood—procedural or conceptual—these rules and roles will vary. The ways language is used within the science learning community is an important point of variation in the "practice" of science.

Using Language

Choices in language influence the ways students learn science. Participation frameworks, "the rights and obligations of participants with respect to who can say what, when, and to whom" (Cazden, 1986, p. 437), may be particularly useful for examining the roles that emerge in classroom activities. People use their understandings of what type of activity they are doing and what their role is to limit what they do and say. This has direct implications for the meaning that they take away from their experience (Erickson & Mohatt, 1982). For example, a classroom may be organized in a traditional manner where the rights, roles, and responsibilities of the students are very limited, such as in the "recitation script." In this common form of classroom discourse, teachers ask questions which can be answered briefly, quickly and without outside resources, the student responds, and the teacher evaluates the response as correct or incorrect (Mehan, 1985). Students and teachers are quick to recognize when they are in this language game and regulate their conversational turns accordingly. Research has shown that participation frameworks, such as this Initiate-Respond-Evaluate recitation script, have direct implications for what the students understand their activity to be about and for what they understand the discipline to be about. In the case of the recitation script in science classrooms, it encourages students to adopt an understanding of science that is equated with memorizing a static body of established facts that are either true or false.

Erickson and Mohatt (1982) point out the importance of reconsidering how power relations between teachers and students play out in interaction to shape what students do, say, and learn. We need to re-define, establish, and negotiate what knowledge is legitimate, how that legitimacy is determined, and who can make knowledge claims. These reflections on the legitimacy of knowledge are part of our becoming conscious of the culture of power in classrooms. In our analysis of the power relations and participation frameworks of the science classroom, we will consider the ways in which the teacher positions herself and the students, paying particular attention to how language is used to achieve these positionings.

"Positioning" refers to the roles participants take on, or are assigned, during everyday interaction that establish the relevance and meaning of the person's actions within the activity (Ritchie, 2002). For example, a student can be positioned as a class clown. Having been assigned this role, his or her behavior in class is interpreted in a particular way. Further, the child may use his/her understanding of what a class clown is and does to control his/her own behavior. A person can position him/herself or be positioned by others. Either way, these positions fit within pre-established roles, power relationships, and expectations that influence what they take away from that experience.

In a science classroom, a teacher has pre-assigned roles as a planner, leader, and evaluator. In these roles, teachers are expected to plan lessons, lead the lessons, and evaluate student performance. These roles, as well as the numerous additional roles that teachers embody, may be fulfilled in numerous ways. Some teachers interact with students in a coach role as they encourage students by making suggestions to improve and to "get ahead." Other teachers are primarily a lecturer passing their own knowledge on to their students. Each of these roles is achieved through different kinds of teacher–student positioning and these positionings are enacted through particular kinds of language use. Students are likely to approach science understanding in very different ways depending on the roles, positioning, and language use that the teachers and students take up.

One question that gets at teacher–student roles is "What is expected of students' talk and behavior?" Cobb's (1999) research in mathematics education offers a way to more clearly frame this question for science classrooms. In his work he differentiates three levels of social organization that pertain to how students grow into their classroom culture or community: classroom norms, socio-mathematical (or in this case, socio-scientific) norms, and mathematical (or in this case, scientific) practices.

At the broadest level, classroom norms involve the expectations and normative purpose for overall participation. That is, teachers establish what is motivating student activity, whether it be grades, finishing a project, or publicly explaining their claims and their reasoning so that everyone can understand (e.g., Cobb, 1999).

Cobb's second level refers to how particular disciplinary content areas have specialized ways of talking that are intimately tied to the big concepts within that domain (1999). This level of norms addresses the fact that some answers and justifications are more acceptable for a given community. For example, what counts as an adequate explanation of seasonal variations in temperature during a dinner conversation may not be an acceptable answer during science class. In environmental science, empirical data and models of the casual relations are critical, whereas at dinnertime, including these elements might make people think you are showing off. This is what Cobb means when he refers to this level of social organization in mathematics classroom as socio-mathematical norms. We translate the term for our purposes as socio-scientific norms.

At the most specific level, we differentiate socio-scientific norms from the specific ways in which tools and procedures are used to achieve scientific goals; this is the level of scientific practices. It is this level that addresses methods and tools for measurement of data, such as a procedure for calculating the amount of water in a soil sample. This scientific practice may be framed by the socio-scientific norm that the calculation is a tool for achieving a scientific goal of inquiry into soil and water. Rather than calculating in isolation, the calculations are framed by the classroom expectation that students must invent and thereby understand their own procedures and methods.

Ms. Cook: A Case Study

In this chapter we describe one teacher's approach to teaching science, including her learning objectives and the classroom culture she helped create to attain those objectives. In our previous work we have compared this teacher's implementation of a science unit to that of another teacher implementing the same unit (Enyedy & Goldberg, 2004; Enyedy, Goldberg, & Muir Welsh, 2006). Although the two teachers had exactly the same lesson plans from the same curriculum, they differed in their focus of science content and types of classroom communities they created. For example, Ms. Cook used inclusive language to encourage a community of co-inquirers; whereas, the other teacher's language and interaction created a boundary between teacher and student-learners. The differences eventually contributed to differential learning gains for the two groups of students, with Ms. Cook's students out performing the others. Based on pre-test and post-tests, Ms. Cook's students outperformed the students from the other classroom, particularly on items involving the interpretation of scientific data and measurement.

Setting and Context

The data in this chapter comes from a study in which we examined how, over a three-month period, two teachers and fifty-four students in a combined elementary and middle school in an urban area of Los Angeles implemented an environmental science curriculum, Global Learning and Observations to Benefit the Environment [GLOBE] (www.globe.gov). GLOBE is an international K-12 curriculum funded by the National Aeronautics and Space Administration and the National Science Foundation to promote learning environmental science and student engagement in authentic science activities. GLOBE links students, teachers, and scientists in a coordinated effort to learn about the earth's environment through observation, data collection, and analysis. Students collect environmental data in their local area and transmit their data via the Internet to an international database. Students then have access to data displays that are based on the combination of their data and the data collected by other schools around the world.

The students in this school are predominately Latino/a (97%) with a high percentage of ELLs (61%). The school itself is located within an industrial area, sandwiched between a paint factory, and abrasive treatment facility, and a cemetery. Because of its location, it is not a community school—most students are bussed in from neighboring residential areas.

The two participating teachers in the original study were both very experienced, each having nearly twenty years of classroom experience, much of it at their present school. Both teachers spoke Spanish as a second language fluently, and in interviews spoke of being dedicated to the community and to building personal relationships with their students. The teachers were committed to act as agents of change and encouraged their students to do the same. From our interviews with the teachers, we also learned that both teachers had similar pedagogical values and beliefs that could be fairly described as social–constructivist—with an emphasis on active learners and reflection through dialogue. Both teachers were also new to the GLOBE curriculum having gone together to the same four-day GLOBE training session. After the training, both teachers expressed concerns about how well they really understood the science content of GLOBE. In response, we provided both teachers the same amount of additional support (weekly meetings prior to and during the unit to plan the activities, practice the protocols, and reflect on the progress and difficulties the teachers and students were encountering), as well as in-class technical assistance with the computers and lab equipment. Despite these similarities, the teachers implemented the curriculum in different ways creating very distinct classroom communities.

Prior to the implementation of the GLOBE curriculum and at the end of the unit, the students were given tests that assessed their understanding of environmental science and their understanding of the tools and procedures of inquiry (e.g., graphing, reading maps, measurement error, etc.). Their pre- and post-test performance was analyzed to test our emergent hypothesis that the different classroom communities established in the two classrooms would influence student-learning outcomes. Students were assessed both on their ability to choose or produce the correct response and on their ability to explain their reasoning.

Student scores in both classrooms improved on the post-tests. A t-test showed that the mean score on the pre-test differed significantly from the mean score on the post-test for both classes suggesting that the students learned some amount of environmental science (t = 8.45, p < 0.01). As an additional method of viewing improvement, we calculated gain scores (the difference between the pre-test and post-test scores). In Ms. Cook's classroom, the mean gain score was 7.4 points versus only 2.6 points in the other classroom. This demonstrated a greater improvement in Ms. Cook's classroom (for a more thorough reporting of these results see Enyedy & Goldberg, 2004) and interested us in the positive learning environment created in her classroom through the different patterns of language use we had noted.

In this chapter, we present our analysis of Ms. Cook's classroom community, focusing on the relationship among her science instruction goals, teacher–student roles, and classroom discourse, including an examination of Spanish use in this bilingual classroom. Our analysis of this classroom makes apparent two of Ms. Cook's underlying science instruction goals: deep understanding of science (including the content and the process of scientific inquiry) and the importance of community. Through her classroom discourse, Ms. Cook positions herself and her students with respect to these goals and cultivates the following values: co-inquiring in a community of learners and connecting science and the "real" world. Ms. Cook's teaching practices and discourse choices support these underlying goals and values, although they do not always appear explicitly on written lesson plans and are not included in the official GLOBE curriculum. Part of these practices and choices involves her use of Spanish to support student learning. In what follows we detail how Ms. Cook made language choices that aligned with her values of co-inquiring in a community of learners and connecting science and the "real" world, and then we document the way she used Spanish to assist her in enacting these values.

Co-Inquiring in a Community of Learners

Ms. Cook applied her socio-constructivist beliefs about knowledge, teaching, and learning by changing the traditional power structures and roles in her classroom. Ms. Cook wanted to be perceived as a learner and as just one of many "teachers" in the classroom. From our video analysis (Enyedy & Goldberg, 2004), we termed this desired role that of "co-inquirer". For example, Ms. Cook took the position that, "I have just as much to learn from them as they do from me and they have just as much to learn from each other" (Conversation 1, 11/14/01). This statement exemplifies the value Ms. Cook placed on the co-inquirer role.

Importantly, recognizing students as co-inquirers was not limited to talking about science concepts. Ms. Cook established a relationship with students where they were comfortable talking to her about her teaching practices as well, even if that meant critiquing her co-inquirer stance. In reflecting on her role in the classroom, Ms. Cook stated:

Students have told me that what I need to do is, if somebody's struggling, sit down next to them and help them out. Before I didn't do that that much. I mean it sounds so dumb that I didn't do that, but I just—I don't know, my philosophy was a little different. And it's okay to be the teacher.

(Interview 1A, 5/30/01)

As the quotation above indicates Ms. Cook believed her students had something to teach her about her own teaching. They made her aware that sometimes it was necessary for her to shift out of her role as a co-inquirer. Her role should be contingent on the particular goals and situation. At times, Ms. Cook did "step-out" (see Rittenhouse, 1998) of her co-inquirer role to help students as a more knowledgeable participant, such as providing content information or directions for an activity. Yet she continually returned to her default role in the classroom as a model inquirer and critical questioner, "because I want them to start asking themselves questions."

Ms. Cook stated that she didn't "want to be the center of attention," because she wanted her students to become independent learners. To this end, she felt it was important to develop relationships with her students that took her out of the role of being the "fountain of knowledge" in the classroom. She wanted to instill in students a sense of belonging to a community in which all members work together to discover new things and solve problems. Her role in this community was still privileged over the students, but she avoided being the sole authority for what counted as knowledge in the classroom. For example, when students had the floor, such as when they were presenting their "Science Talks," Ms. Cook would not answer other students' questions. Instead, she insisted, as will be described below, that all questions be directed at and answered by the presenters themselves.

Science Talks were verbal presentations of a "family project" done mostly at home. Typically the students each chose their topic and invited their parents to the presentation. During a Science Talk the presenter is regarded as the teacher. Accordingly, Ms. Cook positioned herself as a learner or co-inquirer with the students. In one case when a student, Walter, had finished his presentation and was asking if there were any questions, a fellow student directed a question to Ms. Cook. She replied, "¡No soy experto! [I'm not an expert!] Ask Walter!" In ways like this, Ms. Cook played out her role of being both a teacher and a learner.

Additionally, Ms. Cook's pedagogical approach was premised on the assumption that there are multiple paths to answers and that there is no one right way. She did not want to adopt an authoritative position in the classroom. She thought it was important for her students "to try their ways, and for their ways to be validated. Questioned also, but validated definitely." For Ms. Cook, the bigger lesson to be learned is about science as discipline. As she stated:

It's not how to measure the soil and water.... What I would hope that they would walk away with from there is that there's a lot of ways to do things. That all ways are valid, and there's always a better way. And we can always learn from each other. That's basically it, really, when you come down to it.

(Interview 2A, 6/13/01)

Through other perhaps more subtle, yet powerful, ways, Ms. Cook's instructional discourse cultivated a community of learners within the classroom. When discussing science topics, Ms. Cook frequently used inclusive pronouns, such as *we*, *us*, and *our*. Through pronoun usage Ms. Cook helped unite the classroom community while still providing direction, such as in this segment from a lesson on soil moisture (inclusive pronouns are in italics and non-inclusive or exclusive pronouns are underlined):

> What *we're* gonna do today is the soil that you brought in from home. All of you brought in soil. *We* put it in the kiln.... *Our* idea is to see how much moisture is in the soil. So after *we* did all of that experimenting with the baggies, Tony, with the sponges, *we* decided as a group that *we* were going to weigh the wet soil.
>
> (Classroom Video, 5/17/01)

Ms. Cook's use of inclusive pronouns helps establish a sense of community while reserving exclusive pronouns for specific, individual tasks.

The heavy use of inclusive pronouns continues throughout the lesson, such as in the following excerpt from the discussion of the moisture formula:

> *Our* idea is to see how much moisture is in the soil. So after *we* did all of that experimenting with the baggies, Armando, with the sponges, *we decided as a group* that *we* were going to weigh the wet soil. And *we* weighed the can by itself. Correct? Everybody weighed the can and weighed their wet soil.
>
> (Classroom Video, 5/17/01)

By using inclusive pronouns, Ms. Cook helps students feel they are an integral part of the activity and working together as a class unit. When exclusive pronouns were used, they tended to be reserved for specific, individual tasks or specific feedback to students.

Photo 6.1 Ms. Cook displays a soil sample during whole class discussion.

Another tool that helped students fill the role of co-inquirer involved organizing activities so that there was frequent movement between whole class and small group work, encouraging students to interact in a variety of ways while exploring a concept. Students were expected to communicate with their peers and teacher about topics. Ms. Cook regularly encouraged students to talk it out. For example, when introducing a new lesson on humidity and the use of a sling-psychrometer, she asked the students, "I guess my question to the group is this: Is there a big difference between wet and dry? ... This is what I want you to discuss right now. Try to figure it out, start talking" (Classroom Video, 4/27/01).

As these examples make clear, there are multiple paths to cultivating a community of learners through language use. First, Ms. Cook spoke directly with her students about the roles of teacher and students during classroom interaction and monitored these when, as in Walter's example from the Science Talks, she noticed students shying away from their "teacher" responsibilities. Second, she used pronouns to emphasize the co-inquirer role she wanted students to assume with her in the classroom community. And third, Ms. Cook organized activities to involve student talk in multiple venues, including both small group and whole class discussions. In the section that follows we explore how during class discussions and activities, Ms. Cook helped students make connections between science concepts and previous lessons and to the "real" world.

Connecting Science and the 'Real' World

Students' backgrounds also influence the way they approach science content. Ms. Cook summarized the challenge her students' backgrounds posed to her:

> Science is often a difficult one for me because of the myths and cultural beliefs that my students bring with them—some of which are hundreds of generations old. Who am I to say, "You, your parents, and your ancestors

Photo 6.2 Students in small groups use the sling psychrometer to measure humidity.

are wrong. Scientifically what you believe is not true." I would never do that. I do not even want the students to question their beliefs—or better said—I do not want the students to see science in conflict with their beliefs.

(MSE definition, 5/2002)

In reflecting on teaching practices and student learning, Ms. Cook stressed the importance of connecting science concepts to student backgrounds.

During the implementation of GLOBE activities, Ms. Cook often connected science concepts. This was a critical feature of talk about science in this classroom. Topics were connected to other science concepts as well as to familiar situations. In fact, Ms. Cook often made explicit connections between science and students' lives beyond the school walls. She did this intentionally and frequently discussed the importance of these connections during interviews. This focus became clear within analyses of classroom interaction of both teacher and student talk.

For example, in one lesson, students investigated the temperatures of water and soil at various depths. In the introduction, Ms. Cook had students imagine the temperature of sand and the ocean at a beach: "We're going to spend the whole day at the beach ... I mean we go there in the morning and you're all hot and you're playing ... And you have to walk on the sand like this and it gets wet." She had them imagine that they run down to the water and put their towels down on the sand so that it doesn't hurt their feet. During her monologue, student comments connected her description to their own experiences. For example, a student connects the beach discussion to being at the fair, "When I go buy a snack, you know, it [the sand] burns." She made these seemingly side, or peripheral, comments legitimate by building her imagery on them: "And then at high tide you make a fire and you taste marshmallows and hot dogs" (Classroom Video, 4/25/01).

In this lesson, Ms. Cook directed students to, "Talk to your friends about a time that you have been to the beach and about how, how the ground felt, and how the water, how the water felt" (Classroom Video, 4/25/01). This directive communicates to students that they would be discussing this topic with peers and connects it to a familiar activity for these students—going to the beach. Students are given the opportunity at this point to learn with their peers (as co-inquirers) and to continue to build on the imagery as they make their own connections to the temperature of sand and water.

In the classroom community, the process of discovery was stressed as this teacher and her students explored science concepts together. Ms. Cook often began lessons with students by engaging in some sort of free exploration with them or even "goofing around to see how it might work." This was followed by students generating their own conjectures, which they must put in writing prior to engaging in any formal inquiry. The constructivist nature of the classroom activities coupled with frequent connections to previous lessons and students' home lives helped students better understand science concepts as co-inquirers.

Speaking Spanish and English in the Classroom

Another dimension of the classroom culture and discourse that emerged in Ms. Cook's classroom was the use of multiple national languages (Spanish and English).[2] The fact that most students in a classroom are bilingual does not necessarily mean that it is a bilingual classroom. The teacher's ability to understand and speak Spanish and the politics of the school district and the state can greatly influence how Spanish, or any other national language, is perceived and used in the classroom.

Officially, this classroom was not designated as a bilingual classroom. However, this school, though located in a state that mandated instruction only in English, was exempted so that teachers could use multiple languages. Even though Spanish was not recognized as an official language of instruction, Ms. Cook and her students sometimes chose Spanish as the preferred mode of instruction. Both English and Spanish were used as tools for better understanding science concepts. In Ms. Cook's classroom Spanish was common, but the frequency varied with the social context and the activity. This is to be expected because, in a bilingual classroom, multiple languages often have different roles.

As in many U.S. classrooms, even bilingual ones, English is the dominant language spoken during public, whole class science discussions. There are many reasons for this practice, including the desire to model academic English for students who are still acquiring that register. Yet even when Ms. Cook speaks in English, students often respond to her in Spanish. Their comfort level in choosing either language may be due to the way Ms. Cook has positioned them as co-inquirers and because of the way in which Ms. Cook reacts to languages as they are spoken. When Spanish is spoken in whole class discussions, Ms. Cook typically does not comment on the fact that Spanish is spoken and she does not translate or mark the statement as different in any way. Instead, any reasonable contribution is accepted and responded to regardless of the language in which it is made. This further strengthens the inclusive nature of the classroom community.

Photo 6.3 Whole class discussion of atmospheric readings.

For example, each day the students take atmospheric readings at solar noon, which is at a different and predictable time each day. In a discussion about how to determine when it is solar noon, Ms. Cook mentioned that she forgot to take yesterday's measurements for a group of students who were unable to make their own. A student responds that she remembered to take the measurement, and the teacher verifies that the student measured at the right time by asking her if she had seen her shadow. Ms. Cook's question did not ask about the calculation, but got at the concept of what solar noon means—the time when the sun is directly overhead. Notice that in the excerpt below Ms. Cook responds to, but does not translate the contributions made in Spanish. This may send a message to students that it is acceptable to speak in Spanish in the classroom, and at the same time models how to talk science in English.

MS. COOK: Please. I forgot to do it yesterday I was too busy eating my lunch [laughs]. I forgot to go out at 12:51.
STUDENT 1: I *Fui* [I went]. Ms Cook. I did it—I did it.
MS. COOK: Did you have a shadow?
STUDENT 1: No.
MS. COOK: No shadow? Cool.
STUDENT 2: There was no sun.
MS. COOK: Yes there was, yesterday there was.
STUDENT 1: *El sol estaba arriba, arriba, arriba y aparecí y no vi nada.* [The sun was up, up, up and I showed up and did not see anything.]
MS. COOK: Oh but there were a lot of clouds yesterday—
STUDENT 1: —*por eso usted no podría ver nada.* [that's why you couldn't see anything]
MS. COOK: OK, well you shouldn't see it—hopefully today there won't be.
(Classroom Video, 4/27/01)

In this exchange, Spanish and English are both positioned as legitimate tools for students to use to help explore science concepts. Both the teacher and the student begin the exchange in English, and part way through the student switches to Spanish when narrating her actions from the day before. The teacher continues to reply in English, but also responds to the student's contributions in Spanish. Imagine a different case. What if Ms. Cook had translated the turn into English before responding? What if she had made the student rephrase her contribution in English? If this were to happen, Spanish would have been positioned as being unacceptable as a way to talk science in this classroom. Because the Spanish contributions would be treated differently than contributions in English (which are not marked as different by translating them), we would expect that eventually Spanish would disappear from the public talk in the classroom. Furthermore, the focus would shift from understanding science concepts to emphasize language choices. Obviously, this would run counter to Ms. Cook's goals of being inclusive and honoring multiple ways to learn science.

A similar balance between English and Spanish was achieved during the students' formal presentations. In this classroom, oral presentations were made exclusively in English. However, when students were asked questions, they felt free to answer them in Spanish. For example, during a student group's presentation about surface level ozone (Classroom Video, 5/4/01), Ms. Cook commented on a student's picture of a typical traffic jam on a Los Angeles freeway: "You have to explain your picture because I love it." The first student responded in English saying, simply, "freeways." A second student elaborated in Spanish saying, "*las llantas queman y huelen malo* [the tires burn and it smells bad]." Even though the science content is not correct in this example—ozone is a secondary chemical reaction between light energy and carbon dioxide from the car's exhaust, and has nothing to do with the car's tires—the student is still actively participating in the class' dialogue. In the long term, participation is more important than this one conceptual error. If the student continues to actively participate and attempt to make sense of the science concepts he is learning there is always the opportunity for the teacher, his peers, or for him to correct the misconception. The point to note here is that when students were asked to elaborate and go further, this exchange showed that they were enabled to do so by the fact that they were allowed to communicate in Spanish and did not have to worry about struggling with both the science content and its expression in English at the same time. Concurrently, Ms. Cook's responses in English provide a model for how to talk science in English.

To summarize, most of the formal, public talk during whole class discussions and formal presentations was in English. Students occasionally spoke Spanish, and when they did so Spanish was respected and responded to without translating it or marking it as unacceptable. However, Ms. Cook herself consistently spoke English in these whole class discussions—modeling how to talk science in English as she did so.

In private interactions with individual students and during small group work, the language dominance was reversed. In these situations, Spanish and code switching between Spanish and English were the norm. Students communicated

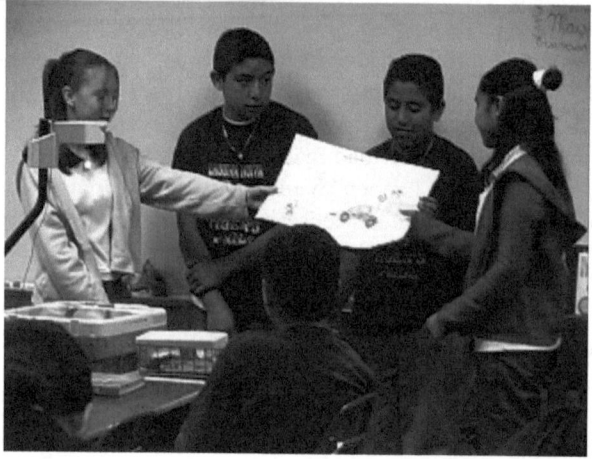

Photo 6.4 Small group presentation of their smog poster.

in whatever language they found most comfortable and helpful. The emphasis was on making sense of the experiments and the data and in these cases Ms. Cook usually responded to Spanish in Spanish.

Small group work comprised a significant portion of the "science time" in this classroom. As a result, Ms. Cook spent much of her time interacting with students in this small group context. One result of responding to Spanish with Spanish and English with English was that it relinquished some of the agency and power over to the students. During the most critical parts of science class—the time when students were trying to interpret and make sense of the science concepts and connect them to what was happening in their experiments—the students were able to choose the language that they thought would be the most useful to them at the time.

For example, during an experiment where students were investigating the difference between the rate of temperature change between water and soil, students used both English and Spanish to work out what was happening with their data. Specifically, the students reported their results and their conjectures to each other in English, but explained their reasoning to each other in Spanish.

In the exchange below, the students have just recorded the temperature for both their soil and water samples at 8 cm, 1 cm, and 1 cm above (i.e., the air temperature). In comparing their readings they discovered that the air temperature 1 cm above the water was cooler than the air temperature directly above the soil. With no teacher present they spontaneously began to discuss why this was the case. Note that this exchange begins and ends with Spanish utterances.

STUDENT 1: Oh, my gosh *(???) sube nada* [nothing is going up]

[Thirty second pause while they record their measurements.]

STUDENT 1: Hey you know what I think. It's not the soil it's the air. That's hot.

STUDENT 2: The air?

STUDENT 1: Yeah.

STUDENT 2: Let me see.

STUDENT 1: (Eight) centimeters.

STUDENT 2: No eight centimeters above is twenty-one. Eight below is twenty-two. One centimeter above is thirty [referring to the soil temperatures].

STUDENT 1: Damn. Mine's one centimeter above is nineteen. I think the water's cold [referring to the water temperatures].

STUDENT 2: I think it is the air that is cold.

STUDENT 1: You think so.

STUDENT 2: For the water let me see.

STUDENT 1: *No va más bajo que esto* [It doesn't go lower than this]. It's just like that, cause for this. *Su no, su no el mismo* [Its not, its not the same].

(Classroom Video, 4/27/01)

If we focus on Student 1 and examine the exchange in terms of the function of the different utterances, we see an initial observation which surprises the students and prompts them to try to make sense of the data. This precipitating observation was made in Spanish. It is followed by a hypothesis (in English), and some measurements (also reported in English). Finally, after Student 2 states her conclusion, "that it is the air that is cold," Student 1 disagrees and counters in Spanish. Significantly, it is in this counter-point where the data is connected to a conclusion to form a basic scientific argument. This kind of active sense-making and scientific argumentation is at the heart of a constructivist approach to science instruction, and it was important to Ms. Cook that the students engage in this type of thinking and talk regardless of the language used.

One implication is that Spanish emerged as an important resource for meaning making in this classroom. Spanish, in both public and private spaces was seen as legitimate—no one (teacher or students) commented on the fact that students spoke in Spanish or code switched back and forth between Spanish and English. However, language choice did follow patterns based on the roles of Spanish and English in this classroom. During small group work, Spanish was dominant during student sense making. Since new activities started with some time for small group exploration, Spanish played a key role in figuring out the task, reflecting on what they already knew about the concept, and for making conjectures and hypotheses to test. In the whole class discussions, Spanish was still legitimate, but Ms. Cook paid more attention to modeling English and the students seemed to respond by speaking English more frequently.

Discussion

When implementing a science curriculum, educators focus on content that meets prescribed standards while considering activities that will help students learn the particular scientific content. As activities unfold in everyday classroom interaction, additional classroom goals emerge, as roles are co-created between teachers and their students. In Ms. Cook's classroom, her language choices supported her over-arching goals of developing in students a deep understanding of scientific ideas and a high regard for the importance of community. To help students attain these goals, Ms. Cook positioned herself as a learner alongside her students as they co-inquired about science. Connections were made to the "real world" so that science was meaningful beyond the school walls. Both English and Spanish were used as legitimate resources for students to help make sense of the content.

There are multiple, legitimate ways to organize any classroom. Each organization is well adapted to certain goals while less effective for others. Ms. Cook chose to structure lessons that supported the goals and roles described above. One teaching practice that helped to cultivate the goals and roles of this classroom involved small group exploratory activities before beginning prescribed experiments. As we saw in the example described previously, when Ms. Cook asked her students about using the sling-psychrometer, "I guess my question to the group is this: Is there a big difference between wet and dry?... This is what I want you to discuss right now. Try to figure out, start talking." These mini-activities and small group discussions encouraged students to work as a community of learners and positioned science as a process of inquiry rather than a prescription to follow step-by-step. Another teaching practice involved making connections, such as in the example of Ms. Cook asking the students to visualize a trip to the beach and then asking students to "talk to your friends about a time that you have been to the beach." This visualization and the following discussion involved sharing explicit connections between the science activities and the students' home lives, as well as connections between various science lessons. Students cannot be silent in Ms. Cook's classroom; they must talk, share, hypothesize, and inquire.

Classroom talk and interaction are fluid and dynamic. There is no way to consistently predict what students will say or how a teacher will respond in various situations, even when lessons are scripted or official classroom rules are posted. Yet clear discourse patterns emerge in science classrooms. In our discussion, we have seen how Ms. Cook's frequent use of inclusive pronouns was a discourse pattern that helped unite the classroom community and further positioned this teacher and her students as co-inquirers.

In a bilingual classroom, an additional layer of complexity is present when analyzing discourse. Official and unofficial rules are generated about which language to use. We have seen how students in Ms. Cook's classroom felt comfortable using both Spanish and English in a variety of settings. English was the dominant language in public discourse, such as during whole class instruction (which was dominated by teacher talk). During small group activities, Spanish emerged as the dominant language and was more often 'controlled' by students. However, in all settings both languages were interspersed and available for students to use to better understand and communicate about science concepts.

Although careful planning is helpful in attaining science goals, consideration must also be given to the roles that emerge in everyday classroom interaction. Teachers and researchers may better understand opportunities that students have to learn by analyzing how talk in the classroom supports or hinders overarching goals. Also, a teacher's beliefs about her or his role further contribute to science-learning and other classroom goals. As stated earlier, Ms. Cook believes strongly that her role is to learn in tandem with the students. All members of the community of learners contribute to the learning of the community. In all settings, but perhaps more so in bilingual settings, issues of power must be considered in terms of goals, roles, and discourse (Darder, 1991, 2002).

The examples we have provided of the ways in which Ms. Cook and her students positioned themselves in the classroom and achieved particular roles are from the end of the school year. The history of this classroom community is tied to these roles. In other words, all of the preceding interactions from the school year have influenced the roles that emerged. It takes time for rules and roles to be cultivated when teachers choose to position themselves in ways that are not already familiar to students. Most students have been socialized to normative, traditional roles (such as waiting to speak until the teacher calls on them). When considering goals and roles of a science classroom, a teacher must consider the background and previous experiences of students and build off of these important components when co-developing a learning community with his/her students. Teachers must work with their students to establish the norms of their classroom environments with each new group of students. Students need to be socialized to their roles in a learning community.

Teachers, like Ms. Cook, struggle with the best ways of presenting science content, particularly in culturally and linguistically diverse settings. Ms. Cook's comments at the beginning of this chapter mention these struggles but that in the end the results are so "absolutely powerful" that teaching science is worth the effort. Although there are no quick fixes, teachers can move beyond simplistic notions of lesson plans and focus on their overarching goals for their classroom. From our analysis of Ms. Cook's classroom, we are convinced that teachers' language choices play an important role in representing teacher and student roles. Teachers dedicated to their convictions about the power of science, as is Ms. Cook, should begin to understand their language use as a way to begin to better understand the relationships that exist in these complex classroom settings.

Targeting Enduring Understandings

1   What are the connections you see between the discussion in the Goldberg team's chapter and the four elements of culturally-responsive instruction, as described by Villegas and Lucas?
2   How would you describe the "work" that language is doing in this multi-cultural science context?

Deepening the Reflection

1   Describe the overarching goals for your science classroom. How are they similar to or different than Ms. Cook's?
2   Reflect on a science classroom you have recently visited. How were the teacher and the student positioned with respect to one another? How did these positionings influence the way in which activities were conducted? How were issues of power interwoven into these positionings? How was this power reflected in language use?
3   How would you explain the relationship between Ms. Cook's approach to the use of Spanish in her classroom and her other goals?

Encouraging Engagement

1   Arrange to visit a bilingual middle school science classroom, similar to Ms. Cook's. In your observation, collect information to answer the following questions:
  a   What are the classroom norms about talking? For example, who dominates talk during whole class or small group interaction? When a student wants to share an idea, is there an expectation that the student will raise his or her hand first?
  b   What language is the dominant one, spoken during various activities (whole class and small group interaction)? Does there seem to be rules about which language is to be spoken during which time?
  c   Who determines which language to use? Is it the teacher or the students, or both? What are some factors that seem to influence language choice?

    d   What is the relationship between the use of the non-English language(s) and science content development?

    e   From your observation, make a list of lessons you learned about language use in bilingual science classrooms and come prepared to share those with your peers.

2   Interview a science teacher in your local district about their philosophy and experiences working with ELL students. In what ways is their approach similar to or different from Ms. Cook's? What do you learn from the interview about this teacher's understanding of the relationship between language, culture, and science?

Notes

1 All names used in this chapter have been changed to provide anonymity to study participants.
2 We use the term "national language" to separate this level of analysis from an analysis of the technical register of science discourse, the specialized vocabulary and grammar of science, which is distinct from everyday talk regardless of whether the technical language is spoken in Spanish or English.

References

Boaler, J., & Greeno, J. (2000). Identity, agency and knowing in mathematical worlds. In J. Boaler (Ed.), *Multiple perspectives on mathematics teaching and learning* (pp. 171–200). Westport, CT: Ablex.

Bransford, J. D., & Donovan, M. S. (2005). Scientific inquiry and how people learn. In M. S. Donovan & J. D. Bransford (Eds.), *How students learn science in the classroom* (pp. 397–420). Washington, DC: National Academy Press.

Brown, B. A. (2004). Discursive identity: Assimilation into the culture of science classroom and its implications for minority students. *Journal of Research in Science Teaching, 41*, 810–835.

Cazden, C. B. (1986). Classroom discourse. In M. C. Wittrock (Ed.), *Handbook of research on teaching* (3rd ed., pp. 432–463). New York: Macmillan.

Cobb, P. (1999). Individual and collective mathematical development: The case of statistical data analysis. *Mathematical Thinking and Learning, 1*, 5–43.

Darder, A. (1991). *Culture and power in the classroom.* Westport, CT: Bergin & Garvey.

Darder, A. (2002). *Reinventing Paulo Freire: A pedagogy of love.* Cambridge, MA: Westview.

Duschl, R. A. (1990). *Restructuring science education: The importance of theories and their development.* New York: Teachers College Press.

Enyedy, N., & Goldberg, J. S. (2004). Developing classroom communities for understanding through social interaction. *Journal of Research in Science Teaching, 41*, 905–935.

Enyedy, N., Goldberg, J. S., & Muir Welsh, K. (2006). Complex dilemmas of identity and practice. *Science Education, 90*, 68–93.

Erickson, F., & Mohatt, G. (1982). Cultural organization of participation structures in two classrooms of Indian students. In G. Spindler (Ed.), *Doing the ethnography of schooling* (pp. 133–174). New York: Holt, Rinehart & Winston.

Fairbrother, R., Hacking, M., & Cowan, E. (1997). Is this the right answer? *International Journal of Science Education, 19*, 887–894.

Greeno, J. G., Benke, G., Engle, R. A., Lachapelle, C., & Wiebe, M. (1998). Considering conceptual growth as change in discourse practices. In M. A. Gernsbacher & S. J. Derry (Eds.), *Proceedings of the twentieth annual conference of the Cognitive Science Society* (pp. 442–447). Mahwah, NJ: Lawrence Erlbaum Associates.

Hammer, D. (1997). Discovery learning and discovery teaching. *Cognition and Instruction, 15,* 485–529.

Lave, J., & Wenger, E. (1991). *Situated learning: Legitimate peripheral participation.* Cambridge, UK: Cambridge University Press.

Lemke, J. L. (1990). *Talking science: Language, learning, and values.* Norwood, NJ: Ablex.

Linehan, C., & McCarthy, J. (2001). Reviewing the "community of practice" metaphor: An analysis of control relations in a primary school classroom. *Mind Culture and Activity, 8,* 129–147.

Mehan, H. (1985). The structure of classroom discourse. In T. A. Van Dijk (Ed.), *Handbook of discourse analysis* (Vol. 3, pp. 119–131). London: Academic Press.

Ritchie, S. M. (2002). Student positioning within groups during science activities. *Research in Science Education, 32,* 35–54.

Rittenhouse, P. (1998). The teacher's role in mathematical conversation: Stepping in and stepping out. In M. Lampert & M. Blunk (eds), *Talking mathematics in school.* New York: Cambridge University Press.

Wells, G. (1999). *Dialogic inquiry: Towards a sociocultural practice and theory of education.* New York: Cambridge University Press.

7 Locating Time in Science Classroom Activity

Adaptation as a Theory of Learning and Change

Jorge L. Solís, Shlomy Kattan, and Patricia Baquedano-López

Introduction

In this era of accountability and high-stakes testing, there is more pressure than ever on teachers to manage how they use their time. Teachers are expected to cover a certain amount of material over the course of the academic year and student progress is measured through a routine of time-marked assessments. The school day itself is, and traditionally has been, broken up into time-based segments and student performance measured by completion of assignments within certain time-based frames. Have you ever considered what the broader implications of such a time-driven view of teaching and learning may be?

The unilinear, forward-looking orientation to time that characterizes schooling is so pervasive to the school environment that it's simply something we all take for granted. Yet in this chapter, Solís and his colleagues suggest that we have much to learn by paying attention to time orientations in schooling and, more specifically, by understanding the work that language does to reinforce conventional, and construct alternative, temporal realities. The tension created by the discontinuities that arise between conventional school routine and dynamic teacher–student engagement drives an adaptive process that is productive for learning. Understanding culturally-responsive science pedagogy as the process of negotiating between student-generated alternative temporal frames and the official timeline frames of schooling sensitizes us to the hidden challenges of such pedagogy while alerting us to the presence, in science classrooms, of larger issues of knowledge and power.

As you read Solís and his colleagues' chapter, be aware of how your taken-for-granted notions about time in classrooms are being confronted. How does the tension you read about between a white science teacher's, Ms. Anna's, use of the scientific (i.e., timeless) word "solid" and that of Kevin, her African-American student, who uses it in a dated, vernacular sense of "solidarity," make you think differently about how the construct of time construes teacher–student interaction in science teaching and learning and how, further, that construct is mediated through language? How does an Asian-American teacher's, Mr. Pepp's, verbalized emphasis on timely completion of assignments, and his Latina students', Emily and Claire's, work to meet his timeline clarify for you how science classrooms, like all school contexts, discipline students into a kind of time-based work-place ethic?

As a science teacher, you will find yourself in many situations like those encountered by Ms. Anna, Kevin, Mr. Pepp, Emily, and Claire. This chapter is sure to make you think differently about science teaching, specifically the work that language-construed notions of time perform in the classroom, and the knowledge and power those notions invoke, whether you realize it or not.

In linguistically and culturally heterogeneous classrooms teachers and students employ a number of strategies to identify and respond to the naturally arising tensions between and among their different backgrounds and expectations. Such tensions take place when the organization and goals of classroom learning activity are no longer transparent or when there is a break in the continuity of classroom events. In previous work (Baquedano-López, Solís, & Kattan, 2005) we have discussed the ways in which the reactions of teachers and students to acts that counter expected behavior in the classroom, such as students not being called upon when raising their hands, combine to create educationally constructive tensions. In this chapter we focus on the ways teachers and students negotiate and orient to tensions inherent in institutional and local understandings and constructions of *time*, analyzing how such negotiations organize everyday normative classroom interaction. Drawing on data collected as part of a longitudinal study of urban elementary science classrooms, we argue that features of scientific discourse (e.g., predicting, estimating, generalizing) have an implicit "time" perspective that organizes scientific learning activity and shapes and reproduces scientific knowledge and epistemologies, or ways of thinking about that knowledge. While such time encoded understandings of learning science tend to reproduce power differentials in ways that support long-standing notions of a unilinear, irreversible learning process (Valsiner, 2002), teachers and students also continually re-negotiate and thus adapt the temporal parameters of their learning activities. The explicit management of temporal experience as coherent and sequentially ordered events or shared knowledge thus appears as a focal theme of tension.

We begin this chapter by introducing our methodological approach and our adaptation framework for analyzing classroom discourse (Baquedano-López et al. 2005). This framework examines the nature of tensions and discontinuities that emerge in competing interpretations of content or subject matter, the different social and cultural histories of participants, and the different expectations they hold for their learning activities. Such tensions and their ensuing negotiations, while largely emerging from ongoing activity, also project alternative temporal possibilities for talk and action against an unfolding curricular timeline. That is, classroom interaction is encoded with but also contests institutional timelines for learning. While schooling imposes work timelines on teachers and students based on generic, shared social histories, our discussion of adaptations uncovers the prosaic socio-cultural discontinuities in how time works to assert power relations and produce knowledge in the classroom. We discuss a number of features of scientific discourse (in particular hypotheticals and conditionals) and reexamine them in relation to more recent understandings of time across disciplinary fields. We also provide a critical review of learning theories and their implicit understandings of time. More specifically, we critically examine the ways "time" has been analyzed in the educational, social science and language-related literature. Through examples of data collected in 3rd- and 4th-grade science classrooms we examine classroom interactions that illustrate an emergent inter-actional grammar for understanding constructions of time in learning. This interactional grammar is significant because it signals and constitutes collabora-tive academic disciplinary engagements of a range of socio-cultural expectations in the classrooms. Awareness for how teachers and students prioritize schooling expectations may afford more responsive pedagogical stances, or what Villegas & Lucas refer to as "gaining sociocultural consciousness" of the unequal reproduc-tive powers of schooling.

A Cross Disciplinary Perspective

Data Sources

The framework for understanding time in learning advanced here evolved from analysis of data collected as part of a three-year long research project studying a science curriculum implementation, Science Instruction for All (SIFA), which was carried out at school sites in Florida, Arizona, and California. From 2001–2004 our university research team collaborated with teachers to imple-ment a science curriculum at six elementary schools in the greater San Fran-cisco metropolitan area in Northern California. Designed in response to a national call for more equitable and inclusive science education at the elemen-tary school level (see Lee & Fradd, 1998; Rosebery, Warren, & Conant, 1992), the SIFA project included local instructional practices and goals in the design and implementation of a highly demanding science curriculum. Our project data include student test-scores, students' written work, ethnographic field notes of classroom observations, pre- and post-interviews with teachers, and video and audio recordings of naturally occurring classroom interaction. A total of fifty-two lessons in ten 3rd- and 4th-grade elementary school classrooms were video recorded. Consistent with findings across the three research teams, quantitative data based on pre- and post-test scores indicated an upward pro-gression and a closing of the achievement gap for all students (Ku, Garcia, & Corkins, 2005).[1] Our ethnographic perspective documented student perform-ance and the implementation of the SIFA curriculum across the duration of the project.

Theory and Method

Starting with the sociocultural premise that language is the primary means through which shared meanings are constructed, mediated, and reproduced into experience and knowledge (Cole & Engestrom, 1993; Wertsch, 1991), we focused on routine classroom activities as units of analysis and thus the central organizing principles of our data collection (Duranti, 1997). Drawing more specifically on the analytical methods of Conversation Analysis (Goodwin & Heritage, 1990; Sacks, Schlegloff, & Jefferson, 1974), as well as on theoretical insights from ethnomethodology and interactional sociology (Garfinkel, 1967; Goffman, 1981; Goodman, 1984) we recognized the primacy of face-to-face encounters in the reproduction and contestation of the social order (see Baquedano-López et al., 2005, Schegloff, Jefferson, & Sacks, 1977). We employ the notion of *participant framework* (Goffman, 1981; Goodwin, 1990; O'Connor & Michaels, 1996) as a way to describe how participants in ongoing activity occupy different roles that are improvisationally shaped through talk across different social settings, including classrooms. Participant frameworks are the ways in which participants in activities create, reproduce, or even challenge social alignments among themselves and others primarily through talk. We recognize, however, the inherent temporalizing bias underlying these approaches to the analysis of talk in interaction. This bias rests primarily in the linearity for noting and analyzing talk (Ochs, 1979; Varenne, 1998). As Ochs (1979) and Varenne (1998) have duly noted, methods for transcribing, and thus analyzing talk, that is the way speech is physically organized and represented on the written page, defaults toward, and is dependent upon, a view of time as unfolding in a unidirectional manner. In this regard, our analysis of temporalizing discourses, is to a certain extent, limited in that our current methods for studying classroom interaction also enforce a perspective of "time" along an irreversible and successive chain of past, present and future activity. Inasmuch as we share this limitation with many approaches to the study of talk and classroom interaction, we hope to contribute to the development of more dynamic ways for analyzing talk, interaction, and learning in classrooms.

Adaptation in Classroom Learning

The ideas that we advance in this chapter are based on our ethnographic observations of the ways in which teachers and students continuously pushed routine classroom science activities in ways that triggered new temporal connections, curricular tensions, and forms of participation in the classroom. Building on our previously developed framework for analyzing tensions and discontinuities in science classroom activities, we align ourselves with literature that examines how learning occurs through collaborative engagements where the potential for conflict and tension exists (Gutiérrez, Baquedano-López, & Tejeda, 1999; Gutiérrez, Rymes, & Larson, 1995; Matusov, 1996). Taking as a starting point the widely accepted idea that learning is the result of shifts in participation over time (Rogoff, 1990; Lave & Wenger, 1991), a perspective that is most central in sociocultural theories of learning, we expand this notion to also include as part of these shifts the accompanying reorganization of social and cognitive parameters during collaborative learning activity (Baquedano-López et al., 2005). We believe that sociocultural perspectives on learning must be expanded to include a discussion of the social and discursive nature of time to fully capture the ways teachers and students accomplish learning goals in relation to a recognizable (yet often implicit) curricular and institutional timeline. Teachers and students construct alternative temporalities to adapt to institutional curricular timelines. Institutionally constructed temporalities organize and impose expected forms of participation and other orderings (academic year curricula, lessons, units, grade level expectations, etc.). More precisely, time-coded expectations highlight tensions underlying every classroom encounter. For example, every instance that classroom participants perform tasks like taking attendance or other initiating classroom routine (e.g., entering the classroom through the door, passing out materials, sitting silently), classroom participants engage potential breaches in executing expected time-coded norms.

Our study of adaptation strategies originated with the observation that teachers repeatedly incorporated new materials to those provided to them by the SIFA project. This trend reflects a persistent teacher inclination to integrate students' online experiences into the curriculum and thereby, build on and affirm students' understandings of science. Moreover, this occurrence motivated us to rethink planned curricular activity. We argued that the changes that teachers make as they teach is not a necessary cause to think that their deviations are problematic, but rather that it is important to understand variation as inherent to curricular implementation. Such variation, in the form of curricular modifications, became the focus of our study of adaptations as unpredictable and unplanned, yet systematic and patterned. That is, while it is true that teachers will often look to modify the received curriculum to make it more suited to their needs, there are, in addition, changes that take place which arise during instruction and yet which are nonetheless confined to occur within the discursive and cultural strategies already available to participants. While related to the notion of constructivist pedagogy, adaptations point out that learning draws from self-organizing interactional dispositions. Teachers certainly can take on the role of distributing a kind of social dialect of knowledge-making[2] (embracing constructivist stances) that push for more inclusive student sharing of knowledge production and by connecting familiar and new experiences. Adaptations, however, operate at a different level and allow us to examine both how social dialects of learning like constructivist pedagogy are used and how other kinds of sociocultural resources are used to achieve these and other ends.

Adaptations are a productive feature of classroom interaction in that they reorganize both the cognitive and social domains of learning activity. For example, shifts in discursive genres from question to narrative, or from small-group to class-wide participation, indicate not merely changes in how participants employ symbolic and material tools (such as language), but also constitute changes in terms of what counts as knowledge on the individual and group level (cf. Solís, 2004). Adaptations, in this sense, are responses to tensions and breaches (see Garfinkel, 1967) and promote new forms of cognitive and social engagement for those participating in classroom learning. Diverging from previous accounts of curricular adaptation (Erickson, 1983; Hunt, 1981; C. D. Lee, 2001; Linn, Clark, & Slotta, 2003; O'Donoghue & Chalmers, 2000), or even the larger anthropological notion attached to it (Alland, 1975),[3] it is our position that adaptation is not a teacher-driven activity based on the modification of curricular materials to accommodate student needs (adapt the curriculum to the student), nor is it emblematic of an assimilationist orientation (adapt the student to the curriculum). Rather, adaptation is a ubiquitous, improvisational, strategic, discursive phenomenon by which all participants make sense of ongoing classroom events. In this sense, adaptation drives learning.

The process of adaptation we describe is most noticeable through what has been identified in ethnomethodology as a *breach* in ongoing interaction (Garfinkel, 1967). A breach, in essence, is a visible violation or disruption to what is agreed-upon and expected in routine activity. Breaches are common in classroom interaction, as illustrated, for example, when students do not respond to a teacher's question or when peers refuse to work in a group. Adaptation strategies might include a reformulation of the teacher's questions (by the teacher or a student), a complaint, or a reorganization of group activity. While in previous work we examined breaches to shared social norms and classroom expectations (Baquedano-López et al., 2005), in this chapter we consider in more detail breaches to the temporal ordering of those expectations. In so doing, we expand our analytical focus to include a discussion of interactional and grammatical resources that create, or as we contend, elaborate on established constructs of time. These discursive devices include, among others, transpositions, quotations, and projections, and often draw attention to breaches and tensions in interaction and reveal challenges to established temporal expectations for knowledge-making and learning. These discursive strategies are amply discussed in the literature on human interaction (Bühler, 1990; Hanks, 1990; Haviland, 1996; Volosinov, 1973) and, to some extent, in the education literature (Cazden, 2001; Michaels, 1981). While we explicate these terms in greater detail below, suffice it to mention at this point that we have found that science activities promote the use of many of these discursive devices. Before illustrating how these devices are employed in the classrooms we observed, we discuss a general perspective on the ways in which time has been operationalized in extant theories of learning.

Time and Learning: Developmental and Sociological Perspectives

While there have been critiques of the use of a notion of time as a constant against which curricular and even institutional educational activities are measured (Orellana & Thorne, 1998; Slattery, 1995), there are few comprehensive descriptions of the relationship between time and learning in educational literature. Lemke's (2000, 2002) notion of "timescales" serves as a prime example of how subjective notions of learning (that is, the different ways individuals learn) are theorized as occurring across objective temporal orderings or timescales viewed as physical, biological, or natural. Timescales are units of time that order different lengths or durations of natural and material time, that in turn, bound and organized expected social activities and phenomena. These phenomena range from utterances, dialogues, and units to semesters, lifetimes, and historical epochs. When applied to learning, however, this view rests on a notion of time as an objectified and naturalized factor along which learning takes place. That is, learning is seen as an internal or subjective process, yet the time across which learning takes place is posited as objective, uniformly measurable, and largely independent of social activity.

A related perspective has been offered by scholars working within the paradigm of Cultural Historical Activity Theory (CHAT) and the larger sociocultural approach on which it draws. In the CHAT perspective, learning is oriented to future activity and is therefore regarded as development through proximal, incremental growth that in turn posits a natural, inevitable time-boundedness (Ochs & Schieffelin, 1986; Rogoff, 1990; Vygotsky, 1978). The CHAT distinction between microgenetic (immediate), ontogenetic (individual life-span), sociogenetic (socio-historical), and phylogenetic (species) time is akin to Lemke's timescales in that it offers various levels and scales of continuity and change in human development (Cole, 1996). Whereas Lemke's timescales merely categorize at what scale an event or activity is influential, the microgenetic, ontogenetic, sociogenetic, and phylogenetic divisions of the CHAT perspective serve as indicators of developmental, proleptic (i.e., future-oriented stances based on past experience) trajectories. Yet, while CHAT is concerned with providing an account for the weight of history on practice, activities, and actors (i.e., the ways past activities influence present events), it too considers time as an objective, physical, measurable and linear unit against which development is diachronically measured.

In contrast to the developmental learning theories described above, the notion of social time in the sociological literature presents an understanding of time as discursively and pragmatically configured by subjects to reflect the social order, rather than as a biological or astronomical constant (Sorokin & Merton, 1937). For example, it is common in schools to break the day up into periods in which different subjects are taught. Within the community of the school, then, periods or subjects can become measurements of time in and of themselves. While the measurement of those class periods is no doubt derived from broader Western astronomical notions of time, within classroom culture they can take on their own meaning. From this perspective, every event and conversation is an example of temporal articulation and reorganization. When social occasions are temporalized they become distinguishable from each other and thus identifiable across socially constructed periods of time (Zerubavel, 2003). We extrapolate this understanding to include classroom interactions as temporalizing and also as temporalized events. In displacing physical time with social time, a sociological perspective identifies social time as subjective rather than objective, more historical and less organic, therefore within the purview of human agency. While also relying on historical antecedents to understanding cultural practice, social time theories are different from timescales or CHAT perspectives in that social time is an actual unit of analysis (e.g., time-reckoning practices across social groups) and not merely a residue of past occurrence in the present or a measurable and projectable variable, independent of social activity. In our analysis of time in classroom interaction we draw attention not only to the time units of classrooms, but the ways in which participants discursively orient to and construct those temporalizing devices. That is, returning to the above-mentioned example of class periods and subjects as time units in schools, we examine how it is that students and teachers talk about and thus reproduce and contest such time units.

Strategies for Constructing Temporal Locations: Towards an Interactional Grammar of Time

Consistent with other forms of cultural practice, participants in classrooms employ resources that shift the temporal parameters and expectations of their ongoing activity while also generating others. Such resources have been the subject of intense investigation in analyses of talk in space and time (Bakhtin, 1981; Gingrich, Ochs, & Swedlund, 2002; Hanks, 1990; Ochs Keenan, 1977; Schieffelin, 2002). Additionally, temporal shifts have been examined in cognitive science as significant devices that allow for talk about not just what is, but also what may be (Fauconnier & Sweetser, 1996). During conversational exchanges, for example, participants shift their alignment along points of reference and perspectives of time and space, participation frameworks, and indexical reference. These shifts are conditioned and generated by the participants' knowledge and shared frames of reference. When these frames are not shared there are breaches to expected understandings and responses so that speakers may produce a number of strategies to provide an alternative frame of reference for their interlocutors (Hanks, 1990; Haviland, 1996). We now turn to our analyses of how it is that teachers and students distribute roles, produce knowledge, and mediate temporal frames of reference during scientific inquiry activities.

In Excerpt 1 below (Figure 7.1), which has been discussed in greater detail elsewhere (Baquedano-López et al., 2005), a third-grade class had just reviewed a lesson on the states of matter (liquid, solid, and gas). Ms. Anna,[4] a Caucasian teacher, had engaged the entire class in a question and answer exchange while the students were seated at their desks. As Ms. Anna directed the class to move to the rug area for a demonstration activity, Kevin, an African-American student, offered a definition of solid that drew on socio-historical usage of the word rather than its scientific meaning. In Kevin's usage, "solid" is an African-American Vernacular expression of solidarity, rather than a state of matter in which particles move slowly. By the time Kevin offered this alternative meaning of solid, the rest of the class had already transitioned to the rug area. As the exchange between Kevin and Ms. Anna continued, she asked him to rejoin the group on the rug. In a transcribed representation[5] of their exchange below, Kevin has arrived at the rug area but remains standing as the teacher continues to elicit an explanation for Kevin's definition of "solid" as something that African-Americans said in the past.

Excerpt 1 (Modified from Baquedano-López et al., 05)

```
1   T   What did that mean to them?
2   K   >Best friends forever.<
        ((shrugs left shoulder, takes a step towards rug area))
3   T   >Best friends for↑ever<
4   K   uh huh, that's kind of like (what they did)
5   T   So what would be a time that somebody might say that
6       Like if I said (.5) I think-I think I know what you're talking about
7       So like (.5) uh if I go up to Maya (.8)
        and Maya says::  and I say (.5)
        and Maya says (.5)
        "hey:: u::h Miss Anna (.) you wanna go get a
                      [ice cream after schoo:l?"
                      [((points with her index finger))
8       And I uh [maybe I     would     say]
9   K   [Oh yeah (you say some)] (.2) °so-
10  T   [Solid?
11  K   [°uh huh you say solid°
        ((nods))
```

Figure 7.1 Excerpt 1.

Ms. Anna and Kevin negotiated a reformulation of the meaning of solid that required Kevin to shift perspective beyond the here and now of the classroom ("what did that mean to them?" in line 1). This shift in perspective was projected away from a common indexical ground (Bühler, 1990; Hanks, 1990), the common, shared starting point of reference. This indexical ground is the time and space relationship articulated by Ms. Anna and Kevin and which constituted a "here and now." The reformulation of the meaning of solid, however, is both objectified and grounded in a projected time-space location (Black people a long time ago in a particular social setting) where Kevin is positioned as an intermediary voice that can speak across temporal frames.

This exchange is consistent with the observation that classrooms include transpositions of perspectives and participant frameworks that are fluid and changeable. Transpositions (Haviland, 1996) are primarily discursive shifts in temporality and spatiality between a speaker and a referential point or object. A type of transposition, which Haviland identifies as "projection," illustrates the ways the use of various communicative modalities (narratives, reported speech, revoicing, and gestures, among others) display and reframe present perspectives in relation to both ongoing expectations as well as potential and imagined actions. We find projections, and the more general category of transpositions, to be ubiquitous in classroom interaction where everyday learning activities build on curricular, cognitive, or social expectations and the potential realizations of those expectations.

In classroom activity, transpositions are realized through the use of hypotheticals, analogies, and metaphors. As participants in classroom interaction negotiate meaning, they break temporal frames. These are examples that make visible existing, ongoing and emerging connections between participants and science learning. For example, in turns 5 and 6, Ms. Anna, ostensibly in order to clarify Kevin's use of solid, proposes a hypothetical scenario that includes her and another African-American student, Maya, as characters in an imagined (if unlikely) conversational exchange. This hypothetical scenario is made possible through the construction of a series of conditional phrases ("like if I said") and modal auxiliaries ("would be a time," "might say that") that changes the location of the original narrative from an undisclosed past to a potential local occurrence. This is an example of how participants break the purported linearity and irreversibility of time and reorganize participant roles.

While we acknowledge the conflation of space and time in Haviland's analysis of transpositions, it remains a useful concept for advancing a more nuanced understanding of temporalizing discourse in classrooms. Besides encompassing time and space configurations, including the "here and now," transpositions also offer distinctions that have to do with degrees of specificity or extension of the present context to other spatio-temporal contexts. Indeed, while a complete separation between space and time relations is difficult, we have tried to focus our analysis on the language of temporalizing discourse. The use of hypotheticals and proposed imaginary scenarios in classroom discourse are in essence types of transpositions that serve as tools for teaching purposes. Considered a pervasive feature of the genre of scientific talk in schools (Halliday & Martin, 1993; Lemke, 1990), hypothetical and imaginary scenarios enable collective scaffolding with attention to preferred modes of reasoning and action. Imaginary scenarios in classroom discourse tend to be future-oriented or revisionist and as such they are meant to approximate shared expectations of competence and knowledge. This type of activity is not individually centered or internal, rather, it is carried out by individuals in collaboration with each other. While collective imagining presupposes a collaborative shared referent, the process naturally creates tension because symbolic references are hardly ever grasped in their totality and individuals inevitably orient to shared referents differently. Thus imaginative work requires careful referential positioning to avoid potential miscommunication and misunderstanding in classroom discourse.

Another discursive strategy that reveals tensions in the production of knowledge, goals, and participation is the use of quotative speech. When Miss Anna animates Maya, that is, speaks for her (Goffman, 1981), in a hypothetical scenario where they are both getting ice cream in "hey, uh, Miss Anna you wanna go get ice cream after school?" (line 7), it triggers the emergence of a new transposition. In this case, Ms. Anna authors and animates a fictive character based on Maya and in so doing constructs a temporal alternative. Quotatives, part of direct and indirect discourse, can be conceived as transpositions since they necessarily project other temporal contexts and indexical grounds (Bühler, 1990; Clark & Gerrig 1990; Haviland, 1996). In the context of classroom activity, quotatives are part of prototypical scaffolding strategies that generalize through the reconceptualization of familiar themes in knowledge-making (Lemke, 1990; Cazden, 2001). In this regard quotatives recreate and organize interactional structures that afford alternative roles, activities, and goals.

We find Wortham's (1992) examination of participant examples important to discuss in relation to the creation of alternative temporal locations for classroom participants. While Wortham speaks of participant examples as the mapping of existing social hierarchies onto hypothetical enactments, we note that the realignment of power positions illustrated in Excerpt 1 is different. That is, rather than a parallel mapping of relationships between two temporal frames, the teacher further distances participants to locations outside the present ground of reference. In the same way that Kevin's solid example distances him by locating him in a frame of reference in a faraway past, the hypothetical ice cream scenario likewise dissociates the meaning of solid pushed for here from the meaning employed in the science class by locating it in an imaginary time and location. The teacher appears to create, rather, a false equalization of student and teacher roles in this imagined ice-cream narrative. Moreover, this enactment highlights the relative power of each participant through quoted speech (Volosinov, 1973).[6] When people quote others, they appropriate not only the words but also the voice of other speakers (Bakhtin, 1981: 293; Volosinov, 1973: 118, 145; Goffman, 1974: 504; Clark & Gerrig, 1990). As such, quotations like the ones exemplified in Excerpt 1 function as symbolic means for maintaining existing power differentials in the classroom (Bourdieu & Passeron, 1977, 1990). Additionally, this strategy served to also preserve the one meaning that was originally intended for the class to learn—a state of matter, not the history of African-American experience in the U.S.

While it may appear that knowledge is bridged and scaffolded for the benefit of all students (e.g., scientific vs. historical notion of solid), our analysis demonstrates that learning is fraught with sometimes irresolvable breaches. In the example above, a student's potentially derailing contribution to the ongoing activity is reauthored and repositioned within the frame of science. In the remainder of this chapter we analyze data from two other classrooms to illustrate similar dynamics in the organization of participant frameworks and resulting power relations, as well as to show how participant frameworks are also related to knowledge production. While intertwined, that is we cannot conceive of power relations separate from knowledge making in classrooms, we offer an analysis of the micro-processes of talk and interaction that are constitutive of each.

Participant Frameworks and the Local Organization of Power

In our study of elementary science classrooms we observed teachers and students engaged in scaffolding activities that juxtaposed alternative temporalities making use of hypothetical and conditional constructions. Generally, a distinction can be made between hypotheticals and conditionals on the basis of their relative use of factual and truth claims, temporal locations indexed, and the use of the probability of realization of the actions they suggest. To clarify their use in this chapter, we refer to hypotheticals as the situations that are created based on collaborative supposition. We use the term conditional to refer to the actual grammatical constructions that capture factual or probability tensions. For example, when a teacher asks his class to imagine that they are giants walking a topographical map of California drawn on the board, we refer to such projected (and fantastical) activity as hypothetical, but when the teacher tells his class that if heat is trapped in a valley surrounded by mountains, it will evaporate and will form clouds, the construction of "if/then" clauses to make such statements constitutes a conditional construction. Conditionals do not always have truth-value, as counterfactual conditionals clearly illustrate. The major function of counterfactuals is not in describing truth claims but rather projecting an epistemic stance from one claim to another. Consider the following example that constructs an analogical relationship between statements: "If we had been in San Francisco today in the evening, we would have seen the wind cooling the Embarcadero." The essential functions of counterfactuals are to set up an imaginary situation which differs from the actual one (Fauconnier & Sweetser, 1996). This imaginary scenario (seeing the wind cooling the Embarcadero) depends on changing the condition of the antecedent claim (not actually having been in San Francisco in the evening).

The purposes and goals of alternative temporal and imaginary discourse in classroom activity are varied. As we saw in Excerpt 1, Ms. Anna appears to dehistoricize certain types of knowledge (African-American uses of solid) in order to adapt or make this concept relevant to the current meanings of solid that are most closely related to cognates in science. In a fourth-grade classroom a Caucasian male teacher, Mr. Gus, employs other forms of imaginary discourse to also include students' experiences in collective reasoning. In Excerpt 2 below (Figure 7.2), Mr. Gus and his students had started a lesson on weather patterns. The main point of this lesson was to learn to simulate changes in temperature resulting in movement of air and wind. As the lesson progressed, the exchanges between Mr. Gus and his class changed from adherence to the curricular materials to adaptations that included examples of students' personal experiences as residents of California who experience the heat and wind patterns talked about in their lesson. After a student had read aloud from their workbook a section pertaining to these weather patterns, Mr. Gus began to draw a topographical map of the Northern California landscape on the blackboard to illustrate the patterns. In the following excerpt, note the way Mr. Gus' explanations of the relationship between wind and heat provided a viewpoint that served to generalize generic references into "truths" occurring in the natural and social environment of his students.

In this excerpt Mr. Gus recontextualized the reading from a common text into a discussion and demonstration that created a new text—a map—on the board. This adaptation was prefaced by spatial-temporal shifters from an existing to a projected ground. We note, for example, that the adverb "now" (line 2) both marks and bounds the sequence and organization of activities that precede and follow it. The ongoing activity shifts and it is adapted to changes in goals, participant roles, texts, and tasks. In this exchange, for example, there are shifts from text to board, from reading to listening, from understanding the generic world of scientific discourse to a reading of lived experience. We note too the ways in which transpositions and hypothetical reasoning ("now," "let's say," "as if") are common in scientific classroom discourse. This exchange is similar to Ms. Anna's reasoning about the meaning of "best friends forever" (Excerpt 1) introduced with the expression "like if I said" as she initiated a hypothetical narrative about getting ice cream with a student. In this case, Mr. Gus introduced a new

Excerpt 2

```
1    T     Okay (.5) >very good<
                ((turning toward the board and putting down workbook))
2          Now I want to explain to you:::  (.5)
3          How this sort of affects (.5) you::r >life<
                ((looking at hands holding chalk))
4          Every summer this affects your l↑ife
                ((facing chalkboard))
5    Ss    [(unintilligible)
6    T     [Let's say this is the ocean right here ()
                l↑ah l↑ah l↑ah l↑a::h
                ((draws horizontal line on the board))
7          This is kind of a side view of (.) California
                ((turns back to classroom))
8          As if you're::: a::h kind of a giant
                ((looking back at chalk box))
9          Standing (.5) and looking over
10         A:nd (1.5) >let me put this in color< (3)
                ((poking chalk box))
11         And then here's San Francisco: (2)
                ((facing board))
12         Right here::: (2)
                ((draws white line on board))
13         Here's the top of the golden gate bri:::dge >sh: sh: sh:<
                ((drawing on board))
14         This is sort of a sideways view
15   S     (unintelligible)
16   T     Here's twin peaks
17         >(unintelligible)pah pahs<
                ((as he draws twin peaks towers))
18   S     [what about bugs bunny
19   T     [some skyscrapers >hey< (3)
```

Figure 7.2 Excerpt 2.

representation on the board with a phrase that keyed a hypothetical construction in "Let's say" (line 6). This hypothetical construction transposes an alternative imaginary social world, while simultaneously laying out the boundaries of what constitutes objective and relevant scientific knowledge. The use of "let's say" combines a first person imperative adverb "let's" with "say," a common verb used to introduce indirect and direct, reported discourse (Clark & Gerrig 1990; Quirk, Greenbaum, Leech, & Svartik, 1983).

The use of "let's" in this exchange also introduced a discontinuity in a recognized and sequential temporal order of activities. In fact, it is this temporal disjuncture into hypothetical discourse that afforded a new participant framework for Mr. Gus and his class. In this case, the modifying object of the phrase "the ocean right here," afforded a collective positioning animated by the teacher for the entire class. This created a prospective and provisional knowledge-base. More specifically, drawing on the accepted methods of scientific inquiry the teacher built on existing givens to construct a list of topographical characteristics of the Bay Area landscape. By universalizing this knowledge-base in the present tense ("Every summer this affects your life" in line 4 and "Now yeah every summer this is exactly what happens" in line 28), the teacher not only generalized but also added a timeless property to these weather events and experiences related to them (Lewis & Weigert, 1981).

We note that Mr. Gus' map incrementally added topographical features through declarative propositions that build an epistemic stance consistent with general truths about the landscape of the northern California coast and valley. With every increment there were specifics of imagined features transposed on the board. In fact, the drawing of this map in conjunction with the teacher walking across the board (and traveling metaphorically across the landscape being drawn) animated and embodied a collective view for the class "as if you were a kind of a giant" (line 8).[7]

A parallel that typifies this process of incremental disclosure can be made between the special perspective-giving the teacher was offering and the language of science or "professional vision" that is expected in any domain specific subject-matter (Goodwin, 1990), in this case scientific inquiry at the elementary level. This perspective is already an embodied form of scientific work that socializes students to ways of seeing their subject-matter and to relate to it in particular ways, for example, through graphic representations that project alternative time-space locations (Ochs, Gonzales, & Jacoby, 1996).

We have already indicated that hypothetical transpositions afford the reorganization of participant frameworks in ways that equalize potentially inherent differences in past-lived experience among those participating in collaborative reasoning. The collective imagining of having "a giant's perspective" affords everyone in the classroom the same vantage point. However, the construction of this fantastical view, that is the drawing of a map and the rules for what goes into this map, are designed by the teacher and remain largely uncontestable. The activity projected in the hypothetical scenario of a giant's view is made possible by both the relative power the teacher has to change the parameters of this class' collective activity and by the construction of an alternative temporal frame that asks those present to suspend their established knowledge sources (including past experiences or experiences outside this classroom). Under the assumption of collective positioning as a giant walking the topographical details of the San Francisco Bay Area, the teacher in fact controls the locus of information, interpretation, and the relevant and applicable analytical concepts at hand (see Figure 7.3).

In Excerpt 1, we pointed out that Ms. Anna appeared to equalize power hierarchies in her class by providing a scenario where she could participate in student behavior, such as getting ice cream after class by quoting and animating a hypothetical exchange with another African-American student. Hypothetical transpositions afford for such temporally equalizing position for all members in Mr. Gus' class, that is, everyone can have the same perspective, and therefore, the same knowledge-referent. But these hypothetical transpositions in classrooms are almost always engineered by teachers in efforts to relate content material to knowledge, of solid in Ms. Anna's class, and of wind and air patterns in Mr. Gus' class.

In another 3rd-grade classroom, an Asian-American teacher, Mr. Pepp, introduced the last lesson of a unit on the changing states of matter which focused on

Excerpt 3

```
20   T    Then there's San Francisco Bay again
          ((draw blue sketch))
21        And then there's East Ba:y (1)
22        And then that turns into some moⁱuntains (1)
          ((draws spiked lines))
23        And then (.5) some <bigger  hills>
24        And then there's the Central Valley over here (1.5)
          ((extends line))
25        We'll say Sacramento (1)
26   Ss   I'm gonna have bad dreams right here xx () stomach (2)
27   S    (inintelligible) Sacramen[to
28   T    [Now yeah every summer this is exactly what happens (1.5)
          ((turns back to classroom))
```

Figure 7.3 Excerpt 3.

boiling point temperature and the accompanying conversion of liquid into vapor. In this class it was customary for students to work with their neighboring peers for completing tasks. After reviewing past inquiry activities where students carried out a boiling experiment that required the recording of temperature changes, the teacher instructed students to complete all the lesson activities of their science workbook. In the example we present next we focus not so much on the power differentials constructed by participants (although they are present in the discourse of this class) but rather on the ways in which participant frameworks temporally organize and are organized temporally in knowledge-making.

Classroom Timelines: Knowledge-Making and Learning

We note that in Mr. Pepp's class the timely completion of tasks underlies learning expectations for all participants. We argue, based on our observations of several elementary schools, that this is generally the case in many classrooms. In addition to student progress being measured by the number of tasks completed, we observed that timely completion of assignments often stood as a proxy for learning in Mr. Pepp's class. That is, completion of an assignment was an indication that scientific knowledge covered in the assignments was now learned. This is vividly illustrated in Excerpt 4 (Figure 7.4) where the interactions between Claire and Emily, two bilingual Spanish-English speaking Latina students are captured. Both students were sitting together in the back of the room working on their workbooks. Their interaction was carried out in English and Spanish and took place against a background with the teacher monitoring student work and keeping track of time for them. In accordance with the knowledge expectations of this class, Claire had been helping Emily with the graphing of the

Excerpt 4

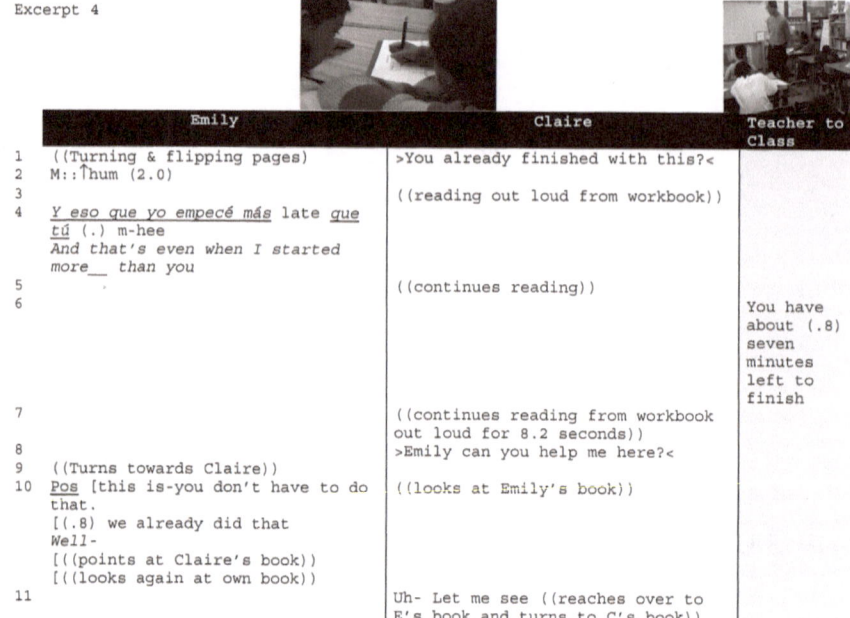

	Emily	Claire	Teacher to Class
1	((Turning & flipping pages)	>You already finished with this?<	
2	M::↑hum (2.0)		
3		((reading out loud from workbook))	
4	*Y eso que yo empecé más* late *que tú* (.) m-hee And that's even when I started more__ than you		
5		((continues reading))	
6			You have about (.8) seven minutes left to finish
7		((continues reading from workbook out loud for 8.2 seconds))	
8		>Emily can you help me here?<	
9	((Turns towards Claire))		
10	Pos [this is-you don't have to do that. [(.8) we already did that Well- [((points at Claire's book)) [((looks again at own book))	((looks at Emily's book))	
11		Uh- Let me see ((reaches over to E's book and turns to C's book))	

Figure 7.4 Excerpt 4.

temperature points recorded during a past experiment. A few minutes into their interaction, Emily began to publicly take inventory of the completed pages of her notebook by rapidly flipping its pages. Claire (sitting to the left of Emily) stopped doing her own work and turned to address Emily. Note that we have reproduced their interaction in a transcript with columns to indicate the exchanges between Emily and Claire with the teacher speaking to all students.

Claire's question to Emily in line 1 ("you already finished with this?") revealed an orientation toward task completion based on a shared task timeline. Claire's question can be interpreted as an unexpected response to the recognition that both students were pursuing different timelines. This is further accentuated by Emily's high-pitched affirmative response that acknowledged she was ahead of Claire. Emily's statement in line 4, "and that's even when I started [after] you did," upgraded a positive assessment of her own time-efficiency and thus, potentially, of science competence. Moreover, it suggested an organization in this classroom that ordered where students were supposed to be in relation to each other and to the teacher's expectations. Emily's response regulated and even chastised Claire's timeline for completing the assignment—illustrated in the explicit mention of the fact that Emily had started the assignment after Claire had done so. As if to emphasize Emily's point, Mr. Pepp's voice was heard in the background telling his students exactly how many minutes were left to complete the assignment.

We note that there is a collective orientation in this class to a practice that enforces a particular understanding of time as a continued timeline against

Excerpt 5

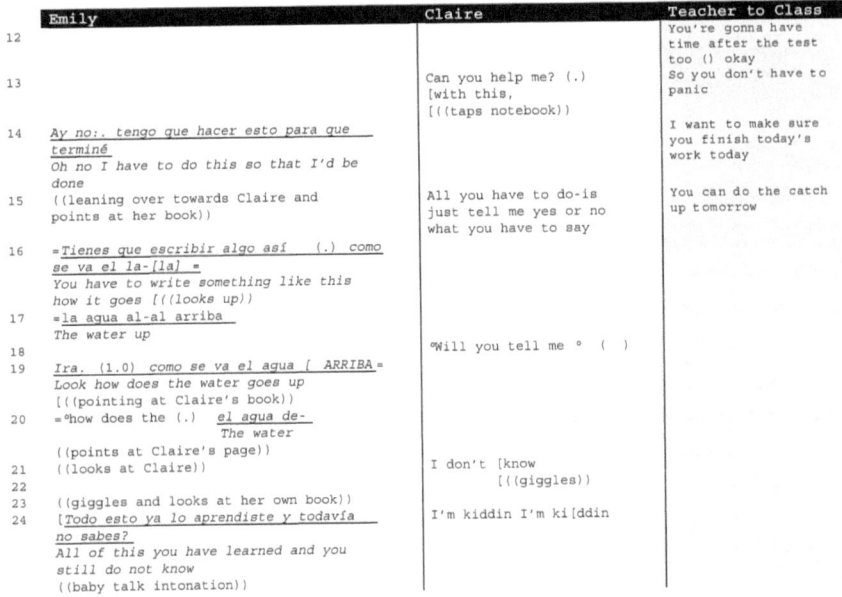

Emily	Claire	Teacher to Class
12		You're gonna have time after the test too () okay
13	Can you help me? (.) [with this, [((taps notebook))	So you don't have to panic
14 *Ay no:. tengo que hacer esto para que terminé* Oh no I have to do this so that I'd be done		I want to make sure you finish today's work today
15 ((leaning over towards Claire and points at her book))	All you have to do-is just tell me yes or no what you have to say	You can do the catch up tomorrow
16 =*Tienes que escribir algo así (.) como se va el la-[la]* = You have to write something like this how it goes [((looks up))		
17 =*la agua al-al arriba* The water up		
18	°Will you tell me ° ( )	
19 *Ira. (1.0) como se va el agua [ARRIBA* = Look how does the water goes up [((pointing at Claire's book))		
20 =°*how does the (.) el agua de-* The water ((points at Claire's page))		
21 ((looks at Claire))	I don't [know	
22	[((giggles))	
23 ((giggles and looks at her own book))	I'm kiddin I'm ki[ddin	
24 [*Todo esto ya lo aprendiste y todavía no sabes?* All of this you have learned and you still do not know ((baby talk intonation))		

Figure 7.5 Excerpt 5.

which work, and therefore learning, is measured. Indeed, the interaction between Emily and Claire is also revealing of this task-completing-deadline adherence. In Excerpt 5 (Figure 7.5) when Claire asked Emily for help, Emily responded with another regulatory explanation of what needed to be accomplished. Her efforts tried to align Claire's timeline with the expected work schedule outlined by the teacher. Note how the teacher explains the timeline for completing the work.

Emily's use of "tengo que" (I have to) and "tienes que" (you have to) in lines 14 and 16 to describe ongoing classroom work referenced timelines tied to normative expectations of student competencies. We note that Emily complied and provided assistance to Claire by scaffolding for her the scientific process of evaporation as it occurred in a previous classroom experiment. She not only gave Claire specific instructions for how to complete the assignment (Photo 7.1, left) she drew with her hands the movement of vapor as an embodied representation of vapor escaping from the drawing of a pot on the page (Photo 7.1, right).

Our analysis of Emily's reasoning exemplifies how notions of generic time are constructed in science. Using a timeless construction, Emily generalizes how water interacts with heat as a scientific fact ("how does water go up" in line 19). Like Ms. Anna and Mr. Gus, Emily scaffolds for Claire, only she does so through gesture. This is initiated when Emily proffered "Ira" (line 19, "look") which highlighted a temporal shift in their interaction, much like Ms. Anna's "Like if I said" and Mr. Gus' "Let's say." All these shifters were offered as a prelude to the impending adaptation that ensued as the participants reordered the cognitive

Photo 7.1 Two bilingual students talk about evaporation.

and social parameters of their activity. We noted in our analysis of the excerpts from Ms. Anna's and Mr. Gus' classrooms that a great deal of knowledge-making activity involves the reorganization of participant frameworks. The interaction between Emily and Claire also demonstrates this point.

Consistent with our earlier analysis of classroom interactions, knowledge-making activities include a temporal component that reorganizes participant frameworks. As Excerpt 1 and Excerpt 2 illustrated, these reorganizations can occur in such instances as when teachers construct examples that require students to imagine and be transposed into distant temporal locations, when a student becomes an emissary to a past cultural frame (Excerpt 1), or when they adopt fantastical identities as giants viewing from on top wind patterns in a local geographical area (Excerpt 2). In Mr. Pepp's class a student scaffolded knowledge for another student by placing her hands on the page and using her fingers to draw the direction of vapor rising from an illustration of a container with boiling water. This gesture is polysemous (with multiple meanings) and pluritemporal (invoking multiple time frames) (Brockmeier, 1995; Ricoeur, 1985). The gesture is a quotative of past scientific activities carried out in class during a boiling experiment in the preceding weeks. It also projects a futural expectation for completion of a task, and thus, learning as defined in this class. As we noted earlier, quotatives are devices that function to advance the reporter's intentions and are also displays of positions of power. In this case, Emily's recollection of the rising steam animates the teacher's expectations and displays her expert knowledge vis-à-vis Claire's current understanding of the task. Moreover, Emily positions herself ahead of Claire in the teacher's work timeline for the members of this class. That is, students shift roles as more expert others in scientific-knowledge while they enforce and reproduce time-marked forms of classroom activity.

Another notable aspect of this exchange is that the parceling out of learning into measurable tasks represents an economy of classroom work schedules (Thompson, 1967) and illustrates how students are disciplined into work ethic dispositions (Bourdieu & Passeron, 1977; Elias, 1992). More specifically, students are being socialized to a work discipline system through time-marked periods of activity (e.g., completing unit lessons). This economy of educational labor creates a false democratic system in which all members of this class appear to have the same opportunity for accomplishing expected classroom tasks. But as the interaction between these two students demonstrated (and as we know from other conventional time-keeping practices), these time-marked classroom schedules that promote standard roles and expectations for all do not always lead to equal outcomes for all students. While classrooms may enable collective work-schedule scripts where power asymmetries continue to be reinforced, participants will inevitably produce and adapt their present circumstances in accord with their emerging goals. In all the interactions examined in this chapter, participants augment and elaborate on the temporal complexity of the tasks they collaboratively engage.

A focus on the discourse of teachers and students allows for an examination of how they negotiate the goals of their ongoing learning activities, including the construction of alternative temporal parameters. A much-studied interactional pattern in studies of classroom discourse is captured in the "scaffolding" metaphor and represents the support of authorized knowledge, or that information which is deemed by the community and the institution to be legitimate. Across the literature on classroom discourse, scaffolding is seen as a mediational activity that supports learning (Bruner & Garton, 1978; Stone, 1993; Tharp & Gallimore, 1988). From a sociocultural perspective, scaffolding involves understanding activity systems of interaction as properties of ongoing, present cognitive demands. We argue that scaffolding activities also rely on constructions of time as alternatives to ongoing present demands; that is, mediating tasks for engaging ongoing activities may actually locate those very activities in multitemporal locations (e.g., relying on past, projected, or imaginary scenarios) not just in a present time. While impetus remains to view scaffolding as predefined sets of interactional categories such as modeling, feedback, contingency management, and cognitive or task structuring (Tharp & Gallimore, 1988), alternative perspectives on scaffolding are emerging (Pea, 2004). Stone (1993), for example, notes that there is generally little mention of the communicative mechanisms from which transfer of responsibility from more expert others to novices is accomplished or how the bidirectional alternation of expert positions occurs during scaffolding activities. We support a dynamic view that recognizes scaffolding not just as simply mediational, but also as an important cognitive and social activity in its own right. That is, scaffolding is not merely a means by which to achieve the transmission of knowledge from more expert to less expert participants, but is itself an interactional accomplishment (Goodwin, 1994; Ochs et al., 1996). Scaffolding interactions socialize participants to projected, expected academic goals, as well as conditioning participants to use those cultural resources that transmit them (Ochs, 2002).

The use of hypotheticals and conditionals in scaffolding activities generally requires participants to make logical (in the scientific sense) and historico-cultural connections between actual past or present phenomena across temporal and spatial locations (Bruner, 1986). For example, in our science classroom data, while engaging his students on the meaning of humidity, one teacher asked: "If you are in your room and you go into the bathroom when someone just takes a shower, which room has more water vapor?" This use of a hypothetical example illustrates how adaptations in classroom activity can involve temporal reorganization of knowledge and participation. The hypothetical scenario proposed by the teacher as a means to elucidate a comparison of humidity as a scientific concept (in their workbook) in another context (at home) provided an alternative framework from which to act and view the concept being learned.

Conclusion

In this chapter we started out by considering different theories of learning and their implicit temporal orientation as a way to better understand learning in heterogeneous science classrooms. We considered how certain unilinear temporal paradigms dominate educational theories. Current theories of learning are based on an implicit notion of time, which objectifies, and therefore, also reproduces a theory of time. We then provided a synthesis of "time" as an analytical category in sociological literature and suggested that we could benefit from incorporating this perspective when conducting research on learning in classrooms. We used, as a starting point for our analysis of classroom interaction, our adaptation framework for studying classroom discourse. A central aspect of our adaptation framework is the recognition that activity continually involves tensions and subsequent reorganizations of social and cognitive parameters. This reorganization takes place in response to breaches to the expected and ongoing interaction in classroom talk. We described the particulars of an interactional grammar that highlights the salience of time in the construction of knowledge and the organization of participant roles. As part of this interactional grammar we examined classroom discursive strategies that included transpositions of temporal frameworks (reported speech, conditionals, hypotheticals) as the loci for understanding how knowledge-making derives and requires temporal reorganization. Our findings illustrated that these discursive strategies, while they uphold teacher expectations and scientific inquiry protocols, they also reveal a lamination of activity timelines and social positionings not always recognizable in science classrooms. We noted that projections and transpositions can uphold asymmetries in classrooms in ways that privilege the position of the teacher as author of all possible scenarios and the source of scientific knowledge.

We observed that while there were efforts to implement the curriculum in culturally or locally appropriate ways, these efforts inevitably defaulted to the standards of the traditional model of scientific inquiry. Such adherence, however, does not indicate a rigidity of practice or thought. On the contrary, this pedagogical tendency required the constant adaptation to commentary and knowledge being generated during lessons. The classrooms that we observed were dynamic and inclusive of student experience, and their teachers were dedicated and worked hard to incorporate a comprehensive scientific inquiry into a curriculum that would not otherwise include it. Yet, the proclivity to follow a standard generic approach to science inquiry obeyed forces beyond the classroom and our project—among these, the need to adhere to regulated knowledge and the mandate to comply with educational standards and even official time-lines. The teachers' decisions to uphold such standards are not reflective of problematic pedagogy; on the contrary, they sensitize us to the demands placed on them and their students to conform to curricular and school expectations.

Our findings indicate that the construction of alternative temporal frames is generated by the tensions of the present ongoing context and other versions of the present context. When these versions come together as in a hypothetical classroom example provided by a teacher, conflicting truths and provisional frames of interaction emerge. These are not necessarily irresolvable. As Goodman (1984) notes, "the multiple worlds of conflicting true versions are actual worlds, not the merely possible worlds or nonworlds of false versions" (p. 31). The focus on the actual, present moment context is important as a starting point for examining the central role of temporal reference or the indexical ground (Hanks, 1990) from which alternative temporal locations are projected. That is, imagined worlds and other social symbolic mediations are consequential to, and constitutive of, the learning process.

Classroom Implications

In this chapter, we have described three inter-related ways of how time is socially produced through classroom exchanges that: (1) recover past collective, group or individual history, memory, and experience, (2) position participants in unequal relationships, and (3) socialize students and teachers to a dominant school schedule. Time contexts, while seemingly invisible to teaching and learning, are constantly used in the classroom to recast past knowledge, simulate and rehearse new ideas, set common schedules, and exert authority. Our analysis of each excerpt provides vivid classroom examples for describing how classroom roles and expectations are expressed through routine time-coded language in the classroom.

Our analysis of classroom discourse reframes what is often referred to as pedagogy that contextualizes experience (where teachers draw and make connections to existing science knowledge such as family cultural practices, community resources, and completed school lessons of individual and group histories) as attempts to recover past experience or assert a shared potential experience. In Excerpt 1 (with Ms. Anna) and in Excerpts 2 and 3 (with Mr. Gus), we use the concept of *transposition* to show the everyday pull for shared time constructs through multi-modal resources that include the use of hypotheticals, analogies, metaphors, narratives, gesture, and quotative speech. We also find examples of student-to-student scaffolding through gestural transposition in Excerpts 4 and 5 where Claire and Emily work through the notion of evaporation in Mr. Pepp's class. Transpositions of this kind engage multiple and new temporal realities in support of collaborative scientific reasoning in the classroom, made possible through student–teacher and student–student engagements.

While transpositions create contexts for integrating and sharing knowledge, they can also, more problematically, maintain and extend unequal social identities excluding student authority. In Excerpt 1, Kevin is asked to explain the meaning of solid of the past (in the realm of African-American practices) thereby separating his contribution from the ongoing science learning. The teacher by quoting this contribution, that is by revoicing Kevin's meaning, maintains her power as the primary authority in both explaining the meaning of solid and how it relates to Kevin's comment.[8] This analysis is useful for the teacher as classroom activities provide contexts for shared social contexts, but can also reproduce them. There is also the possibility of transformation and change. What would have happened had Ms. Anna accepted Kevin's contribution as a possible new way of understanding science or even as another form of knowledge? In our data from science classrooms we found extensive examples like those in Mr. Gus' class, where the teacher generates an imaginary scenario promoting a common context yet produced it in a way that is not always accessible to student reinvention. Rather, these scenarios tend to be fixed and reproductive of knowledge. Transpositions are productive resources in science pedagogy that can create shared learning contexts, as well as maintain teachers as ultimate authorities of knowledge-making. The awareness of these classroom units of discourse are thus useful for teachers to monitor and plan the presentation of scientific content that is more inclusive and less imposing.

Finally, the way time is constructed across social institutions, and especially in schools socializes teachers and students to a dominant school schedule that is one-dimensional. Schools by and large promote a curricular schedule that equates the passage of time and completion of work with learning. The implication here for classroom teaching is the importance of understanding how the imposition of a unilinear school schedule influences classroom collaborations and the assessment of learning. We provide an example in Excerpts 4 and 5 that describes how at 3rd-grade students are already cognizant of the pressures for enforcing and completing work schedules. We find that all students, including bilingual students like Emily and Claire, can learn to manage the school schedule competently but that there may be opportunities for learning that can be overshadowed by a strict focus on time-schedule adherence.

Through analysis of selected classroom settings that contrast social and cultural dynamics, we find that attempts to locate time in science activity provide important insights into how power and knowledge are constructed and reproduced. Locating time in classroom science activity requires interrogating local notions of what constitutes learning and who possesses the authority to enforce that learning. Notions of time and work timelines are persistent themes in our observations of science classrooms participants who socialize each other to act in imagined, remembered, and projected social worlds. In addition, teachers and students constantly engage time notions in how they represent scientific authority as residing in absolute or constructed facts. While the analysis of the science classroom settings provided here addresses routine topics such as how classroom participants negotiate non-school knowledge (Excerpt 1) or the construction of emerging scientific logic (Excerpts 2 and 3), we argue that there is an underlying complexity and tension in these interactions that often goes unexamined. We offer a framework for understanding this complexity by focusing on how participants adapt their ongoing interactional circumstances and how they reorganize the social and cognitive domains of joint attention.

In this chapter we have described how time is the fulcrum through which cultural and linguistic diversity are mediated and remediated. That is, teachers and students in classrooms construct notions of time to accomplish interactional goals, including negotiation of meanings and understandings that emerge from the tensions of the linguistic and cultural diversity that they bring to their interaction. In closing, we propose that a morphology of time in science classroom learning is needed to identify how this mediation occurs linguistically and what kinds of social and cognitive reorderings it generates. By morphology we mean the study of the form and function of the grammatical construction of socially based, temporal worlds (even dominant timelines) through projections, conditionals, and hypotheticals that would enable us to see how 'time' is located in classroom learning and how the construction of such temporal worlds mediate knowledge and power in classroom learning.

Acknowledgments

We thank the teachers and students who participated in the SIFA project and who welcomed us into their classrooms. We learned valuable lessons from their commitment to science education. The study reported here was made possible by a grant from the National Science Foundation and collaborations with Dr. Eugene E. García and Dr. Okhee Lee. We also thank Discourse Lab colleagues in the Graduate School of Education including Gabino Arredondo and Jennifer Collett for their assistance and helpful commentary. Finally, we want to acknowledge productive conversations we've had as participants of UC Berkeley's Linguistic Anthropology Working Group.

Targeting Enduring Understandings

1   What are the connections you see between the discussion in the Solís team's chapter and the four elements of culturally-responsive instruction, as described by Villegas and Lucas?
2   How would you describe the "work" that language is doing in this multi-cultural science context?

Deepening the Reflection

1   How do ideas about time influence the organization of classroom lessons, such as notions like time on-task, wait time, or grade-level skills? What assumptions are usually made about the connection between length of time and learning or comprehension?
2   When we think about tapping into previous students cultural knowledge or experience, what does that mean in terms of linking those experiences to science concepts? How are those past experiences usually characterized in the classroom? Are they relevant in the past only? Are they historical? Are they local? Can these experiences ever be scientific too?
3   Examine a lesson or chapter in a science text and identify how time is

represented. What is the role of time in the scientific phenomena being discussed? How do you think this kind of time is different from or similar to the forms of time students themselves may use?

Encouraging Engagement

1   In follow-up to Reflection Question #3, now observe a group of students working together on a science inquiry activity. Record how students use different aspects of time, timelines, and representations of time to engage with their assignment. How do they imagine and remember what they've learned, are learning, and expect to learn?

2   Observe a science lesson and note the instances when changes in the direction of the expected course of the lesson occurred. Are these modifications in the lesson student- or teacher-initiated? How did student and teacher roles change? For how long? Was new knowledge offered? How was it used?

Notes

1 Across the three research projects students were given both project-specific assessments as well as NAEP and TIMMS referenced items. The findings here include quantitative data from Arizona and San Francisco student performance on these assessments.

2 Social dialect is appropriate here due to the uneven acquisition and familiarity of this code in U.S. classrooms characterized by student-directed, inquiry-focused, and reflective activity.

3 In a review of the use of the notion of adaptation in its biological and anthropological senses, Alland (1975) contends that an understanding of adaptation as a variable in the transgenerational propagation and change of social structure can lead to a better understanding of culture and society. While our use of the term adaptation is meant to capture the idea of simultaneous continuity and change, in a break from traditional uses of the term in anthropology, we examine adaptation as sets of local strategies that are enacted in moment-to-moment interaction, rather than across generational lines.

4 The names of teachers and students are pseudonyms.

5 Transcription conventions follow those developed by Jefferson (in Atkinson & Heritage, 1984). A period indicates a falling tone, not necessarily the end of a sentence; a comma indicates a slight rising inflection, such as enumerating or listing; a question mark indicates rising intonation, not necessarily a question; an exclamation point indicates an animated tone; colons indicate sound elongations, square brackets indicate overlapped speech; underlining indicates relative emphasis; double parentheses are used to include descriptions of gestures; empty single parentheses indicate unintelligible speech for an approximate length of time; parentheses around a word or phrase indicate a good guess at what was said (sometimes this is marked as xx); parentheses with a number inside indicate a length of pause in seconds; fast speech is indicated by the following convention: >fast talk<; speech slower than the surrounding talk <slow talk>; a dash indicates a cut-off or a sudden stop in the flow of talk; an equals sign indicates latched speech; a degree signs indicates relatively quieter speech; an up arrow indicates a rise in tone. Stills are used to show physical behavior not readily describable in the transcript or in order to further illustrate gestures and eye-gaze direction. Still boxes inside transcripts are connected by lines to their corresponding turns. Utterances in Spanish are both italicized and underlined.

6 In the words of Volosinov: "Between reported speech and the reporting context, dynamic relations of high complexity and tension are in force. A failure to take these

into account makes it impossible to understand any form of reported speech" (1973: 118–119).

7 We note that the pronoun "you" can be interpreted as a second plural "you" to refer to the students or a more generic inclusive "you" of the teacher and the students together.

8 For a detailed discussion of the exchange between Kevin and Ms. Anna and the way floating terms such as "solid" are framed as scientific or non-scientific, see Baquedano-López et al., 2005.

References

Alland, A. (1975). Adaptation. *Annual Review of Anthropology, 4*, 59–73.

Atkinson, J. M., & Heritage, J. (1984). *Structures of social action: Studies in conversation analysis.* New York: Cambridge University Press.

Bakhtin, M. M. (1981). *The dialogic imagination: Four essays* (C. Emerson & M. Holquist, Trans.). Austin, TX: University of Texas Press.

Baquedano-López, P., Solís, J. L., & Kattan, S. (2005). Adaptation: The language of classroom learning. *Linguistics and Education, 6*, 1–26.

Bourdieu, P., & Passeron, J. C. (1977). *Reproduction in education, society and culture.* London: Sage.

Bourdieu, P., & Passeron, J. C. (1990). *Reproduction in education, society, and culture* (1990 ed.). London: Sage & Teesside Polytechnic.

Brockmeier, J. (1995). The language of human temporality: Narrative schemes and cultural meanings of time. *Mind, Culture and Activity, 2*, 102–118.

Bruner, J. S. (1986). *Actual minds, possible worlds.* London: Harvard University Press.

Bruner, J. S., & Garton, A. (1978). *Human growth and development.* Oxford: Clarendon Press.

Bühler, K. (1990). *Theory of language: The representational function of language.* Amsterdam: J. Benjamins Pub. Co.

Cazden, C. B. (2001). *Classroom discourse: The language of teaching and learning* (2nd ed.). Portsmouth, NH: Heinemann Publishers.

Clark, H. H., & Gerrig, R. J. (1990). Quotations as demonstrations. *Language, 66*, 764–805.

Cole, M. (1996). *Cultural psychology: A once and future discipline.* New York: Belknap & Harvard Press.

Cole, M., & Engeström, Y. (1993). A cultural-historical approach to distributed cognition. In G. Salomon (Ed.), *Distributed cognitions: Psychological and educational considerations* (pp. 1–46). New York: Cambridge University Press.

Duranti, A. (1997). *Linguistic anthropology.* Cambridge: Cambridge University Press.

Elias, N. (1992). *Time: An essay.* Cambridge: Basil Blackwell.

Erickson, F. (1983). Classroom discourse as improvisation: Relationships between academic task structure and social participation structure in lessons. In L. C. Wilkinson (Ed.), *Communicating in the classroom* (pp. 153–181). New York: Academic Press.

Fauconnier, G., & Sweetser, E. (1996). *Spaces, worlds, and grammar.* Chicago, IL: University of Chicago Press.

Garfinkel, H. (1967). *Studies in ethnomethodology.* Englewood Cliffs, NJ: Prentice-Hall.

Gingrich, A., Ochs, E., & Swedlund, A. (2002). Repertoires of time-keeping in anthropology. *Current Anthropology, 43*, S1–S137.

Goffman, E. (1974). *Frame analysis: An essay on the organization of experience.* Cambridge, MA: Harvard University Press.

Goffman, E. (1981). *Forms of talk.* Philadelphia, PA: University of Pennsylvania Press.

Goodman, N. (1984). *Of mind and other matters.* Cambridge, MA: Harvard University Press.

Goodwin, C. (1994). Professional vision. *American Anthropologist, 96*(3), 606–633.

Goodwin, C., & Heritage, J. (1990). Conversation analysis. *Annual Review of Anthropology, 19*, 283–307.

Goodwin, M. H. (1990). *He-said-she-said: Talk as social organization among black children.* Bloomington, IN: Indiana University Press.

Gutiérrez, K., Baquedano-López, P., & Tejeda, C. (1999). Rethinking diversity: Hybridity and hybrid language practices in the third space. *Mind, Culture & Activity, 6*, 286–303.

Gutiérrez, K. D., Rymes, B., & Larson, J. (1995). Script, counterscript, and underlife in the classroom: James Brown versus *Brown v. Board of Education. Harvard Educational Review, 65*, 445–471.

Halliday, M. A. K., & Martin, J. R. (1993). *Writing science: Literacy and discursive power.* Pittsburgh, PA: University of Pittsburgh Press.

Hanks, W. F. (1990). *Referential practice: Language and lived space among the Maya.* Chicago, IL: University of Chicago Press.

Haviland, J. B. (1996). Projections, transpositions, and relativity. In J. J. Gumperz & S. C. Levinson (Eds.), *Rethinking linguistic relativity* (pp. viii, 488). Cambridge: Cambridge University Press.

Hunt, D. E. (1981). Teachers' adaptation: "Reading" and "Flexing" to students. In B. R. Joyce, C. C. Brown, & L. Peck (Eds.), *Flexibility in teaching: An excursion into the nature of teaching and training* (pp. 59–72). New York: Longman.

Ku, Y. M., Garcia, E. E., & Corkins, J. (2005). *Impact of the instructional intervention on science achievement of culturally and linguistically diverse students.* Paper presented at the American Educational Research Association, Montreal, Canada.

Lave, J., & Wenger, E. (1991). *Situated learning: Legitimate peripheral participation.* Cambridge: Cambridge University Press.

Lee, C. D. (2001). Is October brown Chinese? A cultural modeling activity system for underachieving students. *American Educational Research Journal, 38*, 97–141.

Lee, O., & Fradd, S. H. (1998). Science for all, including students from non-English language backgrounds. *Educational Researcher, 27*(4), 12–21.

Lemke, J. (1990). *Talking science: Language, learning and values.* Norwood, NJ: Ablex.

Lemke, J. (2000). Across the scales of time: Artifacts, activities, and meanings in ecosocial systems. *Mind, Culture and Activity, 7*, 273–290.

Lemke, J. (2002). Language development and identity: Multiple timescales in the social ecology of learning! In C. Kramsch (Ed.), *Language acquisition and language socialization* (pp. 68–87). London: Continuum.

Lewis, J. D., & Weigert, A. J. (1981). The structures and meanings of social time. *Social Forces, 60*, 432–462.

Linn, M. C., Clark, D., & Slotta, J. D. (2003). WISE design for knowledge integration. *Science Education, 87*, 517–538.

Matusov, E. (1996). Intersubjectivity without agreement. *Mind, Culture, and Activity, 3*, 25–45.

Michaels, S. (1981). "Sharing time": Children's narrative styles and differential access to literacy. *Language in Society, 10*, 423–442.

O'Connor, M. C., & Michaels, S. (1996). Shifting participant frameworks: Orchestrating thinking practices in group discussion. In D. Hicks (Ed.), *Discourse, learning and schooling* (pp. 63–103). New York: Cambridge University Press.

Ochs, E. (1979). Transcription as theory. In E. Ochs & B. B. Schieffelin (Eds.), *Developmental pragmatics* (pp. 43–72). New York: Academic Press.

Ochs, E. (2002). Becoming a speaker of culture. In C. Kramsch (Ed.), *Language acquisition and socialization* (pp. 99-120). London: Continuum.

Ochs, E., & Schieffelin, B. B. (eds). (1986). *Language socialization across cultures* (Studies in the social and cultural foundations of language). New York: Cambridge University Press.

Ochs, E., Gonzales, P., & Jacoby, S. (1996). "When I come own I'm in the domain state": Grammar and graphic representation in the interpretive activity of physics. In E. Ochs, E. A. Schegloff & S. Thompson (Eds.), *Interaction and grammar* (pp. 328–369). Cambridge, UK: Cambridge University Press.

(Ochs) Keenan, E. (1977). Why look at unplanned and planned discourse. In E. O. Keenan & T. L. Bennett (Eds.), *Discourse across time and space* (pp. 1–41). Los Angles, CA: Department of Linguistics, USC.

O'Donoghue, T. A., & Chalmers, R. (2000). How teachers manage their work in inclusive classrooms. *Teaching and Teacher Education, 16*, 889–904.

Orellana, M. F., & Thorne, B. (1998). Year-round schools and the politics of time. *Anthropology and Education Quarterly, 29*(4), 1–27.

Pea, R. D. (2004). The social and technological dimensions of scaffolding and related theoretical concepts for learning, education, and human activity. *The Journal of the Learning Sciences, 13*, 423–451.

Quirk, R., Greenbaum, S., Leech, G. N., & Svartik, J. (1983). *Studies in English linguistics for Randolph Quirk.* London: Longman.

Ricoeur, P. (1985). *Time and narrative* (Vol. 1). Chicago, IL: University of Chicago Press.

Rogoff, B. (1990). *Apprenticeship in thinking: Cognitive development in social context.* New York: Oxford University Press.

Rosebery, A., Warren, B., & Conant, F. (1992). *Appropriating scientific discourse: Findings from language minority classrooms.* Cambridge, MA: TERC.

Sacks, H., Schlegloff, E. A., & Jefferson, G. (1974). A simplest systematics for the organization of turn-taking for conversation. *Language, 50*, 696–735.

Schegloff, E. A., Jefferson, G., & Sacks, H. (1977). The preference for self-correction in the organization of repair in conversation. *Language, 53*, 361–382.

Schieffelin, B. B. (2002). Marking time. *Current Anthropology, 43*, 5–17.

Slattery, P. (1995). A postmodern vision of time and learning: A response to the national education commission report "prisoners of time." *Harvard Educational Review, 65*, 612–633.

Solís, J. L. (2004). Locating student classroom participation in science inquiry and literacy activities. In J. Cohen, K. McAlister, K. Rolstad, & J. MacSwan (Eds.), *ISB4: Proceedings of the 4th International Symposium on Bilingualism.* Somerville, MA: Cascadilla Press.

Sorokin, P. A., & Merton, R. K. (1937). Social time: A methodological and functional analysis. *American Journal of Sociology, 42*, 615–629.

Stone, C. A. (1993). What is missing in the metaphor of scaffolding? In C. A. Stone (Ed.), *Contexts for learning: Sociocultural dynamics in children's development* (pp. 169–183). New York: Oxford University Press.

Tharp, R. G., & Gallimore, R. (1988). *Rousing minds to life: Teaching, learning and schooling in social context.* New York: Cambridge University Press.

Thompson, E. P. (1967). Time, work-discipline, and industrial capitalism. *Past and Present, 38*, 56–97.

Valsiner, J. (2002). The irreversibility of time and ontopotentiality of signs. *Estudios de Psicología, 23*, 49–59.

Varenne, H. (1998). Local construction and educational facts. In H. Varenne & R. McDermott (Eds.), *Successful failure: The school America builds* (pp. 183–206). Boulder, CO: Westview Press.

Volosinov, V. N. (1973). *Marxism and the philosophy of language* (L. Matejka & I. R. Titunik, Trans.). New York: Seminar Press.

Vygotsky, L. S. (1978). *Mind in society: The development of higher psychological processes* (M. Cole, Trans.). Cambridge, MA: Harvard University Press.

Wertsch, J. V. (1991). *Voices of the mind: A sociocultural approach to mediated action.* Cambridge, MA: Harvard University Press.

Wortham, S. (1992). Participant examples and classroom interaction. *Linguistics and Education, 4,* 195–217.

Zerubavel, E. (2003). *Time maps: Collective memory and the social shape of the past.* Chicago, IL: University of Chicago Press.

8 "You're Magmatic Now"

Language Play, Linguistic Biliteracy, and the Science Crossing of Adolescent Mexican Newcomer Youth

Katherine Richardson Bruna

Introduction

In this chapter, Richardson Bruna illustrates the cognitive, linguistic, and affiliative role of language play in the schooling lives of young Mexican adolescents. While all students forge academic identities out of the relationships they discover or create between school-, community-, and family-based knowledges, immigrant youth do so while also learning a new culture and language. Richardson Bruna takes up the notion of border crossing to relate the literal movement Mexican immigrant youth have made across the physical Mexico–U.S. border to their continued, more figurative, movement between the linguistic borders of Spanish and English. Framing this linguistic border-crossing with ideas of literacy development, oral culture, humor, and power, Richardson Bruna regards Mexican adolescents' playful language use as an important display of their evolving biliteracy, one that, for them, is historically culturally relevant and, most strategically within science, makes space for their voices in a place where those voices are traditionally seldom heard.

The discourse about teaching science to English Learner students, like those featured in Richardson Bruna's chapter, is often dominated by a lengthy characterization of what these students "lack"; they lack knowledge of the content area from inadequate previous schooling, they lack exposure to formal scientific understandings and expressions from non-scientific folk and religious culture in the home, they lack motivation to be strong science students and future science professionals due to lack of opportunity in and access to the field, and, of course, they lack English. Richardson Bruna, in contrast, highlights what these students bring to their science learning. In the classroom she describes here, we see how Mexican newcomer adolescents actively engage their linguistic biliteracy within the context of their science learning. In a form that would likely go unrecognized and be misread by many science teachers as off-task talk, these students' movement across Spanish and English in their language play shows how closely instead they are attending to and engaging with the linguistic context of the science classroom. Their language play is a tool, on the one hand, that they use to participate in the discourse community of that particular classroom, as well as a lens, on the other, through which we can view these students' experiences with science learning in a new cultural and linguistic context.

Richardson Bruna states that, if affirmed, language play offers teachers themselves opportunities to make cognitive, linguistic, and affiliative connections with their students. As you read this chapter think about how you, as a teacher, could use a student-based resource, like language play, in the classroom. What is the best way to learn about and build on, in your teaching, this and other resources students and their families bring into your classroom, school, and community?

We remember the funny things, you know. We were, like, tired and we just tried not to be depressed in the middle of the walk and we were, like, joking.
(Captainville High School student, Jesús Romero,[1] on crossing, with his sister, the Mexico–U.S. border, Interview 11/16/06)

When marginalized communities go from passive objects of ridicule to active agents of mischief, they are definitely subverting the social order (Latorre, 2003, p. 4).

This chapter documents how adolescent Mexican newcomer youth make meaning of their science education through the strategy of language play. Language play, I argue, is an important cognitive and affiliative resource. Just as joking helps Jesús, in the quote above, cope with the stresses of his, more literal, border crossing, language play helps make Mexican newcomer youth "at ease" as they navigate the borders between their home culture and that of the school science setting (Phelan, Davidson, & Cao, 1998; Aikenhead & Jegede, 1999, p. 276). It is an act of "mischief" that subverts the social order in the science classroom while providing a strategic window through which to view these students' experiences with literacy development in the science setting.

Drawing on the sociocultural theoretical framework modeled by Elmesky (2003) in her work on the structural resonance between the strategies employed by youth in their community and classroom contexts, I will describe language play as a point of structural resonance between adolescent Mexican newcomer youths' lives and the sociolinguistic context of an "English Learner Science" course in a demographically-transitioning Midwestern meatpacking community. Like Elmesky, I will make the argument that this strategy could and, indeed, should be utilized by teachers to assist these youth with their border crossing into school science (Phelan et al., 1998; Aikenhead & Jegede, 1999). While science education scholars have provocatively used this metaphor of "crossing borders" to talk about the transitions students, in general, negotiate as they confront the culture and language of science (Saul, 2004), my work applies the explanatory power of this metaphor to those for whom the identity of a "border crosser" remains most marginalizing—Mexican newcomers in U.S. society and schools.

The New Latino Diaspora and Its Implications for Science Education

My interest in Mexican newcomers, and more particularly, my vantage point in the Midwest, is neither haphazard nor trivial. Over the last three decades, immigration from Mexico has come to account for almost forty percent of the total national immigration increase, up from below 800,000 to a swelling eight million (Camarota, 2001). Accompanying this dramatic increase in the number of Mexicans immigrating to the U.S. is a similarly dramatic increase in the numbers of those who, in immigrating, bypass the usual gateway states like California and Texas and take up residence instead in traditionally non-settlement communities like the rural Midwest (Millard & Chapa, 2004; Lyman, 2006). The resulting dispersion of Mexicans into new host communities is a central feature of what scholars term "The New Latino Diaspora" (Wortham, Murillo, & Hamann, 2002).

These sites of rapid ethnic diversification provide intriguing opportunities for educational anthropologists to examine the emergence of new cultural identities and social relationships (Levinson, 2002). Schools, as significant identity-mediating institutions (Goode, Schneider, & Blanc, 1992) which typically serve to reproduce existing economic and cultural status hierarchies (Bowles & Gintis, 1976; Bourdieu & Passeron, 1977), are a critical context in which these new identities and relationships emerge. In particular, classrooms in these demographically-transitioning communities are characterized by hybrid discourse practices through which teachers and students work together by "juxtaposing forms of talk, social interaction, and material practices" from different cultural and linguistic contexts (Kamberelis, 2001, p. 86). For science education scholars, this means that real changes in business-as-usual-thinking about science instruction are afoot. Students and teachers talk science (Lemke, 1990), position themselves with respect to science teaching and learning, and utilize resources in science instruction in ways that challenge conventional cultural narratives of science teaching in the U.S.

In my research on a demographically-transitioning community of the New Latino Diaspora, I have seen how an understanding of science education as the development of conceptual mastery is tempered by students' non- or limited-English-proficient speaking designation. In a special English Learner Science course, Mexican newcomers learn science vocabulary in a curriculum emptied of thorough explanations of the scientific processes to which such vocabulary refers (Richardson Bruna, Vann, & Perales Escudero, 2007). I have also seen how an understanding of science education as the preparation of students to enter science-related professional fields is troubled by students' undocumented status. Mexican newcomers are expected to go work the line at the local meatpacking plant (Richardson Bruna & Vann, 2007). Finally, I have seen, as I will demonstrate here, how an understanding of science education as the acquisition of scientific literacy is transmuted by students' orality. Mexican newcomers bring an affinity for oral tradition that, unrecognized through the literacy-privileging[2] lens of modern science, is, nonetheless a part of their meaning-making mechanisms in science instruction. Their language play provides an important window on literacy development while also serving, in the science classroom I studied, as a structurally-resonant crossing strategy in science. Here I aim to document these students' language play, describe its structural resonance in their English Learner Science classroom, and discuss its possibility for the future of science education.

A Possible Orality of Science?

Rampal (1992) asks us to consider a possible orality of science. She makes her case by describing how the history of scientific knowledge is, in fact, the result of a language project, an explicit campaign "to forge language as a precise instrument for unambiguous representation" (p. 228). The move towards objectivity, so characteristic of thinking about science, was reflected in language that became "detached and disembodied" (p. 229), as well as in a personal stance toward scientific undertakings that "shed all allusions even to the natural affective manifestations of scientists, including their sense of wonder, excitement..." (p. 231). Today, Rampal asserts, "the very spirit of science—of exploration and curiosity about everyday life—is conspicuously absent" (p. 233). She suggests that we need to reclaim personal engagement, redirect the evolution of science discourse, and embrace the wonder and excitement that scientific knowledge and its language has left behind. In effect, we need to consider what it would take to "incorporate the pyschodynamics of orality into processes of learning science" (p. 241).

Rampal points out how the experiences of children are characterized by rich expressions of oral knowledge that do not meet science's expectation for abstraction and objectivity (p. 236). Her account resonates with that of others who remind us that before children become literate, their universe is an oral one, full of storytelling, metaphor, rhyme and rhythm, jokes, and humor. These provide powerful meaning-making tools and nurture important cognitive abilities (Egan, 1986, 1988, 1997). In "First World" environments, as children progress through school and acquire literacy, many of these tools and abilities are lost to the unimaginative routine of schooling, curriculum, and instruction. In "Third World" environments, however, where communities have histories of illiteracy or low-levels of literacy, many of these tools maintain their powerful grip on the collective imagination throughout adulthood (Rampal, 1992).

The Mexican newcomer populations of the New Latino Diaspora come from such communities—small, poor, rural spaces in Mexico in which, for example, nearly seventy-five percent of families are in the lowest four percent of the nation's income distribution (Farm Foundation, 2006). The need for these communities to go *norte* (north) to escape from such poverty and provide for their families, as well as the gain enjoyed by U.S. industries as a result of their low-wage labor, is part of a shared U.S.–Mexico history.

That history has generated an oral tradition of Mexican–American humor. This genre centers on the bilingual and bicultural situation which is caricatured, as de Caro (1972) writes, in the "jokes which circulate among Mexican-Americans [and] involve the figure of the unacculturated Mexican, the immigrant, the bracero farm laborer, and his absurd misunderstandings of a foreign culture, *particularly in regard to language*" (de Caro, 1972, emphasis mine). The tradition of Mexican–American humor serves, as much humor does, to allow the individual "to take a more playful perspective on a stressful event and thereby reduce the often deleterious emotional consequences" (Dixon, 1980, as cited in Martin, 1996). In this chapter, my point of departure is that adolescent Mexican newcomer students bring their socialization in the oral mode into the science classroom, as well as the tool of humor as a coping strategy to deal with their marginalization in U.S. schools.

Students' Multiple Worlds: Language Play as a Structurally Resonant Crossing Strategy

Efforts to understand the experience of Mexican newcomers in science education and the strategies they use in the crossing from their home to classroom contexts must be informed by a model that acknowledges the multiple worlds that they inhabit and attends to how they integrate their meanings (Phelan et al., 1998). Phelan, Davidson, & Cao supply such a model from their work, as seen in Figure 8.1.

Informed by this model, Phelan and her colleagues found that similarity between the cultural values and norms of students' multiple worlds, and/or students' employment of strategies to put themselves at ease despite the differences,

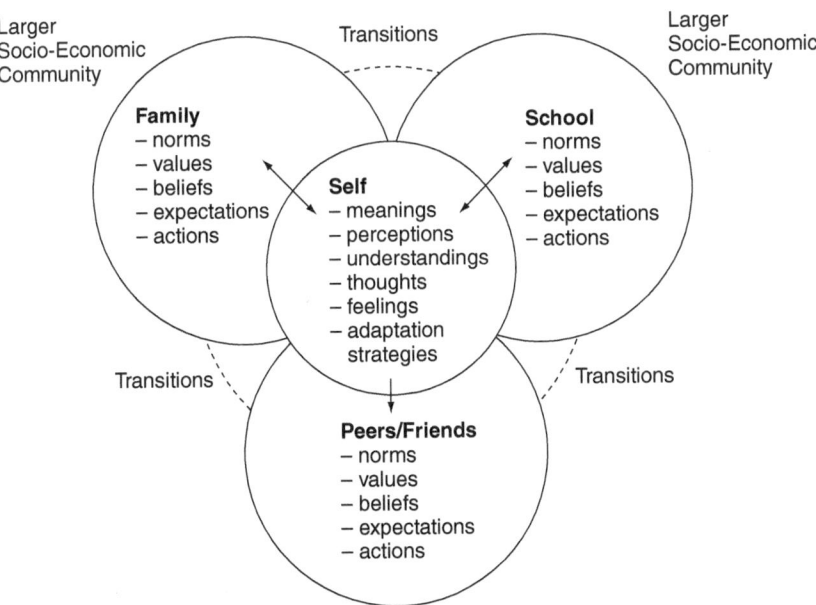

Figure 8.1 A model of the interrelationships between students' family, peer, and school worlds. (Reprinted by permission of the Publisher. From Patricia Phelan, Ann Locke Davidson, and Hanh Cao Yu, *Adolescents' Worlds: Negotiating Family, Peers, and School.* New York: Teachers College Press, © 1998 by Teachers College Press, Columbia University. All rights reserved.)

were key factors related to successful transitions (1998). Aikenhead & Jegede (1999) took up Phelan et al.'s work with respect to science education, more particularly, and concur with the assertion that it is possible and desirable "to identify institutional structures that operate to facilitate boundary crossing strategies and do not require students to give up or hide important features of their lives" (p. 246). They talk about this work of facilitation as helping students feel at ease (p. 276) and cite Lugones' (1996) experientially-rich account of her own at-ease making, a prominent feature of which is the adoption of an attitude Lugones describes as "playful." She writes: "The attitude that carries us through the activity, a playful attitude, turns the activity into play" (431). The concern is not competence but about being creatively and actively in the moment.

To the extent that structural elements of a classroom facilitate playfulness, resonance may be said to exist between the students' own needs and strategies for at-ease making and the processes of instruction (Elmesky, 2003). These strategies, as Elmesky (2003) illustrates in her work on the science education experiences of inner city youth in Philadelphia, are largely unconscious, a product of the interaction between an individual and the cultural field which s/he inhabits (p. 32). When students, crossing from one (neighborhood) world to another (science classroom) world, carry with them shared cultural strategies, they inhabit and structure that world in a way that reflects the ideological, material, and human resources embodied by those strategies. Elmesky describes how African American youth, for example, enact the strategy of confrontation within the field of the inner city neighborhood and science classroom. Importantly, she illustrates how, contrary to dominant school culture's and society's perceptions of such behavior as destructive (p. 33) a student's argumentative stance in a science classroom may be, in fact, intimately related to her productive development of science conceptual mastery (p. 49). She presses the science education community to understand students' use of such strategies in the classroom, asserting that "youth sometimes utilize such mannerisms agentically in the process of building a science identity and becoming scientifically literate citizens" (p. 33). In the next section, I extend the ethnic, linguistic, and geographical reach of Elmesky's ideas to the experience of adolescent Mexican newcomer students in the rural Midwest. Instead of the strategy of confrontation, I examine the strategy of language play as a window onto these students' language development and social positioning in the science classroom.

Language Play and Linguistic Biliteracy

The importance of play on children's cognitive development is well-documented (Bergen & Coscia, 2001; Bergen, 1998, as cited in Bergen, 2002). One kind of play in which children engage is language play (Dunn, 1988, as cited in Berko Gleason, 2005). "Children treat language as they would any other object," writes Garvey "as a rich source of material that can be playfully exploited" (Garvey, 1977, pp. 397–398, as cited in Berko Gleason, 2005, p. 397). Metalinguistic awareness, what scholars refer to as linguistic literacy, is a central feature of language play. Linguistic literacy is, as Ravid & Tolchinksy (2002) define it, "the ability to consciously access one's own linguistic knowledge and to view language from various perspectives" (p. 418). Children exhibit linguistic literacy when they demonstrate the ability to, for instance, adapt their "pronunciation, morpho(phono)logy, choice of lexical items and syntactic structures depending on geography and social motivation and also for situationally-determined reasons" (p. 421). Making these kinds of adjustments entails being aware of one's own linguistic identity, the linguistic identities of others and being conscious of the linguistic features that distinguish both (p. 421). An examination of linguistic literacy is important because, as Ravid & Tolchinsky describe it, "what children know or think they know at any step in their development functions as an interpretive system of what they are currently engaged in" (p. 419).

In the analysis that follows, the type of language play I am interested in is a specific form of linguistic literacy that relies on bilingual competency (what I will call linguistic *bi*literacy). My adaption of the term is necessary to extend it beyond the monolingual norm implicit in many of its definitions.[3] By documenting examples of students' linguistic biliteracy and situating it as a strategy within a content-learning context, I am beginning to fill in the blind spots which scholars lament exist in the scholarly record. As Ravid & Tolchinsky point out, "School-age children and *adolescents* continue to acquire new linguistic constructions and new functions for existing constructs within contexts that have so far not been of central concern in developmental psycholinguistic inquiry" (p. 441, emphasis mine). Adolescent Mexican newcomers learning English in a science setting is one of these unexplored contexts that, given demographic trends, merits attention.

The Research Context

Captainville is a rural town in central Iowa that, over the last fifteen years, has become a principal receiving community for Latino immigrants because it houses Bensen, one of the world's largest pork plants. Mexican residents who came as part of the original settlement wave easily recite the changes they've seen in the size of the K-Mart, the Hyvee (the local grocery chain), and, as the number of Mexican employees at Bensen increased, the number of pigs packed per day on the plant floor. School enrollments, they also boast, are bursting (Parent Interview with Adela Vargas, June 16, 2006). Work at the plant allows them to provide for themselves, as well as their families back in Mexico, and it affords their children the opportunity for a "quality" education and, more specifically, the chance to learn English. As one Mexican newcomer parent told me, she said to her teenage son: "*Sí, vas a ir a la escuela. ¿Sabes por qué? Porque aquí en este país si tú no sabes hablar inglés no somos nada* (Yes, you're going to go to school. Do you know why? Because here in this country if you don't know how to speak English, we aren't anything)" (Parent Interview with Angela Manso, October 18, 2005).

Its changing demographic context provides Captainville schools with a daunting responsibility. As Latino enrollment climbs, it does so against a societal backdrop of lagging achievement, high drop out rates, and low teacher expectations for this population (Ruiz-de-Velasco & Fix, 2000; Waggoner, 1999). Developing students' English proficiency is cited as necessary to improving school achievement. But since, as research shows, students may take four to ten years to acquire the kind of academic English needed for success in school (Cummins, 1981; Thomas & Collier, 2002), withholding content instruction until they have acquired the requisite English language skills is not a viable option as this only serves to widen the already large achievement gap that exists between them and their native English-speaking peers (Thomas & Collier, 2002). Schools offer, instead, "sheltered" courses, like English Learner Science, in which content is taught with language development support (Echevarria, Vogt, & Short, 2004).

The English Learner Science classroom of concern here was taught by Linda Crabtree, a white, middle-aged, native-English-speaking woman who had acquired limited proficiency in Spanish due to missionary work in South America. Linda was generally well-liked by her students. As one student expressed, "hace la clase divertida/she makes the class fun" (Aalia Rodríguez, May 3, 2005). In particular, the setting had three features relevant to the following analysis: a metalinguistic environment, a bilingual medium of instruction, and, as just noted, a humorous tone to teacher–student interaction. As part of my description of the research context, I explain these features in more detail below and provide brief examples.

The first feature, the metalinguistic ("talk about talk") environment of this course, was fostered by its "English Learner" designation, reflecting its purpose, in the words of Linda, to "use the vocabulary and help them to prepare for going into the regular classroom" (Teacher Interview with Linda Crabtree, April 7, 2004). This objective resulted in particular kinds of discourse practices that called explicit attention to language as a representational system, i.e., "What's that word?" (Richardson Bruna et al., 2007). The second feature, the bilingual medium of instruction, was a result of Linda being willing and (partially) able to conduct the class in two languages. This was not done as part of a school-wide model of transitional bilingual education, but rather because Linda simply thought switching back and forth between English and Spanish, which she did quite erratically, would help her students. Her attempts to speak to her students in their own language often resulted in requests to her students for help with unknown vocabulary ("¿Cómo se dice?/How do you say…?)", or unsolicited corrections of her Spanish by the students (Linda: "We need to explain how they form *como ellos form … las rocas forma…*" S: [Providing a correction] "*Se forman*").

The third feature, the humorous tone to teacher–student interaction, was an outcome of Linda's wanting to build relationships with the students and also, perhaps, related to her self-consciousness and perceived vulnerability as an imperfect speaker of Spanish. Linda used humor in two central ways: to lighten her admonitions of individual misbehaving students, and to establish general classroom rapport. Her admonitions were often humorous teases in which she poked fun at particular aspects of a student's behavior (his neglect to bring his books to class, i.e., "If your book's in your locker, it's smarter than you are" or his reluctance to do the pig dissection, i.e., "Are you a man or a mouse?"). The way she established general classroom rapport was often through self-deprecating references to her size ("You guys thought I was fat. I just have four stomachs") which would make the students laugh and garner their empathetic responses. These three features, humor, bilingualism, and metalinguistic awareness, permeated the hybrid discourse community of the English Learner Science classroom and provided points of structural resonance for students' strategy of language play, as I will describe in an upcoming section.

Methodology

The teacher and student interview data and classroom observation data reported on here were initially collected as part of a larger exploratory study examining explicit academic language instruction in science classrooms (Richardson Bruna et al., 2007). Videotaped observations of this particular classroom took place over a one-month period at the end of the school year (April–May). These observations documented the teacher and her students engaging in one lesson of a rock cycle unit and four lessons of a body systems unit, two of which involved the dissection of fetal pigs. My interpretation of these data reported has been informed by my subsequent longer-term ethnographic work in this school and community. The student interview quote that opens this chapter, the newcomer parent interview data reported in the previous section, and a teacher interview quote (from the mainstream Earth Science teacher) in an upcoming discussion come from this larger ethnographic record.

Having reviewed the transcripts of this particular English Learner Science setting to document other phenomena (Richardson Bruna & Vann, 2007; Richardson Bruna et al., 2007), I have returned to these transcripts, for this chapter, to collect examples of students' language play that I had previously noted, but disregarded as irrelevant, in my initial analyses. In my return to these transcripts, I coded any student-initiated humorous utterance that relied on linguistic biliteracy as language play. I purposefully excluded the teacher and students' routine direct or indirect corrections of each others' languages during instruction, as this was a predictable outcome of the metalingustic environment of the class as an "English Learner" setting and of the bilingual nature of instruction. I also discarded the teacher's own enactments of language play. This process resulted in the identification of nine clearly-audible utterances that fit my criteria. (Because students were not wearing microphones, these data only represent what one microphone on a single videocamera was able to pick up in a very noisy classroom.) I coded these utterances based on the nature of the linguistic biliteracy they displayed: phonological (an awareness of the sound systems of languages and the rules that govern pronunciation—two such utterances), semantic (an awareness of word meaning—four such utterances), and morphological (an awareness of word formation—three such utterances). I then used the coded transcripts to identify segments to return to in the video. I watched these video segments to gather multi-modal contextual data, like student or teacher interaction preceding or following the utterance or accompanying gestures (Kress, Jewitt, Ogborn, & Tsatsarelis, 2001) that may provide important interpretive information. (In some cases, the playful utterance was uttered when the student was off-camera, so limited multimodal data is available.) Finally, I applied to each example an analysis informed by Fairclough's (1989) three levels of critical discourse analysis. I describe the text of the humorous utterance itself (the micro level); I account for its purpose by interpreting the relationship between the utterance and the interaction in which it was embedded (the meso level); and I attempt to explain its meaning, exploring the relationship between the interaction and the larger classroom, community, and social context (the macro level) (Fairclough, 1989, p. 26).

Learning from Students' Language Play

Here I provide three examples of language play, one each in the categories of phonological, semantic, and morphological. The first two examples come from the pig dissection unit which, importantly, Linda has introduced by talking about its relevance to their presumed future work at the pork plant (Richardson Bruna & Vann, 2007). The last example comes from the preceding rock cycle unit.

Two Pizzas!: An Example of Phonological Language Play

In what I consider to be phonological language play, students make humorous utterances that reveal their implicit knowledge of the Spanish and English sound systems. The example I have chosen of phonological language play comes from the second day of the pig dissection lesson. Before the weekend, the students had viewed a glass-encased model of a fetal pig to identify its body parts, as they corresponded to a handout Linda had provided. On this day, the students actually are given fetal pigs and asked to tie them against dissecting trays so their legs are splayed backwards, facilitating the center cut and subsequent examination of body parts.

During the first lesson, the students expressed reluctance over the activity, concerned, as I have noted elsewhere (Richardson Bruna & Vann, 2007), for ethical reasons—the origin of the fetal pigs and, more specifically, what became of their mother. On this day, their reluctance is physical. As the pigs come out of the large plastic tub, the sight and smell of their small, limp, gray bodies and the smell of the chemical in which they have been preserved, produces revulsion. This is expressed by heightened activity and noise level. Students vocalize their disgust—"uuhhh"—and hold their noses, turning their faces away from the sight. Linda works to call the students to order and begins to show them how to get the pigs ready in the trays and how to tie their legs back with string. The expressions of revulsion turn into an uneasy laughter at the sight of Linda tying back the pigs' legs. The following excerpt begins with Linda's admonition for them to regard the activity as work and not play:

LINDA: [Standing at the left-hand side of a row of desks that have been pushed together to form a table, behind a male student] *Este es una trabajo . Este no es un juego*[4] . This is work . This is not a game.... You're going . EXCUSE ME . Gentlemen . [Holding up a ball of string and cutting while she is talking] You're gonna take two . long . pieces of string . With the first one . you're gonna tie it around . You're gonna tie it to the front leg . and then you're gonna tie . tight . tight [using student's pig to demonstrate] under . See guys? . Then you're gonna do the same . to the back legs . [giving student more string] ok? Pedro [addressing a student sitting in front of her and handing him some string], you're ready . *mirar que yo hice* (look at what I did) . So you want two pieces ...

ANONYMOUS MALE STUDENT: [off-camera] Two *pizzas!*

LINDA: Not pizzas [shaking head]

STUDENTS: [laughter]

LINDA: Take two halves...

Here we see how a student makes a spontaneous realization that the phonology of the English word "pieces" and the Spanish pronunciation of "pizzas" are similar. Since the first syllable of both words consists of the high, front, spread, tense vowel [i] (as in "seat") and since, in Spanish, "z" is pronounced as [s], these words differ only in the vowel of their second syllables: the English word "pieces" contains the mid, back, spread, lax vowel [ʌ] (as in "suds"), whereas the Spanish word *"pizzas"* contains the low, back, spread, lax vowel sound [a] (as in "sod"). Seeing both words in phonemic notation clarifies how it is awareness of the vowel sound of the second syllable, and how that differentiates word meaning between two otherwise very similar utterances, that achieves this particular act of language play:

[pisʌs] = pieces
[pisas] = pizzas (Spanish pronunciation)

What is immediately clear from the exchange is that the student uses his linguistic biliteracy to play with the fact that the teacher has said they want "pieces" when, it seems, he would rather have "pizzas." But another purpose emerges from examining the context of the interaction. This interaction is the first humorous one following Linda's admonition of the students' noisy behavior, so it serves to lighten the tone of classroom interaction and diffuse the power of her remark. It is, significantly, immediately followed by another student's remark describing the activity as "la pasión del puerco" (the passion of the pig), a reference to Mel Gibson's film, "The Passion of Christ," which had just been released and was causing a controversy in the weeks leading up to Easter, the time in which the pig dissection took place. In this way, although Linda has made clear that she regards the activity as work and not play, the students continue to be playful, but in a way that is less disruptive to the collective completion of the activity.

Contrary to Linda's perception that the pig dissection would be interesting to the students because of their families' work at the plant, the students are not eager dissectors. On a physical level, they are noticeably disgusted and, on an emotional level, they are confused about why Linda is "wasting pigs" (Richardson Bruna & Vann, 2007). Being forced to do an activity that they find repellant and, further, doing it under the guise that it is appropriate (to their presumed futures as low-status workers in the community), positions these students in powerless social roles (both within the classroom and without). Through this example of language play, a student exerts power in the ways that he can. He finds humor in the situation and thereby puts himself and his peers more at ease.

Es un Puerco: An Example of Semantic Language Play

At the semantic level, students express their linguistic biliteracy by playing with word meanings in Spanish and English to produce humorous statements. This play sometimes relies on their awareness of Linda's limited Spanish proficiency, particularly her inappropriate vocabulary usage, of which she was often completely unaware.

The semantic example I have chosen also comes from the first day of the pig dissection activity. Linda, as noted above, is trying to call the students to order (*Orden! Orden!*/Order! Order!). The students, as described, have placed their fetal pigs in trays and Linda is showing them how they are to measure their pigs (so they can make predictions about age and whether bigger pigs will have bigger intestines). She is behind a pair of boys, one of whom is Augusto, on the left-hand-side of the table. She has just asked the students what the appendage coming out of the pigs' belly is (wanting them to see that it is the umbilical cord). Augusto, taking up her interest in details of the pigs' bodies, asks her, pointing to his pig, "Can you notice that this is female?" What follows is, in addition to a subtle language lesson, a powerful juxtapositioning of Linda's institutional authority as the teacher of the pig dissection lesson with Augusto's genuine authority, given his farm experience in Mexico, as a breeder of pigs:

LINDA: How can you tell if that's a female?

AUGUSTO: [Says something inaudible while holding up his fetal pigs' legs with one hand in the same way one would do it to change a baby's diaper. He is shaking his head and gesturing with his hand in a way that indicates the obviousness of what he's saying. As he's talking, Linda is looking at another fetal pig in a tray to her left.]

LINDA: [Pointing to the pig to her left] OK. What about this?

AUGUSTO: [Pointing to the different pigs around the table before picking up the legs of the pig across from him to point more closely] This is a female, that's a male, that's a male, that's a male.

LINDA: [Stepping back from the table and calling out to students at the front of the table who, presumably, she thinks may not have been paying attention] Do you understand which is which? *¿Es hombre o mujer?* (Is it a man or a woman)? *¿Es hombre o mujer?* (Is it a man or a woman)?

ANONYMOUS MALE STUDENT: [Off-camera] *Es un puerco.*

STUDENTS: [Laughter]

LINDA: [Asking the same question to different sets of students around the table] *¿Es hombre o mujer?* (Is it a man or a woman)?

The humor of this student's semantic commentary is in the way it draws attention to the fact that one of the important semantic features of the words *"hombre"* ("man") and *"mujer"* ("woman") is the feature "human," a semantic feature not shared by the word "pig" (*"puerco"*) (see Table 8.1).

Because "human" is not a semantic feature of pigs, Linda's use of the words *"hombre"* and *"mujer,"* which carry "human" as a central semantic feature, to describe a pig is semantically odd. The student's remark, *"Es un puerco* (It's a pig),*"* is another way of saying "It's not a man *or* a woman, so the question you have asked doesn't make sense." If Linda had chosen to ask the question a different way by, for example, indicating gender through inflected word-endings (*"puerco;" "puerca"*) or gender-specialized terms (*"hembra"* for female animal; *"macho"* for male animal), the joke would not have been possible. The joke, then, relies on and points out the limitations of Linda's Spanish usage and demarcates group membership between the Spanish-speaking students who get the joke and Linda who, as we can see because she continues her use of the inappropriate terms, doesn't.

Beyond the mere linguistic power over Linda the students in this example wield, there is evidence as well of the experiential power their lives as farmers in Mexico has given them. While Linda officially occupies the teacher role in the classroom, she abdicates that role to Augusto when it comes to familiarity with fetal pig anatomy. In fact, Augusto, over the course of the pig dissection, makes repeated references to his life on the farm in Mexico in order to contribute information about, for example, the length of pig gestation, how pigs can abort, and what these aborted fetuses look and smell like (Richardson Bruna & Vann, 2007). In these moments, like the one depicted in the excerpt above, the contrast between Linda's perception of the students as workers and their representation of themselves as knowers is quite striking. Language play served to reinforce group affinity by signaling in- and out-group membership but it served another purpose as well: to expose the gap between the institutional authority of the teacher who directs the pig dissection in the school science context and the genuine authority of the students who lived pig science within the context of their daily home lives.

"You're Magmatic Now": An Example of Morphological Language Play

In morphological language play, students use what they know of Spanish and English morphology to form a humorous derivation of a root form. The example I have selected of morphological language play comes from the rock cycle unit

Table 8.1 Important semantic features

	hombre (man)	mujer (woman)	puerco (pig)
[adult]	+	−	+
[male]	+	−	+
[human]	+	+	⊖

that immediately preceded the pig dissection unit (see Richardson Bruna et al., 2007).

Linda is standing at the front of the room at the white board on which she has drawn a picture of a volcano. Her goal is to explain the role of volcanic activity in producing igneous rocks. She is eliciting from the class the name of the substance inside a volcano before it erupts (the desired answer is "magma") and the name of that substance as it comes out of the volcano after it erupts (the desired answer is "lava"). The students are not clear on the meaningfulness of the distinction, believing that "lava" is an appropriate word to describe both substances. Linda tries to elicit the term "magma" again, eventually supplying it herself, with a note of exasperation in her tone. Trying to personalize the concept and do a comprehension check of the students' understanding, as well as referencing the frustration she is exhibiting at the students' confusion, she compares herself to a volcano heating up:

LINDA: [Turning and pointing to the outside surface of her drawing of a volcano on the white board] When it's out of the volcano, it's lava [turns to face front]. When it's down here [turning and pointing to the inside of her volcano] getting hot hot . getting ready to erupt [turning forward and moving hands in circles in front of her], it's magma [turning back to board and pointing to the inside of her volcano] ... [Pointing to herself] So what am I? Am I magma right now [turning and pointing to the inside of the board] . because I haven't erupted [turns forward and moves hands up and down to indicate an explosion] and turned mean?

AUGUSTO: [Off-camera] You? Oh yeah . you're magmatic now.

LINDA: I'm magma . I'm still heating up aren't I? OK. Watch out if I erupt . Alright . so there's gonna be a difference between the magma and the lava.

In this example of language play, Augusto takes up Linda's supposition that she is magma getting ready to explode by describing her as "magmatic." He is apparently using his awareness that -tico/a is a derivational suffix meaning "of the kind of" that turns Greek-based (-ma word—final) nouns into adjectives (as in problema + tico = problemático, esquema + tico = esquemático, so magma + tico = magmático). Translating that as the English derivational suffix -atic, he produces an adjective for Linda that fits the hypothetical scenario she has developed. Linda's exasperation, because she hasn't yet "erupted and turned mean," is exasperation "of the kind of," or like, magma; that is; her exasperation has not yet spilled over:

Interestingly, later in the lesson, Linda has reason to rethink her insistence on the distinction of the terms "magma" and "lava." She is reading from the text to the students and interrupts herself to comment on the fact that what the text has previously called "lava," is now calling "magma" (Linda [Reading from text]: "Recall that an extrusive rock forms from ... uh oh ... In one part of the book it says that this is 'lava' and now in this part of the book it says it's 'magma'"). Underscoring the power dynamics present in the classroom, Augusto says, "You see, I told you it's the same," and adds that "In Mexico, we just call them 'lava.'"

(In truth, both words do exist in Spanish). Linda responds by commenting on English as "the craziest language" because "we have so many names for the same thing." Augusto makes a parallel statement about Spanish by supplying the example that, in Mexico, turkeys can be called both *"guajalote"* and *"pavo."* Linda, again wanting to diffuse the power conflict between her interpretation and the students' and the challenge to her teacher authority and knowledge that the incident points to, makes a joke at Augusto's expense: "Well, here I call turkey 'Augusto.'"

In this example, we again see language play being invoked in the midst of power dynamics between the teacher and her students to lighten the tone of interaction. Linda has, in fact, made explicit the fact that she is beginning to feel frustrated with the students' apparent lack of familiarity with the science terms she is trying to teach and it is this frustration that Augusto's joke dispels. We also see the fragile and arbitrary nature of teacher authority. Linda comes to rethink her insistence on the distinction between the two terms when she encounters a seeming contradiction in the text. This makes clear that it is not Linda herself that possesses ultimate authority and knowledge but the textbook she is using. (Indeed, Linda is not a science teacher; she is an ESL teacher assigned to a science classroom). In the face of having her authoritative position eroded, she rationalizes her confusion by referring to English as a "crazy" language and, ultimately, using a form of language play herself, punning, to call Augusto a "turkey."

This incident clearly highlights the infrastructural challenges Mexican newcomers pose to business-as-usual science teaching. Science teachers are typically not equipped, as part of their professional preparation, with the knowledge and skills to work effectively with English Learners. Therefore, schools turn the science education of these students over to ESL teachers. These ESL teachers are, similarly, typically not equipped, as part of their professional preparation, with the knowledge and skills to effectively teach science. In this way, limited capacity exists in the educational system to adequately meet the needs of newcomer students, like these Mexican adolescents in Captainville. Their increasing presence in science classrooms points to the profound absence of attention culturally- and linguistically-non-dominant communities have been given in science education. This, in turn, reflects the construction, as Rampal (1992) describes, in modern science, of the idealized disembodied observer, one uninhibited by the "trappings" of culture and language.

Language Play as Reality Play

The humorous language play of adolescent Mexican newcomers, bringing as it does, the affective stance of culture and language back into consideration, is a

Table 8.2 Illustration of language play

Magma	+	tico	=	Magmático
Magm(a)	+	atic	=	Magmatic
Root (N)		derivational suffix "of the kind of"		(Adj.)

refusal of a disembodied state. In this way, the work these students do through their language play in the science classroom is, ultimately, critical. Though humor does do the following—it displays a common perspective, develops a deeper interpersonal link, and builds rapport (Keltner, Capps, Kring, Young & Heerey, 2001, as cited in Lampert & Ervin-Tripp, 2006), thus giving it, significantly, a central role in adolescent socialization (Sanford & Eder, 1984)—as Paolucci & Richardson (2006) write, "Humor's critical role ... lies in poking a hole through often undiscussed but official versions of everyday reality, exposing their contradictions and the arbitrary basis of their social power" (p. 334). The language play of these youth is thus inherently transformative because it makes us consider science education from their perspective.

When adolescent Mexican newcomers arrive in U.S. science classrooms, they arrive with cultural knowledge, inherited from their families' and communities' lived experiences with Mexico–U.S. labor history. That cultural knowledge includes awareness of the deficit stereotypes associated with their particular rung on the race hierarchy ladder of U.S. society. As their length of residency in the U.S. increases, they accrue lived experience with how these stereotypes not only exist, but are acted upon. At Captainville High, for example, white students throw ketchup packets at them in the cafeteria, white community members assume they are involved in gangs and, in English Learner Science classes, teachers, who are not science teachers at all, instruct them with the expectation that they are all going to work the plant line (Richardson Bruna & Vann, 2007). This is the official version of everyday reality in which, I argue, these students' language play pokes holes.

In the examples I have provided, we see how they use language play, just momentarily, to subvert the teacher's authority in the classroom. Their use of this particular strategy, which reflects, I suggest, their rich community histories of orality, and, most specifically, the deployment of humor as a coping mechanism, was facilitated by the metalinguistic, bilingual, and humorous environment of the English Learner Science classroom (see Figure 8.2). By drawing on their linguistic biliteracy in this way, these students, to invoke Lugones, carry themselves through their science classroom activities with a playful attitude that "turns the activity into play" (p. 431). The playful attitude of "being creatively and actively in the moment," through their shared histories of orality and humor, served to reinforce group solidarity against the institutional authority of the teacher. This social power, which, given her low estimation of their futures (reflecting the "Mexicans as meatpackers" ideology of the community), her limited Spanish proficiency, and lack of science expertise, must have felt, to them, contradictory and arbitrary indeed.

Reconsidering Connected Science: Orality as a "Stairway" to Literacy

This discussion of the science education experiences of newcomer Mexican youth in an English Learner Science classroom is informed by a particular meaning of the humorous, not, in its popular usage, as something trivially

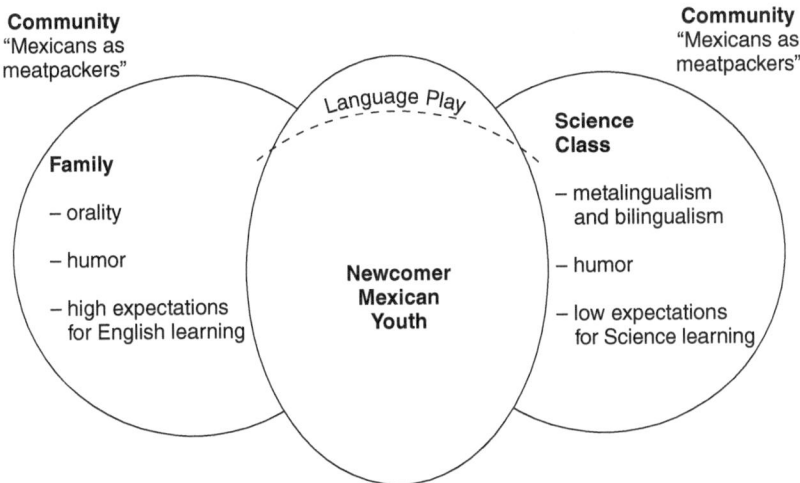

Figure 8.2 Language play as a structurally resonant crossing strategy. (Adapted from Phelan et al., 1998.)

amusing, but, instead, as substantive cognitive and emotional work. Thus, the purpose of this article is not to reinscribe U.S. Mexican students as non-academic "jokesters," but to illuminate the different form that "talking science" (Lemke, 1990) can take among a particular non-dominant student community in U.S. schools. Newcomer Mexican students are travelers along the triple-threat border of, more generally, American culture and the English language, and, more specifically, the American school and science culture, including those specialized languages. The way these adolescents playfully engage with their border identity within the context of science instruction stands to reveal significant information[5] about the cognitive and emotional life engendered by cultural and linguistic hybridity in science teaching and learning and its implications for envisioning a science education agenda that, in a changing America, is truly for all.

Current theorizing about science education with non-dominant populations stresses the idea of connected science. "This way of approaching learning and teaching challenges teachers and researchers alike to assume that children are always connecting in some important way to the discipline and to learn to see these connections unfolding in the life of the classroom" (Warren, Ogonowski, & Pothier, 2005, p. 122). Connected science assumes that "the experiences, ideas, and ways of talking and knowing of children from groups historically placed at risk are productively related to those characteristic of scientific communities" (p. 122). My work, as does Elmesky's, challenges us to extend this approach even to strategies that, on the surface, may be easily misread as "tangential," "disruptive," or "off-task." Such strategies, as Elmesky argues, are an "important part of their cultural toolkit" and the skills behind them are most productively regarded as a "stairway"[6] to participation (p. 50).

How ready are we to expand our thinking about science literacy to these kinds of hybrid discourse practices? If we take up Gutiérrez, Baquedano-López,

Alvarez, & Chiu's (1999) call to regard hybridity as a resource for learning, to understand such practices as "a systematic, strategic, affiliative, and sense-making process among those who share the code, as they strive to achieve mutual understanding" (Gutiérrez et al., 1999, p. 88), we must be prepared to embrace alternative forms of sense-making in science, such as language play, that reflect histories of orality traditionally marginalized in science. As Rampal (1992) writes, "Science curricula, despite claims in the last decade regarding a progressive commitment to learner-centred approaches, have, almost ironically, relegated the predominantly oral universe of children, especially those from non-literate backgrounds, to a distinctly degraded subaltern status" (p. 239). To counter the privileging of literacy that is part of the history of modern science will be, to say the least, difficult, but, if we take the call seriously, doesn't "Science for All" (AAAS, 1989) demand it?

One place to start is in making teachers and teacher educators cognizant of the intimate link between the type of linguistic awareness indicated by, for example, these students' oral language play and literacy. While teachers may recognize Mexican newcomers' playful behavior, they may not fully grasp the work that, in particular, their language play may be doing. An interview with the mainstream Earth Science teacher (students take Earth Science following Linda's "English Learner Science" course) is instructive on this point. He talks about the need for teachers to build a connection with the Mexican newcomer student population and notes that part of that connecting means creating a classroom environment in which these students can "interact with the teacher and be kind of joking *but still* be able to get some understanding, get some learning done at the same time" (Teacher Interview with George Roberts, October 10, 2006, emphasis mine). Embedded in this quote is a tension between play (the students' "joking") and "getting some learning done" that my research attempts to problematize. In the case of language play like that I have described here, the play is not *in* the way; play *is* the way. Phonological play, for instance, has been suggested to hone linguistic skills and correlate to later literacy development (Bryant, Bradley, Maclean & Crossland, 1989, as cited in Berko Gleason, 2005, p. 399). The metalinguistic abilities, then, which underlie such play may be productively regarded as predictors of literacy (Pellegrini & Galda, 1993, pp. 172–173). Thus, while students may use their linguistic biliteracy to play, that play is, in fact, important literacy work; it is preparing them to learn to read and write (p. 432).

An initial suggestion to Linda, then, would be to regard her students' language play, not as noise at the margins of her instruction, but as evidence that they are right with her, attending to the very sound of her science teaching in moments of imaginative engagement (Egan, 1986, 1988, 1990, 1992) resulting in displays of linguistic productivity. She could use these instances as teachable moments, recording them on the board so students can see their graphic representations, comprehend the power of their words and, through analyzing their humor, recognize and feel proud of their linguistic biliteracy—to be bilingual is to them, remember, a most valuable, life-enhancing commodity. Being cognizant of the relationship between orality and literacy, she could purposefully enrich and enliven the oral dimension of her teaching through a storytelling (Montgomery & Kumar, 2000) or poetics (Walders, 2000) of science.

While nurturing a climate of imaginative engagement might provide an anti-dote for the alienation these students may experience in their science learning and for the weight of family and community concerns that many bear daily, importantly, however, it will not make the socio-economic structures that cause this alienation and these concerns to go away. It will not address the economic plight that has brought them to these communities to begin with, or the associated racist and classicist structures upon which globalizing movements of human capital from the "Third World" to the "First" depend. It will only humanize (Freire, 1988; Salazar, 2004) the science classroom as a place of shared wonder and mutual appreciation. Imagine if that were just what these students needed to hang on, against a current of low expectations and limited opportunity, to learn to love and live, on this side of the border or the other, their learning of science.

Targeting Enduring Understandings

1   What are the connections you see between the discussion in Richardson Bruna's chapter and the four elements of culturally-responsive pedagogy, as described by Villegas and Lucas?
2   How would you describe the "work" that language is doing in this multi-cultural science context?

Deepening the Reflection

1   Thinking about your own schooling experience as an adolescent, in what ways can you relate to students' use of language play, as described by Richardson Bruna? What purpose did this kind of playful activity serve you and your peers in the classroom?
    Now, taking up the perspective of a student who doesn't speak the language of instruction, how would you describe some other purposes that such activity might serve?
2   At the end of her chapter, Richardson Bruna suggests that science teachers could take advantage of students' language play for instructional purposes. Think more specifically about what this might look like in the classroom. What would be the advantages of these teachable moments? What might be some disadvantages?
3   Referring to the Multiple Worlds model presented in this chapter (Phelan et al., 1998), reflect on the different family, school, and peer worlds that you cross in and out of every day. Are there changes in your ways of behaving or speaking when you transition from one world to another? Are there aspects of some worlds that are similar to those of others so the transition is easy? Are there differences that make the transition more difficult? Think more particularly about science. What, for you, characterizes the transition into the science-learning world?

Encouraging Engagement

1   Observe a science classroom with students from non-dominant cultural and linguistic groups and note instances of language play or other playful activity. Note the context out of which such play arises. What are the students, the teacher, and other peers doing that elicits the play? What is the relationship between the play and science teaching and learning? Come to class prepared to share your findings.

2   Ask to interview one or two students in the science classroom you have observed. Ideally these students would be ones that have been involved in some kind of playful activity that interests you. Talk to them about and show them the Multiple Worlds model (Phelan et al., 1998). Ask them to describe the different worlds that they move in and out of every day and to explain the difficulty or ease with which they make those transitions. Then ask them about the playful behavior you have observed and how they would make sense of it against the Multiple Worlds model. Is the transition into science a difficult or easy one for them? Why? Share these insights, along with your observations (above), with your classmates.

Notes

1   Student, parent, and teacher names are pseudonyms.

2   While the notion of scientific literacy includes more than just the activities of reading and writing, I take those activities to be central to my understanding of literacy here for, as Halliday and Martin (1993) detail, the very rise of modern science itself is inextricably linked to the production of particular kinds of read and written texts that served as mechanisms for "codifying, transmitting and extending the 'new learning'" (p. 61).

3   For example, in their discussion of user-related differences in a linguistic community, Ravid & Tolchinsky say that such differences are recognizable by language users "as they come into contact with speakers of other dialects, sociolects, ethnic and gender-related variations" (p. 422). The idea of linguistic literacy as being awareness not only of similarities and differences within a language, but between two separate languages is not present. This reveals the monolingual norm that underlies much research on language acquisition (Valdés & Figueroa, 1994).

4   Here and throughout the text, errors in Linda's Spanish have not been corrected.

5   Adolescents are supreme innovators; they are, indeed, the linguistic avant-garde (Gee, Allen, & Clinton, 2001). Paying attention to adolescent register, then, particularly in hybrid discourse communities, will reveal important dynamics of cultural reception, production, and transformation.

6   The "stairway" metaphor is problematic because of the way it seems to imply a tidy progression from spoken to written language, when, in reality, their relationship is best understood as reciprocal. Citing Clark (1997), Berko Gleason states, "...they form a 'virtual loop' where speech and writing constantly feed and modify each other" (p. 430). Perhaps, for this reason, a "moving stairway" metaphor is more apt because it offers the image of a revolving loop in which, at any given time, what is visible on the surface as written language has an invisible correspondence to underlying oral language, and vice versa.

References

Aikenhead, G. S., & Jegede, O. J. (1999). Cross-cultural science education: A cognitive explanation of a cultural phenomenon. *Journal of Research in Science Teaching, 36*(3), 269–287.

American Association for the Advancement of Science (AAAS). (1989). *Science for all Americans.* Washington, DC: Author.

Bergen, D. (1998). Stages of play development. In D. Bergen (Ed.), *Readings from ... Play as a medium for learning and development* (ED421 252; pp. 71–93). Olney, MD: Association for Childhood Education International.

Bergen, D. (2002). The role of pretend play in children's cognitive development. *Early Childhood Research & Practice, 4*(1). Retrieved November 1, 2007, from http://ecrp.uiuc.edu/v4n1/bergen.html.

Bergen, D., & Coscia, J. (2001). *Brain research and childhood education: Implications for educators.* Olney, MD: Association for Childhood Education International.

Berko Gleason, J. (2005). *The development of language.* Boston: Allyn & Bacon.

Bourdieu, P., & Passeron, J.-C. (1977). *Reproduction in education, society, and culture.* London: Sage.

Bowles, S., & Gintis, H. (1976). *Schooling in capitalist America.* New York: Basic Books.

Bryant, P. E., Bradley, L., Maclean, M., & Crossland, J. (1989). Nursery rhymes, phonological skills and reading. *Journal of Child Language, 16*, 407–428.

Camarota, S. (2001). *Immigration for Mexico: Assessing the impact on the United States.* Washington, DC: Center for Immigration Studies. Retrieved August 23, 2006, from www.cis.org/articles/2001/mexico/mexico.pdf.

Clark, A. (1997). *Being there: Putting brain, body and world together again.* Cambridge, MA: MIT Press.

Cummins, J. (1981). The role of primary language development in promoting educational success for language minority students. In *Schooling and language minority students: A theoretical framework* (pp. 3–49). Los Angeles, CA: Evaluation, Dissemination, and Assessment Center, California State University.

de Caro, R. J. (1972). Language loyalty and folklore studies: The Mexican-American. *Western Folklore, 31*(2), 77–86.

Dixon, N. F. (1980). Humor: A cognitive alternative to stress? In I. G. Sarason & C. D. Spielberger (Eds.), *Stress and anxiety* (Vol. 10, pp. 281–289). Washington, DC: Hemisphere.

Dunn, J. (1988). *The beginnings of social understanding.* Cambridge, MA: Harvard University Press.

Echevarria, J., Vogt, M. E., & Short, D. E. (2004). *Making content comprehensible with English learners.* Boston, MA: Allyn & Bacon.

Egan, K. (1986). *Teaching as storytelling.* London, Ontario: The Althouse Press.

Egan, K. (1988). *Primary understanding. Education in early childhood.* New York and London: Routledge.

Egan, K. (1990). *Romantic understanding: The development of rationality and imagination, ages 8–15.* New York and London: Routledge.

Egan, K. (1992). *Imagination in teaching and learning.* Chicago, IL: University of Chicago Press.

Egan, K. (1997). *The educated mind: How cognitive tools shape our understanding.* Chicago, IL: University of Chicago Press.

Elmesky, R. (2003). Crossfire on the streets and into the classroom: Meso/micro understandings of weak cultural boundaries, strategies of action and sense of the game in an inner-city chemistry classroom. *Cybernetics and Human Knowing, 10*(2), 29–50.

Fairclough, N. (1989). *Language and power*. Essex, UK: Longman Group.

Farm Foundation. (2006). *Future of animal agriculture in North America: Chapter 7: Community and Labor*. Retrieved May 31, 2006, from www.farmfoundationorg/projects/documents/CommunityandLabor.pdf.

Freire, P. (1988). *Pedagogy of the oppressed*. New York: Continuum.

Garvey, C. (1977). Play with language and speech. In S. Ervin & C. Mitchell-Kernan (Eds.), *Child discourse* (pp. 27–47). New York: Academic Press.

Gee, J. P., Allen, A.-R., & Clinton, K. (2001). Language, class, and identity: Teenagers fashioning themselves through language. *Linguistics and Education, 12*, 175–194.

Goode, J. G., Schneider, J. A., & Blanc, S. (1992). Transcending boundaries and closing ranks: How schools shape inter-ethnic relations. In L. Lamphere (Ed.), *Structuring diversity: Ethnographic perspectives on the new immigration* (pp. 173–213). Chicago, IL: University of Chicago Press.

Gutiérrez, K. D., Baquedano-López, P., Alvarez, H. H., & Chiu, M. M. (1999). Building a culture of collaboration through hybrid language practices. *Theory Into Practice, 38*(2), 87–93.

Halliday, M. A. K., & Martin, J. R. (1993). *Writing science: Literacy and discursive power*. Pittsburgh, PA: University of Pittsburgh Press.

Kamberelis, G. (2001). Producing heteroglossic classroom (micro)cultures through hybrid discourse practice. *Linguistics and Education, 12*(1), 85–125.

Keltner, D., Capps, L., Kring, A. M., Young, R. C., & Heerey, E. (2001). Just teasing: A conceptual analysis and empirical review. *Psychological Bulletin, 127*, 1231–1247.

Kress, G., Jewitt, C., Ogborn, J., & Tsatsarelis, C. (2001). *Multimodal teaching and learning: The rhetorics of the science classroom*. London: Continuum.

Lampert, M. D., & Ervin-Tripp, S. M. (2006). Risky laughter: Teasing and self-directed joking among male and female friends. *Journal of Pragmatics, 38*, 51–72.

Latorre, G. M. (2003). Humor and hemispheric consciousness: Chicana/o and Native American contemporary art. *Journal of American Studies of Turkey, 18*, 1–13.

Lemke, J. L. (1990). *Talking science: Language, learning, and values*. Westport, CT: Ablex.

Levinson, B. A. U. (2002). Foreword. In S. Wortham, E. G. Murillo, & E. T. Hamann (Eds.), *Education in the new Latino diaspora: Policy and politics of identity* (pp. vii–xi). Westport, CT: Ablex.

Lugones, M. (1996). Playfulness, "world"-traveling, and loving perception. In A. Garry & M. Pearsall (Eds.), *Women, knowledge, and reality: Explorations in feminist philosophy* (pp. 419–434). New York: Routledge.

Lyman, R. (2006, August 15). Census shows growth of immigrants. *New York Times*. Retrieved August 24, 2006, from www.nytimes.com/2006/08/15/us/15census.html?ex=1156564800&en=bcedf603839afde&ei=5070.

Martin, R. A. (1996). Humor as therapeutic play: Stress-moderating effects of sense of humor. *Journal of Leisurability, 23*(4). Retrieved October 26, 2007, from www.lin.ca/resource/html/Vol23/v23n4a3.htm.

Millard, A. V., & Chapa, J. (2004). *Apple pie and enchiladas: Latino newcomers in the rural Midwest*. Austin, TX: University of Texas Press.

Montgomery, S. L., & Kumar, A. (2000). Telling stories: Some remarks on orality in science. *Science as Culture, 9*(3), 391–404.

Paolucci, P., & Richardson, M. (2006). Sociology of humor and a critical dramaturgy. *Symbolic Interaction, 29*(3), 331–348.

Pellegrini, A. D., & Galda, L. (1993). Ten years after: A reexamination of symbolic play and literacy research. *Reading Research Quarterly, 28*(2), 162–175.

Phelan, P., Davidson, A. L., & Yu, H. C. (1998). *Adolescents' worlds: negotiating family, peers, and school.* New York: Teachers College Press.

Rampal, A. (1992). A possible "orality" for science? *Interchange, 23*(3), 227–244.

Ravid, D., & Tolchinksy, L. (2002). Developing linguistic literacy: A comprehensive model. *Journal of Child Language, 29*, 417–447.

Richardson Bruna, K., & Vann, R. (2007). On pigs and packers: Radically contextualizing a practice of science with Mexican immigrant students. *Cultural Studies of Science Education, 2*, 36–54.

Richardson Bruna, K., Vann, R., & Perales Escudero, M. (2007). What's language got to do with it?: A case study of academic language instruction in a high school "English Learner Science" classroom. *Journal of English for Academic Purposes, 6*(1), pp. 36–54.

Ruiz-de-Velasco, J., & Fix, M. (2000). *Overlooked and underserved. Immigrant students in U.S. secondary schools.* Washington, DC: Urban Institute.

Salazar, M. del C. (2004). The transformative potential of humanizing pedagogy: Addressing the diverse needs of Chicano/Mexicano (1) students. *High School Journal*, April.

Sanford, S., & Eder, D. (1984). Adolescent humor during peer interaction. *Social Psychology Quarterly, 47*(3), 235–243.

Saul, E. W. (Ed.) (2004). *Crossing borders in literacy and science instruction: Perspectives on theory and practice.* Newark, DE: International Reading Association.

Thomas, W. P., & Collier, V. P. (2002). *A national study of school effectiveness for language minority students' long-term academic achievement.* Santa Cruz, CA and Washington, DC: Center on Research, Diversity, and Excellence.

Valdés, G., & Figueroa, R. A. (1994). *Bilingualism and testing: A special case of bias.* Westport, CT: Ablex.

Waggoner, D. (1999). Who are secondary newcomer and linguistically different youth? In C. Faltis & P. Wolfe (Eds.), *So much to say: Adolescents, bilingualism, and ESL in the secondary school* (pp. 13–41). New York: Teachers College Press.

Warren, B., Ogonowski, M., & Pothier, S. (2005). "Everyday" and "scientific": Rethinking dichotomies in modes of thinking in science learning. In R. Nemirovsky, A. S. Rosebery, J. Solomon, & B. Warren (Eds.), *Everyday matters in science and mathematics: Studies of complex classroom events* (pp. 119–152). Mahwah, NJ: Lawrence Erlbaum.

Walders, D. (2000). Poetry and science education. *Educational Resources Information Centre* (ED463946). Retrieved February 12, 2007, from www.ericdigests.org/2003–1/poetry.htm.

Wortham, S., Murillo, E. G., & Hamann, E. T. (eds) (2002). *Education in the new Latino diaspora: Policy and politics of identity.* Westport, CT: Ablex.

Part III

The Development of a Science Learner Identity

9 Academic Identity and Scientific Literacy

John M. Reveles

Introduction

Reveles begins his chapter by asking that we consider the question of how students learn what it means to "do" school. Whether we realize it or not, this is one of the things students learn in school; they learn about the "appropriate" and "valued" ways of being a successful student. These ways of being include expectations for writing and talking in classrooms. For example, after reading this chapter you may be expected to write a reflection about what you learned and you will very likely be expected to talk about it with your classmates in relation to other course topics. The way you write a reflection for your professor might be very different than the way you write an email message that you would send to a friend. Additionally, when talking about this chapter with your classmates, you might choose a very different register of speech than you would use during a casual conversation. Scholars, like Reveles, who are interested in how students acquire their academic identities want to know more about how students make particular kinds of choices in learning how to represent themselves in relation to the subjects they are studying. Specifically, Reveles examines how students learn what it means to "do" science, as well as what it means to "be" a scientist by presenting their written work as evidence of such learning.

The evidence he uses in his discussion—(a) drawings of scientists, (b) written reflections on students' perceptions of themselves as scientists and on the work of science, and (c) drawn and written information recorded on science fair posters—reveals a number of interesting insights about identity formation processes as they occur with elementary-aged children. He further explores how these processes unfold against the backdrop of their year-long science learning. These insights help us see the ways in which students can internalize stereotypical images of science activity or of scientists themselves and the potential that science teaching has to influence such images. Given the interest in improving the outcomes of school science learning along with enhancing access to science professions for underrepresented populations, understanding how students' representational choices serve as clues to their affiliation with science is an important endeavor in shaping science instruction. Science instruction that encourages mastery of conventional forms of talking and writing in the science community, as well as the adoption of more inclusive forms of representation. This will be an essential element of culturally-responsive science teaching in multicultural classrooms.

How is it that students learn the cultural practices (i.e., ways of speaking, writing, and interacting) associated with success as they navigate within schooling institutions? And how is it that students realize their full academic potentials as they move from their home culture to the institutional culture of school? Such "ways of being" are learned by different students in different ways as students are acculturated into the institution of school (Halliday & Martin, 1993). Yet, students are often not aware of these ways of being which may be more or less obscured from consciousness among the plethora of school expectations in their daily lives. Early on—even before they enter the classroom—some children learn how to develop varied identities from context to context. However, often it is the case that children who come from ethnic and linguistic minority backgrounds have not had ample opportunities to learn such "ways of being" associated with the practices of the mainstream school culture (Halliday & Martin, 1993). Moreover, some teachers may not attempt to draw upon these students' native cultural knowledge and may further complicate student experience within schooling institutions (Warren, Rosebery, & Conant, 1994). As we consider the idea of students' ability to develop understanding of varied contextual identities, this chapter seeks to connect such learning with the development of students' academic identity[1] and scientific literacy.[2]

The chapter seeks to conceptualize how language, culture, and academic identity are intricately connected to the extent to which scientific activities may be inviting for student participation. Said another way, this study examines how students in a third grade classroom utilized the scientific language and cultural practices of the classroom to inscribe aspects of their academic identities into their written and drawn work. To achieve this goal, the chapter addresses several issues that provide theoretical as well as empirical support for my argument. First, I provide a sketch of some of the accepted characteristics of scientific literacy put forth by national organizations and science education researchers. Second, I review conventional theories of identity development that provide a foundation for connecting my own theoretical perspective on academic identity and scientific literacy inscription. Third, once the theoretical underpinning is presented, I provide the orienting research question, data sources, and educational setting utilized in the study. Fourth, I then highlight the research methodology drawn on in this chapter along with collected data evincing students' academic identity and scientific literacy inscription[3] across an academic year. As a final point, I discuss some of the educational implications that this study has for providing opportunities intended to facilitate meaningful science experiences for *all* students.

Theoretical Framework

The theoretical framework in this chapter builds on sociocultural theories of learning that consider ways scientific concepts are internalized and appropriated by students via spoken and written discourse as they participate in scientific inquiry (Lemke, 1990, 2001; Mortimer & Scott, 2003). Research across disciplines confirms the importance that language plays in the construction and communication of scientific knowledge (Knorr-Cetina, 1999; Latour, 1987). In particular, science is communicated through language, with specific sorts of discourse patterns in written and spoken forms (Yore, Bisanz, & Hand, 2003). However, if the student is unaware of the discourse expectations and requirements necessary to succeed, or is unable to meet such curricular demands, then she or he is at a disadvantage (Lemke, 2000). Furthermore, the language, content, and activity of science require that students learn to appropriate scientific understanding from a specific frame for reference. What this means is that in order for students to become scientifically literate, it is imperative that they be explicitly taught how to use the language of science and engage in the content from a scientific point of view. Therefore, if *all* students are to acquire certain degrees of scientific literacy during their academic experiences, it becomes necessary to examine how it is that the language of science may facilitate literate ability among some student populations and not others. As an interrelated examination, I argue that science education researchers should investigate the implications that the acquisition of scientific literacy has upon students' academic identities as learners of science.

Distinguishing Characteristics of Scientific Literacy

There are numerous perspectives on scientific literacy that have been proposed by professional organizations (AAAS, 1993; NRC, 1996) and researchers (for reviews, see DeBoer, 2000; Eisenhart, Finkel, & Marion, 1996). Past notions of literacy (in general) placed emphasis on students' command of fundamental reading, writing, listening, and speaking abilities. These days, such views of "literacy" in general—and "scientific literacy" in particular—are too restricted to provide an education to children that will help them survive in a complex and increasingly technological world. Thus, scientific literacy has become an important research topic for those concerned with the education of the next generation of citizenry. Moreover, reform efforts in science education have made scientific literacy a central goal for science instruction (AAAS, 1993; NRC, 1996). Therefore, in order to better understand how to promote scientific literacy among *all* student populations, it is essential to examine some widely accepted characteristics that science education researchers and professional organizations have advocated as encompassing varying levels of scientific literacy. For example, some science education researchers have investigated students' ability to construct scientific arguments consistent with those of a scientific community (Bazerman, 1988; Latour & Woolgar, 1986). In a similar manner, other scholars have viewed scientific literacy through the lens of facilitating students' ability to analyze evidence as they participate in the social structures that guide scientific activity (Roth & Lee, 2002; Eisenhart et al., 1996). Still other researchers view scientific literacy as being achieved by students when they are able to acquire an understanding of specific literate practices that underscore scientific endeavors (Halliday & Martin, 1993; Heath, 1983; Norris & Phillips, 2003).

Most recently, ethnic minority students' challenges of assimilation into science classrooms have been framed as a product of linguistic conflict (Lee &

Fradd, 1998), gender and ethnic identity dissonance (Brickhouse, 1994; Gilbert & Yerrick, 2001), and curriculum inappropriateness (Roseberry, Warren, & Conant, 1992). These studies suggest the need for a close examination of the contextual activity within classrooms to help identify ways science can be made more accessible to ethnic and linguistic minority students. At this point, students from diverse backgrounds are simply not afforded equitable access to science content, language, and practices that facilitates an affiliation with science and the development of scientific literate skills. Rather, these students are often pressured toward linguistic and cultural assimilation because they did not arrive at school with officially sanctioned language, culture, and background experiences necessary for academic success (Nieto, 1999). According to these and other theoretical perspectives, students' scientific literacy extends beyond the development of a practical understanding in science. For instance, definitions of scientific literacy have been expanded to include an ability to control the discourse of science for sociocultural purposes (Gee, 1991; Roth & Lee, 2002). From this theoretical standpoint, developing scientific literacy enables access to socially accepted ways of speaking, writing, and interacting. Taking a sociocultural perspective on scientific literacy implies the need to explore the dynamic relationship between students' and teachers' language use and its connection to the everyday co-construction of scientific knowledge.

In this chapter, I argue that in order to understand the role of scientific literacy development for *all* students, educational researchers ought to continue expanding the notion of scientific literacy development to include conversations about the role of the classroom culture, academic identity, and student inscription. As members of a community, students take action and interact with others to construct the contextual knowledge of the classroom. Their learning of and about science, therefore, is inseparable from the surrounding environment in which it takes place. This suggests that issues of student science learning need to be examined over time, as a community jointly constructs ways of speaking, writing, and acting. If the "classroom discourse" can be viewed as a central cultural tool utilized for classroom participation, individuals who recognize such discourse as a property of a cultural group may consciously associate themselves (via their speaking and writing) with a particular group by use of the classroom discourse. The language of the classroom thus becomes a resource for students to draw upon and is manifest in their spoken, written, and enacted classroom participation. In this way, students not only consciously inscribe their academic identities utilizing classroom discourse; they also appropriate specific literate practices in science through their writing. Thus, students' academic identity and scientific literacy inscription are closely tied to the ways they participate in the culture of the class.

Approaches to Understanding Identity

Traditional perspectives on identity development have been examined within psychological domains of research, exploring identity as a set of possible characteristics gradually developing within the continuum of a person's lifespan

(Erikson, 1968; Marcia, 1966; Yoder, 2000). However, contemporary researchers have extended arguments posed by scholars who define identity strictly from psychological perspectives. They have done so by examining the contexts in which identity construction ought to be understood (Holland, Lachiocotte, Skinner, & Cain, 1998; Nasir & Saxe, 2003; Wenger, 1998). Other identity scholars offer descriptions of expansive orienting frameworks that define a person's identity dependent upon the "kind of person" one is recognized as being (Gee, 2002; Hacking, 1994, 1995, 1998). Some researchers have also conceived identity as being closely tied to identity construction within the context of a "community of practice" (Holland et al., 1998; Wenger, 1998). Finally, recent research has cast identity as a dynamic entity that must be examined with respect to the context from which it is being interpreted (Brown, Reveles, & Kelly, 2005; Reveles, Cordova, & Kelly, 2004). Hence, identities viewed in this way are not conceived as static characteristics. Rather, they are constantly being shaped and reshaped as people position and re-position themselves within sociocultural contexts over time. Contemporary models of identity construction look very different than traditional psychological models and are frequently described in more dialogic terms, that make it possible to understand the interactive nature of identity from one social context to another.

Academic Identity Inscription Defined in Science Classrooms

Viewing classrooms as sociocultural contexts and the interaction that makes up the life of the classroom as cultural interaction, educational researchers are able to observe and analyze the actual co-construction of a classroom culture over time (i.e., days, weeks, months, semesters, and academic years). Therefore, the language, activity, and interaction that takes place within a classroom context becomes a means for not only enhancing a student's conceptual knowledge base about a specific topic of interest, but also serves as the catalyst in the construction of her or his academic identity within the classroom culture. From this perspective, academic identities are co-constructed by both the teacher and students through specific "classroom discourse" as it is spoken into existence.

In this chapter, "academic identity" is defined as a recursive contextual co-construction that takes place on a daily basis within classroom contexts everywhere. In other words, as students interact with others (e.g., administrators, teachers, other students, and so forth) in schooling institutions and in classroom contexts, they are active agents in their own academic identity construction. Thus, students' academic identity is characterized as a prescribed set of symbolic expressions that are both ascribed by the other (teacher and/or other members of a classroom context) and continually co-constructed by the self. In this way, students consciously and unconsciously construct and co-construct their own and each other's academic identities as members of a particular classroom community. As a result, academic identity as a theoretical concept is thought of as being tied to the literate practices drawn on to create cultural contexts where the discourse itself is utilized as a resource to frame membership and rules for cultural participation. It follows then, that student academic identity is perceived as a reality experienced by participating members through their interactional engagement in classroom science activities. Over time, students' academic identities become visible through their interactions in the ways they participate in the science activity of the classroom culture. And, as others see in them what they see in themselves, students' academic identities as learners of science are inscribed into the very literate practices that make up the daily life of the class. In this sense, a student's academic identity is constructed in the moment-to-moment interaction, over local time, as well as over a broader social historical context (Nasir & Sax, 2003).

Writing as an Indicator of Academic Identity and Scientific Literacy

While it is true that the spoken language use in a classroom serves as an essential resource for cultural participation, it is not the only indicator of student identity. According to Ivanic, when words are expressed by writers in particular ways, they are aligning themselves with others who use such words and are thus making a statement of identity about themselves as they signal meaning, values, beliefs, and affiliation of particular social groups (Ivanic, 1998, p. 45). As a result, through their writing, students inscribe affiliation with, and participation, in the cultural practices of the classroom. What this indicates is that writing in science often bears relevance in understanding how students appropriate their comprehension about the science content, as well as how they see themselves as members of the classroom community. As students affiliate with community ways of being, over time, they inscribe fundamental aspects of their academic identities and literate practices into their work. For this reason, analyzing student writing in science brings forth insight regarding the ways students' academic identities in a classroom community are connected to the ways they acquire scientific literate abilities.

Orienting Research Question

Utilizing an ethnographic research perspective, this study examined the ways that students' inscribed their academic identities and scientific literacy by addressing the following research question: *How did students inscribe their academic identities and scientific knowledge into classroom artifacts across the school year?*

In order to accomplish this research objective, I completed analyses of four classroom artifacts. As a result, analyses are presented that illustrate trends across four student-produced artifacts indicating academic identity and scientific literacy inscription. In this chapter, I argue that within the context of this particular classroom community, students' scientific literacy development was directly connected to their academic identity and was inscribed in their work during the school year.

Data Sources and Educational Setting

The data collection for this study took place throughout the 2001–2002 academic school year. The data sources were drawn from an extensive year-long ethnographic study. As a participant-observer, the author (referred to in the class as "Mr. Reveles") collected ethnographic data across the entire school year in order to gain an insider perspective of what it meant to learn science within the classroom context. Collected data sources include: (a) videotaped records of classroom interaction (approximately 120 hours of video data, recorded on forty days, spanning nine months), (b) digital photographs of student products (notebooks, projects, journals), (c) fieldnotes compiled during video data collection, and (d) interviews with teachers and students. The data analysis examines developing student academic identities across three artifacts in the first half of the school year and then turns to one artifact in the second half of school year. These artifacts indicate reflections of student academic identities as members of the community inscribed into their science work.

The setting for this study took place in an elementary classroom from a public school located in a small city in southern California. The data were analyzed to examine the ways that the activity system (in this case, a 3rd-grade class) was constructed across different units of analysis following Lemke (2000) and Wortham (2003).

Teacher

The teacher in this classroom engaged in teaching and learning practices that draw upon an ethnographic perspective. Mr. Cordova was both a teacher educator and an elementary classroom teacher during the year of this study. In his own elementary classroom, he sought ways to communicate how disciplinary knowledge is framed with and through the spoken, written, read, and enacted discursive choices of teacher and students. Teaching goals included providing students ways of interacting with and learning from academic content. As such, students had opportunities to develop metacognitive understandings of how their identities "as scientists" were in use as they explained scientific phenomena and engaged in investigations throughout the school year. That is, students were explicitly taught to be cognitively aware of their own thinking as "scientists" while learning science.

Participants

Table 9.1 provides student ethnic and gender distributions within the classroom studied. This table uses district-defined categories and identifies the percentages of students represented within the classroom. Table 9.1 additionally indicates aggregate ethnic and gender distributions of the students in the classroom.

The 3rd-grade classroom from which the data were selected was comprised of eighteen students at the beginning of the school year. However, one student was transferred to a sheltered English immersion classroom because she was new to

Table 9.1 Ethnic and gender distribution of student population

Ethnicity	Percentage	Gender distribution
Hispanic	39	Boys = 9
White	56	Girls = 9
Asian-American	5	
		Total students = 18

the U.S. and spoke virtually no English. The student population (as shown in Table 9.1) was comprised, primarily, of two ethnic groups, defined by the district as "White" (56%) and "Hispanic" (39%), with smaller percentage of one other ethnic group "Asian-American" (5%). The students in this classroom ranged in age from eight to nine years of age with nine females and nine males.

Research Methodology

In this study, ethnography served as the principle research orientation for examining how science, as a socially situated enterprise, was constructed through the discursive interchanges of members of a classroom community of practice (see Reveles, 2005, chapter 3). Employing micro-ethnographic techniques (Erickson, 1992; Hicks, 1995; Kelly, Chen, & Crawford, 1998), I investigated the ways science was accomplished in a 3rd-grade class. In order to examine how science was inscribed into students writing, two sorts of analyses were conducted. These analyses included: (a) video analysis of classroom life and (b) artifact analysis of student inscribed products. The analyses utilized in this work focused specifically upon student written and drawn inscriptions as a point of reference for examining how it was that students within this particular classroom community inscribed aspects of their academic identity and scientific literacy into their science work across the school year.

Video Data Analysis

Following the video analysis methods described by Erickson (1992), I first conducted a review of the video and audio taped records of classroom activities. Notes were taken and compared to the other records such as fieldnotes and artifacts of the classroom science activities across the entire year. After the initial review of data, timelines were constructed to identify the sequences of events and time distributions for each day recorded (for an example of ethnographic timelines, please see Reveles et al., 2004). Since these timelines identified the sequence of science activities recorded, they provided a precise guide for analyzing the corpus of real-time data collected. The timelines indicate the actual days, months, and years when video data included in the dataset were recorded and, therefore, provide a guide for locating specific days and times analyzed.

Artifact Data Analysis of Inscribed Student Science Work

The next level of analyses consisted of examining a range of student products across the entire class. The student-produced artifacts provide the greater part of data analyzed in that they were the tangible products through which students inscribed scientific meaning as they engaged in ongoing activity in the classroom. For each artifact, a domain analysis (Spradley, 1980) was conducted to classify student responses to the task into categories. Tables of students' inscriptions were also constructed to examine in detail how the teacher and students co-constructed science understanding and how students appropriated certain aspects of the science made available in the classroom introduced during the school year.

Data Analyses

Popular Perceptions of a Scientist

Recent science education literature has expressed concern over the gap between typical images that primary and secondary students have of scientists and the images they hold of themselves as a potential factor impacting later interest in science (Noh & Choi, 1996). The popular perception of a scientist has—for a number of years—tended to be that of a white male, with facial hair, wearing spectacles and a lab coat, working at a bench with a chemical apparatus (Chambers, 1983; Mead & Metraux, 1957; Rubin et al., 2003). This first artifact came from student drawings in which they were asked to produce a visual image of what they thought a scientist looked like, as well as the activities they were believed to engage in. The artifact is of interest for analyzing pictorial inscriptions where free expression revealed subtle conceptions held by students. The inscribed images provide a baseline for investigating held student perceptions of typical scientist characteristics. In this way, the findings provide insight for understanding how it was that students in this classroom viewed scientists in relation to their own images near the beginning of science instruction.

Analysis of Artifact I: Picturing a Scientist

The first artifact (Picturing a Scientist) provides evidence that came from systematically analyzing students' perceptions of what they thought a typical scientist looks like. Figure 9.1 represents one student's illustration of her perception of a scientist. The artifact indicates that this student (as well as other students) already held specific ideas of what a typical scientist looks like. Furthermore, the artifact shows that students also held tangible ideas about the types of activities they believed scientists engaged in as professionals.

The artifact illustrates indicators of student academic identity related to inscribed perceptions of scientists in the sense that they were depicting internalized images held at the time. For example, it is interesting to note that the drawing indicates the scientist gendered as female, looking through a

Figure 9.1 Student illustrated example of "Picturing the Scientist".

microscope, and wearing a t-shirt that reads, "I love science." The student's held perceptions indicate a positive affiliation with science at this time. This artifact was collected at the end of October after "The Watermelon Investigation" (an introductory inquiry project conducted during the first days of school) and just before students began participating in weather-related science experiments in November. The exact date of the drawing is not known because of the fact that it was drawn as a cover page for students' science weather journals to be used during upcoming experiments and may not have been completed by all the students on the same day. Table 9.2 demonstrates particular characteristics of the scientists drawn by students in their "Picturing a scientist" illustrations. Table 9.2 was created in order to compare stereotypic vs. not-so-stereotypic features in student illustrations.

Picturing the Scientist Findings

Student perceptions in Table 9.2 indicate trends in student illustrations across the classroom. Analysis of student drawings reveals some stereotypical images of scientists held by several students, as well as some not-so-stereotypical images held by quite a few students at the time. For instance, some of the more universal (Chambers, 1983; Mead & Metraux, 1957) stereotypic images that various students expressed in their drawings were as follows: eleven of fourteen students that completed the drawing pictured the scientist alone, nine out of the fourteen gendered the scientist as male, and seven of fourteen students pictured the scientist handling equipment. Some not-so-stereotypical images students pictured in their drawings were: nine of fourteen illustrated the scientist as smiling, six of fourteen illustrated the scientist standing, and three of fourteen students gendered their scientist as female. Of the students who wrote words directly in the drawings, seven of fourteen students wrote such comments in their pictures as, Amelia—"What is that?" Kathy—"Scientist asks questions and look for answers." "They do it because they are curious and because it is fun." Jade—"I

Table 9.2 Student pictorial representations in "Picturing a Scientist" drawings

Student pseudonym	Yzabel	Cody	Eduardo	Amelia	Jade	Osvaldo	Samuel	Luke	Jacob	Juan	Julie	Kathy	Cameron	Emily	Rodrigo	Alexa	Rosa	Nicholas
Characteristics Pictured:																		
Scientist traits: Gendered		✓	✓	✓	✓	✓	✓	✓	✓			✓	✓	✓	✓		✓	✓
M/F	*	M	I	F	M	M	M	M	M	*	*	M	F	F	I	*	M	M
Verbalization: quote by scientist	*			✓	✓					*	*					*		
written	*		✓	✓						*	*				✓	*		
comment	*	✓	✓	✓	✓		✓			*	*	✓	✓	✓	✓	*	✓	✓
Social context: alone	*	✓		✓	✓		✓			*	*	✓	✓	✓		*		✓
with others	*			✓		✓		✓	✓	*	*				✓	*		
Actions: looking	*			✓	✓	✓			✓	*	*		✓	✓		*		✓
speaking	*			✓	✓					*	*					*		
writing	*									*	*					*		
reading	*									*	*					*		
smiling	*	✓		✓	✓	✓	✓			*	*	✓	✓	✓		*	✓	
handling equipment	*			✓	✓	✓		✓	✓	*	*		✓	✓		*	✓	✓
standing	*	✓					✓	✓		*	*	✓		✓		*		
sitting	*		✓	✓						*	*				✓	*		
outside	*				✓				✓	*	*		✓			*		
Equipment: wearing glasses	*	✓				✓				*	*					*	✓	
test tubes/ flasks	*			✓		✓				*	*			✓		*		
microscope	*						✓			*	*			✓		*		
table/lab bench	*			✓		✓	✓			*	*			✓	✓	*		
lab coat	*	✓								*	*			✓	✓	*		✓
telescope	*									*	*		✓				✓	

Notes
Key: * = missing data, I = indistinguishable.

found a bone of a baby dinosaur." Cody—"They research and study to find answers."

While this first artifact was not intended as an assessment instrument; nor was it a formal class assignment requiring extensive thought and reflection. The pictorial representations do reveal aspects of students' held images (at this time) regarding what a scientist was thought to look like and the typical activities that he or she was thought to engage in. Traditionally held images of a scientist may have resulted from the lack of extensive student experience in understanding the range of career fields that scientists are employed in. Some of the not-so-stereotypical images that particular students held (e.g., gendered the scientist as female) may be associated with the fact that certain students came from households where stereotypical images of scientists were not endorsed. Hence, indicating that some students already held an unambiguous understanding that women are indeed present in science professions.

Another area of analytic interest connected to students' academic identity inscription in this artifact has to do with the timing of the pictorial representations. These drawings were completed in October after students already had a few key science related experiences, which, may have left impressions (a) upon student perceptions of science; (b) upon student perceptions of conducting science investigations; (c) upon student perceptions of scientists; and (d) upon student perceptions of the literate practices associated with doing science. Such school experiences early in the year may have increased the likelihood that some students internalized positive aspects of science engaged in during previous science activities.

Student comments of interest reflecting generally positive attitudes toward science written into some pictures include: Kathy's comment—"Scientist asks questions and look for answers." "They do it because they are curious and because it is fun," and Cody's statement—"They research and study to find answers." These statements reflect a certain degree of understanding that scientists are professionals that predict, explore, and discover phenomena in the world. Additionally, another positive indicator involved a large percentage of students picturing their scientist smiling (nine of fourteen). It is of interest to note that besides picturing the scientist alone (eleven of fourteen students) and gendering the scientists as male (nine of fourteen students), picturing the scientist as smiling was the most common characteristic featured in student drawings. While some of the inscribed images may have been associated with students' knowledge outside the classroom, it is plausible that many of the students in this classroom community were beginning to internalize particular images of scientists and the activities they engage in as they themselves participated in science.

Sources of Student Inscription

Numerous factors can be said to influence students' beliefs about science in school. A number of researchers have corroborated the finding that students' school experiences can influence the images students hold of scientists (Chambers, 1983; Fort & Varney, 1989; Noh & Choi, 1996; Song & Kim, 1999) and that the ways students think of scientists in general does tend to influence their attitudes toward science (Finson, Beaver, & Carmond, 1995) to some degree. In the first artifact, I presented findings of student academic identity inscription related to images held of scientists near the beginning of the school year and discussed possible meaning of student inscriptions. The purpose of this artifact is to show some of the ways students' perceptions of scientists reflected emerging aspects of their academic identities within the classroom community of practice. The artifact presented analysis of students' images at a time when they were still on the periphery of community participation. Before students could become fully participating members of the community, they needed to acquire a certain degree of facility with the ways of being a member of this particular classroom community. In the next artifact, the science writing analyzed was an assignment that required students to think about and write some of the ways they saw themselves as a scientist within the classroom community.

Analysis of Artifact II: I as a Scientist

The second artifact was a self-reflective writing assignment that went into students' inquiry journals entitled: *I as a Scientist.* This artifact was completed on November 2, 2001, just before students began a series of weather-related investigations in the month of November. In the self-reflective comments written, students were beginning to acknowledge their own and each other's academic identities within the classroom community. At this point, students were moving from peripheral participation to increased involvement as members of the classroom community. Student engagement in the social practices of the classroom began to reflect negotiated aspects of community affiliation as manifest in the artifact inscriptions. The analysis for this artifact began with the examination of all student inquiry journals via collected digital photos. This was done to determine which students had completed the assignment. Once students' work was identified, I copied verbatim student text into Table 9.3 to look for trends in student data with a focus on student academic identity inscription.

I as the Scientist Findings

As an illustration of the self-reflective writing indicating students' academic identity inscription, I review several representative examples. For instance, Eduardo wrote, *"I'm a scientist because I study a lot of plants or living things; I predicted how the watermelon weighed, height, and width in an investigation."* Emily indicated, *"I'm a scientist because I did some experiments; I ask questions and get answers; I learn about new things everyday, I do lots of experiments."* Luke stated, *"I do math, I measure, I learn."* And Samuel wrote, *"I'm a scientist because I read like a scientist."* These written statements indicate various ways that students began to view themselves as scientists during their participation in classroom science activities. A number of written statements made by students also reflected an internalization of knowledge gained during earlier classroom science experience. For example, eight students (Cody, Eduardo, Amelia, Osvaldo,

Table 9.3 Students' writing about themselves as scientists

I As The Scientist
Phenomenon: How You See Yourself as a Scientist
How are you a scientist? What kinds of things do you do that are examples that show you are a scientist?

Student	Student's Written Comments
Yzabel	I predict and read.
Cody	In the watermelon investigation we used instruments.
Ed	I'm a scientist because I study a lot of plants and living things. I predicted how much the watermelon weighed, height, and width in an investigation.
Amelia	I'm a scientist because I learn. I did the watermelon investigation and I measured a watermelon.
	I ask a lot of questions everyday. I learn things from my culture.
	This year my class did a watermelon investigation and I observed.
Samuel	I'm a scientist because I read like a scientist.
Luke	I do math, I measure, I learn.
Jacob	I predict lots of things. We did the watermelon investigation. We measured things. I read lots of books.
Juan	I do experiments. I did the watermelon investigation.
Kathy	I think I'm a scientist because I record and read. I did the watermelon investigation. I did earth sciences.
Cameron	I like science. I did a science project.
Emily	I'm a scientist because I did some experiments. I ask questions and get answers. I learn about new things everyday. I do lots of experiments.
Rodrigo	I did the watermelon investigation with a partner. We measured the fruit it weighed $6\frac{1}{2}$ lbs.
	I'm a scientist because when I was in second grade we learned about magnets.
Rosa	I'm a scientist because my cousin tells me a question.
Nicholas	I'm a scientist because in New Jersey I predicted that there were dinosaur bones in Montana. I ask at least five questions a day.

Jacob, Juan, Kathy, and Rodrigo) wrote statements that were directly related to activities engaged in during the Watermelon Investigation. Student inscriptions also reflected a meta-cognitive awareness of internalized scientific literate actions taken. Case in point, Yzabel wrote: *I predict and read.* Luke wrote: *I do math, I measure, I learn.* And Kathy wrote: *I think I'm a scientist because I record and read.* Such student inscriptions are reflective of actions of scientists identified by teacher and students during earlier classroom conversations. The statements made by students suggest an affiliation with established cultural practices of the classroom. Written self-reflective statements, in coordination with identified actions of scientists, confirm students' emergent academic identity affiliation with the classroom community. Moreover, students' inscriptions indicate the beginning of a history of student academic identity affiliation using terms and concepts co-constructed by community members during prior classroom conversations. Students were beginning to internalize certain ideas, images, and scientific literate practices related to their prior classroom experience. As students participated in the scientific literate practices of the classroom, their academic identity positioning was becoming evident in the ways they inscribed their own identities as scientists.

Analysis of Artifact III: "What Counts as Science? What Do We as Scientists Do?"

The third artifact analyzed occurred at the end of November (November 21, 2001) and was a student assignment labeled: *What Counts as Science? What Do We as Scientists Do?* This artifact directly asked students to reflect upon how they viewed themselves "as scientists" in the community. The artifact includes written responses to three questions. Given that this artifact has three questions to it, I first describe how each question was analyzed to provide the logic of my analytic approach. Next, I present data analysis of the third question as an example of how all three artifact questions were analyzed. Afterwards, I identify general findings in student inscriptions.

The coding procedure utilized in the analysis of this artifact resulted in the classification of student responses into three categories. The categories constituted a domain analysis of student inscriptions of "doing science" following Spradley (1980). The semantic relationship identified in the domain analysis constitutes a Means-end relationship, where X is a way to do Y. The emergent categories from this analysis are:

- Students wrote what they had done in terms of *abstract practices.*
- Students wrote statements about *the making of scientific instruments.*
- Students wrote responses in reference to the *watermelon investigation.*

Student written responses were grouped in three domains, under categories AP, MI, and WI where: (a) AP = student reference to an "Abstract Practice," (b) MI = student stating that they "Made an Instrument," and (c) WI = student response in reference to the "Watermelon Investigation." In the analysis of the second question, I examined inscribed data of all students who completed the assignment to look for trends in students' responses. Response codes used in the analysis of question two were categorized as either C = written response regarding a scientific Concept they learned, or P = student reference to a learned scientific Procedure. The responses to this question exhibit an understanding of concepts and procedures engaged in during previous weather investigations. In their responses, students inscribed a degree of understanding about particular weather related phenomena (conceptual responses) studied as well as an understanding regarding how to carry out a scientific investigation (procedural responses).

In the analysis of student written responses to the third question, response codes used in the analysis of the artifact question were categorized as either SC = written response referencing a "Specific Concept" to learn about, or TC = written response referencing a "Theoretical Concept" to learn about.

In analyzing written inscriptions, two types of answers are evident in the responses. Students wrote about what they wanted to learn in terms of specific

Table 9.4 Student written responses to "What Counts as Science" question three

What Counts as Science? What Do We as Scientists Do?			
3. We will learn how to measure the earth's air pressure, and also the earth's rainfall. You will design your own experiment using your weather instruments. What would you like to learn about the earth's weather?			

Domain Analysis: Student Responses

Student Written Responses	Student	SC	TC
Missing data.	Yzabel	*	*
If there is a thermometer more than 23°C?	Cody	✓	
How tornadoes are formed.	Eduardo		✓
I would like to take a rain gauge and put it outside with water in it. Then see if it freezes.	Amelia	✓	
I would like to learn if you can put all the instruments together and make one instrument.	Jade	✓	
How cold should it be to make us die from freezing.			✓
I would like to learn about how to make a tornado.			✓
Missing data.		*	*
I would like to learn about the tornado experiment.	Jacob		✓
No answer.	Juan	*	*
Missing data.		*	*
I would like to learn to put the rain gauge out on the porch for a night, an hour and see how much is in and then.	Cameron	✓	
Can you use all your instruments in one day? If you did what did you learn?	Emily	✓	
How long does it take to stop a tornado?	Rodrigo		✓
I want to learn what would happen if you used the anemometer, thermometer, and the wind vane at the same time.	Alexa	✓	
Missing data.	Rosa	*	*
That if the thermometer can go as high as 200°F.	Nicholas	✓	

concepts and in terms of theoretical concepts. When pointedly asked what they would like to learn about the earth's weather, students wrote several interesting responses that likely drew on past classroom experiences. For instance, Jade's response (*I would like to learn if you can put all the instruments together and make one instrument.*) indicates that she had already learned about certain weather instruments but wanted to expand such knowledge by combining all instruments into one single weather instrument. Her answer reflected a desire to gain specific conceptual understanding regarding the instruments they had already studied. Alexa wrote similar responses about combining instruments. She even named the instruments that would be combined (*I want to learn what would happen if you used the anemometer, thermometer, and the wind vane at the same time.*). Other student responses indicated that they desired to learn theoretical concepts related to the earth's weather. For example, Eduardo expressed a desire to learn how tornadoes were formed, Osvaldo wanted to learn how cold it should be for someone to die from freezing, and Rodrigo wanted to learn how long it would take to stop a tornado.

General Findings Across Artifact III

Drawing on Bakhtin's notion of *"ventriloquation"* in which one voice transmits what another voice has said (Bakhtin, 1973, 1981, 1984), I explain the importance of understanding student academic identity inscription as an interconnected construct for taking up scientific literate practices in this classroom. The written responses in this artifact consistently indicated science literate practices engaged in during past science investigations. Across all three artifact questions, students inscribed their academic identities into their work as they "ventriloquated" the words they used in their writing. Bakhtin made clear an understanding of this phenomenon as he wrote:

> When each member of a collective of speakers takes possession of a word, it is not a neutral word of language, free from aspirations and valuations of others, uninhabited by foreign voices. No, he receives the word from the voice of another, and the word is filled with that voice.
>
> (Bakhtin, 1973, p. 167)

From a Bakhtinian perspective (as is positioned in Bakhtin, 1973, 1981, 1984, & Wertch, 1991), student academic identity inscription and appropriation of scientific understanding is visible in the written responses that students gave to each question. This implies, that, as students were writing answers to the questions posed by the teacher, they were involved in a dialogic process in which they (as language users) made particular language choices obtained from past scientific experiences they had engaged in. Learning science in this classroom, then, was a cultural experience contributing to students' academic identities and was manifest in the ways they wrote about what they did "as scientists" in the community. As students put pen and pencil to paper in order to express themselves, they entered a dynamically charged dialogue that socially marked them as members of the classroom community by the words they chose to use. Thus, the "language of the classroom" (over time) became the "language of students" as they inscribed meaning to their life experiences as members of this community of practice. Analysis across the written responses in artifact three indicate that the words students used to inscribe meaning and express scientific understanding were comprised of the words they had learned in the classroom as they participated in science weather experiments. Thus, students' ventriloquation of the language of the classroom in their written responses reflected student academic identity inscription and appropriation of scientific understanding.

Analysis of Artifact IV: Student Science Fair Posters

The final artifact analyzed is the science fair poster. Students completed five group posters and one individual student poster (the student investigated plant cures which did not fall under any other category). This last artifact analysis lends strength to the argument that students' academic identities changed throughout the year by looking specifically at student identity inscription in their science poster write-ups. The examination of student science fair posters indicates student appropriation of classroom scientific literate practices introduced during the academic year. The science fair poster presentations were born out of students' plant experiments in May and encompassed students' last inquiry-based science activity of the school year. Students worked in collaboration to write-up and put together various components of the science posters (e.g., experiment hypothesis, materials, procedures, findings/conclusions). Student posters were displayed at the science fair on May 10, 2002 and provided an auxiliary perspective into how students inscribed aspects of their academic identities as scientists into their written work. The data presented in Table 9.5 comes from digital photographs of students' science fair posters and presents analysis findings of student research groups in aggregate.

Student Science Poster Findings

The results from the analysis of students' science fair posters reflect engagement in particular scientific literate practices introduced by the teacher and co-constructed by students in the course of the academic year. Evidence suggests that a significant number of the scientific literate practices introduced and engaged in throughout the year were inscribed into student poster write-ups. As students wrote up all aspects of their plant experiments, they were appropriating a number of scientific literate practices that they in union with the teacher had previously co-constructed. Looking across student research groups, it can be seen that all groups exhibited (in the poster write-ups) at least ten of the twenty-three scientific literate practices introduced by the teacher and co-constructed by students earlier in the year. In conducting their plant experiments and poster write-ups, students were drawing on prior science experiences and inscribing meaning to their project posters based on past classroom scientific literate practices. This point is particularly salient in the sense that in doing so, students were again aligning themselves with the culture of the class in their act of writing up their posters. By drawing on specific scientific literate practices used throughout the school year, students were inscribing their academic identities into their poster write-ups as they drew on the language of the classroom community and used it as their own.

In Table 9.5 it can also be seen from the aggregate data that some of the fields representing scientific literate practices introduced during the academic year are completely blank across all student groups. The reason for this is that many of the scientific literate practices introduced and used during earlier science investigations were not germane to these experimental designs. For example, practices such as sketching, defining literate practices, making scientific instruments, and so forth, were not essential elements in student science fair posters; yet were central components during previous weather experiments. Therefore, there is a noticeable gap in the appropriation of scientific literate practices that were not needed to complete this particular inquiry-based science activity.

Shifting analytic focus from overall data trends to student research group findings, it can be seen that some student groups utilized more or less scientific

Table 9.5 Student science fair posters

Student Research Groups	Structure	Sunlight		Reproduction		Water	Photosynthesis		Plant cures									
Student Pseudonym	Eduardo	Kathy	Samuel	Emily	Julie	Amelia	Cameron	Osvaldo	Luke	Cody	Yzabel	Alexa	Jade	Rodrigo	Nicholas	Jacob	Juan	Rosa
Student Scientific Literate Practices evident in posters:																		
Introduced during academic year																		
Design research plan	✓	✓	✓	✓	✓	✓	✓	✓	✓	✓	✓	✓	✓	✓	✓	✓	✓	✓
Pose research questions	✓	✓	✓	✓	✓	✓	✓	✓	✓	✓	✓	✓	✓	✓	✓	✓	✓	✓
Investigate research questions	✓	✓	✓	✓	✓	✓	✓	✓	✓	✓	✓	✓	✓	✓	✓	✓	✓	✓
Record data	✓	✓	✓						✓	✓	✓	✓	✓					
Provide estimates																		
Provide evidence for investigation	✓	✓		✓	✓	✓	✓		✓	✓		✓	✓	✓	✓	✓		✓
Write down estimates																		
Sketch																		
Complete investigation worksheet	✓	✓	✓	✓	✓	✓	✓	✓	✓	✓	✓	✓	✓	✓	✓	✓	✓	✓
Use scientific instruments	✓	✓										✓	✓					
Take measurements	✓	✓									✓	✓	✓					
Record actual measurement data	✓	✓	✓								✓							
Write down research questions for investigation	✓	✓	✓	✓	✓	✓	✓	✓	✓	✓	✓	✓	✓	✓	✓	✓	✓	✓
Compare estimates with evidence																		
Pose investigation hypothesis	✓	✓	✓	✓	✓	✓	✓	✓	✓	✓	✓	✓	✓	✓	✓	✓	✓	✓
Carry out investigation	✓	✓	✓	✓	✓	✓	✓	✓	✓	✓	✓	✓	✓	✓	✓	✓	✓	✓
step-by-step	✓	✓	✓	✓	✓	✓	✓	✓	✓	✓	✓	✓	✓	✓	✓	✓	✓	✓
Record investigation information in table	✓	✓	✓	✓														
Draw schematic representations of science instruments																		
Work in collaborative research groups	✓	✓	✓	✓	✓	✓			✓	✓	✓	✓	✓	✓	✓	✓	✓	✓
Make research observations	✓	✓	✓	✓	✓	✓	✓	✓	✓	✓	✓	✓	✓	✓	✓	✓	✓	✓
Defining disciplinary practices																		
Making scientific instruments																		
Keeping inquiry journals																		

literate practices in their posters than other student groups. Case in point, in group one's poster (the structure group), fifteen out of twenty-three scientific literate practices introduced during the academic year were appropriated by the group in their poster. Students wrote about such literate practices as: designing a research plan, investigating research questions, providing evidence for their investigation, and so forth. In groups two (the sunlight group) and three (the reproduction group) students evinced ten of twenty-three scientific literate practices in their poster write-ups. Students in both groups engaged in several scientific literate practices as they completed their science fair posters. In group four's poster write-up (the water group) thirteen of twenty-three scientific literate practices were inscribed. Group five's poster (the photosynthesis group) evinced ten of twenty-three scientific literate practices and the individual student inscribed ten of twenty-three scientific literate practices into her poster write-up.

These findings indicate that students' inscribed scientific literate practices were not practices acquired merely through completing the assignment. Instead, all student groups were drawing on their past science experiences in the classroom to inscribe specific scientific literate practices into their poster write-ups relevant to this particular plant experiment. Student posters further indicate that the scientific literate practices drawn on were literate practices that students had a significant part in co-constructing throughout the school year. Such co-constructed literate practices occurred in the course of science conversations, activities, and writing that positioned students as members of this classroom community. Hence, the co-constructed science knowledge of this classroom community was a year-long process in which students continually connected prior scientific understanding to new understanding and insight gained during inquiry-based science activities. As students' appropriated scientific literate practices into their writing, they were also inscribing various aspects of their academic identities and scientific literacy into their posters. Although the written text that students produced was removed from the immediacy of classroom spoken conversations, students inscriptions can still be constituted as being dialogically charged. In this sense, students wrote as though they were in a dialogue with a "presumed other" (in this case their teacher) and made written language choices vis-à-vis what they thought he might feel was acceptable for the assignment. Therefore, as students wrote up their science fair posters they were exhibiting aspects of their academic identities in the ways they positioned themselves as scientifically literate members of the classroom community.

Educational Significance

Overall Research Findings

Research has shown that discourse practices in science are not purely related to the ways students talk science (Lemke, 1990), they are also connected to the ways students appropriate semiotic systems for meaning making (e.g., graphs, charts, illustrations, gestures, mathematical formulas, and tables of analytic representations) (Halliday & Martin, 1993; Latour & Woolgar, 1986; Norris & Phillips,

2003). Thus, student writing in science becomes an additional cultural tool for inscribing mental reflections of students' understanding about science. Furthermore, students' writing in science is often a telling indicator of how they view themselves as members of a particular scientific community (e.g., pre-med students writing a research paper for a medical school prerequisite course, honor students submitting an entry to the high-school science fair, or elementary school students keeping classroom science journals). The examination of student writing in science brings forth improved understanding regarding the ways students choose to participate in science. Written and spoken literate practices in science require speakers and listeners to co-construct meaning through interactions that serve to position them as scientific, literate, and competent. These ways of being suggest that learning to engage in the discourse of science requires developing new repertoires for interaction with people, texts, and tools. The consideration of student academic identity thus becomes fundamental for understanding the ways students inscribe meaning in science during their educational experiences. This view proposes that identity is situational, contextualized, and evident in students' spoken and written discourse. As members of groups make decisions about how to position themselves discursively, they draw from a repertoire of ways of interacting that are manifestly expressed in how they speak and write about their experiences.

In this chapter, I examined various ways students in multicultural settings appropriated scientific meaning and inscribed their academic identities in science through their production of written and pictorial representations. The substance of this chapter came from the systematic analysis of four student-produced artifacts. Examining student work, I identified student appropriation of specific scientific literate practices as manifest in their written inscriptions. Each of the artifacts was analyzed within the sequence of its occurrence during the academic year. Within the context of the analyzed artifacts, students made conscious and unconscious choices in their written and drawn inscriptions that signaled an affiliation with classroom literate practices, thus, shaping their academic identities and scientific literacy. As student inscription of meaning was juxtaposed amid students' academic identity inscription from artifact to artifact, classroom interactions were also shaped in patterned ways of interacting that helped define the parameters of this community of practice. The resultant analyses indicate that students' academic identities changed over time as they participated in the literate practices of doing science in this classroom. Furthermore, analysis of four artifacts indicated ways students inscribed their academic identities as they made specific language choices in their artifact inscriptions that echoed the scientific language of the classroom.

Teacher Research Findings

The work in this chapter has educational implications in general, as well as educational implications for teachers in particular. For instance, classroom teachers can utilize the insight gained through the analysis presented in this chapter to structure their own classrooms as communities of practice where students are taught specific ways of talking and writing science. Teachers at all levels of instruction can provide students with opportunities to learn science by actively contributing to the scientific literate practices of their classroom community. As students collaboratively work together during inquiry-based science projects, they can help define collective scientific understanding that can be drawn upon by all members of the community. While students speak and write about common science activities, projects, and lessons, their teachers could document student contributions of "science literate practices" specific to particular classroom communities.

Another implication in this chapter that teachers can draw from is the use of student writing and drawing as tools for science learning. Teachers can implement student science inquiry journals as a way to encourage students to keep a written record of the science content that they learn throughout the school year. The written/drawn inscriptions can be utilized to gain insight into student conceptual understanding as well as scientific misconceptions that might need to be clarified. While students learn to inscribe their understanding about specific science concepts, teachers can also use student inscriptions as performance assessments to gauge student learning. Thus, student inquiry journals can become another important tool that teachers can use to teach science content while simultaneously assessing student understanding.

Lastly, the idea of valuing the cultural knowledge that all students bring with them to the classroom is of paramount importance in encouraging positive student affiliation with school science. If all students are viewed by teachers as bringing cultural knowledge to the classroom that can be drawn upon by community members then students are more likely to develop positive affiliations with the school institution itself. Students' cultural identities are closely tied to the ways that they view themselves in the world. Therefore, if students are to successfully participate in school science, teachers should validate the cultural identities brought by the students as the initial step in helping to shape student academic identities as science learners. Once students are valued for what they bring with them to the classroom community, they are more likely to develop a vested interest in the collective knowledge that is utilized by community members. The three ideas brought forth from this research study—structuring a class as a community of practice, utilizing student inscriptions, and valuing student culture—can serve as additional resources that teachers add to their repertoires when teaching science. If all students are expected to develop increasing levels of scientific literacy, teachers need to continually seek new ways of presenting science content that invites student participation and develops positive academic identities as science learners.

Targeting Enduring Understandings

1   What are the connections you see between the discussion in Reveles' chapter and the four elements of culturally-responsive instruction, as described by Villegas and Lucas?
2   How would you describe the "work" that language is doing in this multicultural science context?

Deepening the Reflection

1   What are some specific approaches to writing-in-science that you have seen teachers use or that you have used yourself to promote scientific literacy? How are they similar to or different than the activities that Reveles describes in this chapter? What kind of science learner identity does each activity encourage?

2   What are some techniques that teachers can use to create a classroom environment that encourages students to view themselves and each other as participants in a community of practice? How does this view of the science classroom alter more conventional understandings of a science learner identity?

3   Reflect back on your own science education experiences. How would you describe your science learner identity? Was it largely positive or negative? What were the most important factors construing this identity? What can you learn from your own experience that may be helpful to you as a science teacher?

Encouraging Engagement

1   In continuing to think back to your own science education experiences, write down a couple of specific instances that you think helped shape your own internalized images of what a scientist looks like, as well as the practices she or he typically engages in. Share these perceptions with your classmates and discuss how you think the internalized images that students have of scientists impact their affiliations with science.

2   Drawing on your discussion, make a class list of some of the most common perceptions that you believe students have of scientists. Next, ask a group of science students in a local classroom to draw their own images of what they think a "scientist at work" looks like. Now, analyze these student-inscribed images utilizing the procedures outlined in Reveles' chapter. Are they what you predicted or not? What surprises did you find?

Notes

1   In this chapter, academic identity will be defined within the realms of science education but will build upon broadly accepted notions and sociocultural theories of identity in order to make the case for examining its social construction in a classroom over time.

2   The term scientific literacy will also be situated among accepted definitions of scientific literacy that have been proposed by professional organizations and science education researchers. However, the term will be expanded to include specific scientific literate practices that students engaged in and inscribed over the course of the study.

3   Inscription is a term used to describe one of the ways students signal understanding in science. Thus, scientific literacy inscription connotes internalized student perceptions that are written and are drawn into their schoolwork about learned scientific practices and content.

References

American Association for the Advancement of Science (AAAS). (1993). *Benchmarks for science literacy.* New York: Oxford University Press.

Bakhtin, M. (1973). *Problems of Dostoevski's poetics.* Ann Arbor: Ardis.

Bakhtin, M. (1981). The dialogic imagination. In M. Holquist (Ed.), *Four essays by M. M. Bakhtin.* Austin, TX: University of Texas Press.

Bakhtin, M. M. (1984). *Problems of Dostoevsky's poetics.* Minneapolis, MN: University of Minneapolis Press.

Bazerman, C. (1988). *Shaping written knowledge: The genre and activity of the experimental article in science.* Madison, WI: University of Wisconsin Press.

Brickhouse, N. (1994). Bringing in the outsiders: Reshaping the sciences of the future. *Journal of Curriculum Studies, 26,* 401–416.

Brown, B. A., Reveles, J. M., & Kelly, G. (2005). Scientific literacy and discursive identity: A theoretical framework for understanding science education. *Science Education, 89,* 779–802.

Chambers, D. W. (1983). Stereotypic images of the scientist: The drawing of a scientist test. *Science Education, 78,* 225–265.

Deboer, G. E. (2000). Scientific literacy: Another look at its historical and contemporary meanings and its relationship to science education reform. *Journal of Research in Science Teaching, 37,* 582–601.

Eisenhart, M., Finkel, E., & Marion, S. F. (1996). Creating the conditions for scientific literacy: A re-examination. *American Educational Research Journal, 33,* 261–295.

Erikson, E. H. (1968). *Identity: Youth and crisis.* New York: Norton.

Erickson, F. (1992). Ethnographic microanalysis of interaction. In M. D. LeCompte, W. L. Milroy, & J. Preissle (Eds.), *The handbook of qualitative research in education* (pp. 202–224). San Diego, CA: Academic.

Finson, K. D., Beaver, J. B., & Carmond, B. L. (1995). Developmental and field test of the checklist for draw-a-scientist test. *School Science and Math, 95,* 195–205.

Fort, D. C., & Varney, H. L. (1989). How students see scientists: Mostly male, mostly white and mostly benevolent. *Science and Children, 26,* 8–13.

Gee, J. (1991) What is literacy? In C. Mitchell & K. Weiler (Eds.), *Rewriting literacy: Culture and the discourse of the other* (pp. 3–11). Westport, CT: Bergin & Garvin.

Gee, J. P. (2002). Identity as an analytic lens for research in education. *Review of Research in Education, 25,* 99–125.

Gilbert, A., & Yerrick, R. (2001) Same school, separate worlds: A sociocultural study of identity, resistance, and negotiation in a rural, lower track science classroom. *Journal of Research in Science Teaching, 38,* 574–598.

Hacking, I. (1994). The looping effects of human kinds. In D. Sperber, D. Premack, & A. J. Premack (Eds.), *Causal cognition: A multidisciplinary approach.* Oxford, UK: Clarendon Press.

Hacking, I. (1995). *Rewriting the soul: Multiple personality and the sciences of memory.* Princeton, NJ: Princeton University Press.

Hacking, I. (1998). *Mad travelers: Reflections on the reality of transient mental illnesses.* Charlottesville, VA: University of Virginia Press.

Halliday, M. A. K., & Martin, J. R. (1993). *Writing science: Literacy and discursive power.* Pittsburgh, PA: University of Pittsburgh Press.

Heath, S. B. (1983). *Ways with words.* Cambridge, UK: Cambridge University Press.

Hicks, D. (1995). Discourse, learning, and teaching. *Review of Research in Education, 21,* 49–95.

Holland, D., Lachicotte, W., Skinner, D., & Cain, C. (1998). *Identity and agency in cultural worlds.* Cambridge, MA: Harvard University Press.

Ivanic, R. (1998). *Writing and identity.* Amsterdam: John Benjamins.

Kelly, G. J., Chen, C., & Crawford T. (1998). Methodological considerations for studying science-in-the-making in educational settings. *Science Education, 28,* 23–49. Special Issue on Science and Technology Studies and Science Education, Wolf-Michael Roth & Cam McRobbie (Eds.).

Knorr-Cetina, K. (1999). *Epistemic cultures: How the sciences make knowledge.* Cambridge, MA: Harvard University Press.

Latour, B. (1987). *Science in action: How to follow scientists and engineers through society.* Cambridge, MA: Harvard University Press.

Latour, B., & Woolgar, S. (1986). *Laboratory Life: The social construction of scientific facts.* Beverly Hills, CA: Sage.

Lee, O., & Fradd, S. (1998). Science for all, including students from non-English-language backgrounds. *Educational Researcher, 27,* 12–21.

Lemke, J. L. (1990). *Talking science: Language, learning and values.* Norwood, NJ: Ablex.

Lemke, J. (2000). Multimedia literacy demands of the scientific curriculum. *Linguistics and Education, 10,* 247–271.

Lemke, J. L. (2001). Articulating communities: Sociocultural perspectives on science education. *Journal of Research in Science Teaching, 38,* 296–316.

Marcia, J. E. (1966). Development and validation of ego-identity status. *Journal of Personality and Social Psychology, 3,* 551–558.

Mead, M., & Metraux, R. (1957). Image of the scientist among high-school students: Pilot study. *Science, 126,* 384–390.

Mortimer, E. F., & Scott, P. H. (2003). *Meaning making in secondary science classrooms.* Maidenhead, UK: Open University Press.

Nasir, N. S., & Saxe, G. B. (2003). Ethnic and academic identities: A cultural practice perspective on emerging tensions and their management in the lives of minority students. *Educational Researcher, 32,* 14–18.

National Research Council (NRC). (1996). *National science education standards.* Washington, DC: National Academy Press.

Nieto, S. (1999). *The light in their eyes: Creating multicultural learning communities.* New York: Teachers College Press.

Noh, T., & Choi, Y. (1996). The differences between the image of scientist and self-image in terms of sex-role and their relationships with science-related attitudes. *Journal of the Korean Association for Research in Science Education (JKARSE), 16,* 286–294 (in Korean with English abstract).

Norris, S. P., & Phillips, L. M. (2003). How literacy in its fundamental sense is central to scientific literacy. *Science Education, 87,* 224–240.

Reveles, J. M. (2005). Scientific literacy and academic identity: Creating a community of practice. *Dissertation Abstracts International, 66–05A,* 332.

Reveles, J. M., Cordova, R., & Kelly, G. J. (2004). Science literacy and academic identity formulation. *Journal of Research in Science Teaching, 41,* 1111–1144.

Rosebery, A., Warren, B., & Conant, F. (1992). Appropriating scientific discourse: Findings from language minority classrooms. *The Journal of Learning Sciences, 2,* 61–94.

Roth, W. M., & Lee, S. (2002). Scientific literacy as collective praxis. *Public Understanding of Science, 11,* 1–24.

Rubin, E., Varda Bar, I., & Cohen, A. (2003). The images of scientists and science among Hebrew- and Arabic-speaking preservice teachers in Israel. *International Journal of Science Education, 25,* 821–846.

Song, J., & Kim, K. S. (1999) How Korean students see scientists: The images of the scientist. *International Journal of Science Education, 21,* 957–977.

Spradley, J. P. (1980). *Participant observation.* New York: Holt, Rinehart and Winston.

Warren, B., Rosebery, A., & Conant, F. (1994). Discourse and social practice: Learning science in language minority classrooms. In D. Spencer (Ed.), *Adult biliteracy in the United States* (pp. 191–210). McHenry, IL: Center for Applied Linguistics and Delta Systems.

Wenger, E. (1998). *Communities of practice: Learning, meaning, and identity.* Cambridge, UK: Cambridge University Press.

Wertsch, J. V. (1991). *Voices of the mind: A sociocultural approach to mediated action.* Cambridge, MA: Harvard University Press.

Wortham, S. (2003). Curriculum as a resource for the development of social identity. *Sociology of Education, 76,* 229–247.

Yoder, A. (2000) Barriers to ego identity status formation: a contextual qualification of Marcia's identity status paradigm. *Journal of Adolescence, 23,* 95–106.

Yore, L. D., Bisanz, G. L., & Hand, B. M. (2003). Examining the literacy component of science literacy: 25 years of language arts and science research. *International Journal of Science Education, 25,* 689–725.

10 The Math Initiative in a 7th-Grade Science Class

How a Daily Routine Results in Academic Participation by ELLs

Holly Hansen-Thomas

Introduction

As the title of this chapter implies, daily routines in the classroom can support the academic achievement of English Language Learners (ELLs). Hansen-Thomas shows us in what ways classroom collaborative problem-solving groups offer ELLs opportunities to participate in problem-solving discourse and the impact of such grouping on interaction and learning. As she illustrates, a math routine designed to support students in problem solving and math/science talk provides students an opportunity to use vocabulary and procedural language with their monolingual as well as bilingual peers, building communities of practice within their larger classroom setting.

However, as Hansen-Thomas illustrates, not all small group collaborative environments benefit ELL students. Much of the opportunity afforded by small group interaction depends on the composition of the group (i.e., monolingual or bilingual) and willingness and ability of group members to include, support, and academically interact with each other around problem solving. In addition, though teacher-organized grouping can be an important tool to increase student participation and learning, teachers must be aware of and reflective about power differentials in science classrooms. To this end, Hansen-Thomas points to the importance of helping teachers develop more linguistic awareness of the characteristics of students' academic and everyday talk. She also cautions us that reforms that call for seemingly "routine" classroom activities may have implications for ELL students that need to be better understood so that the linguistic and social needs of all students in the science classroom are better met.

As you read this chapter think about what you've learned in your teacher education about the relationship between language and collaborative group activities. What have you experienced in your own learning regarding the dynamics of small-group work? What have you been taught about the benefits or challenges of small-group situations particularly with respect to students for whom the language of instruction is a second or additional language? What more do you think teachers need to know in order to provide ELL students with the best opportunities for successful interaction in small-group problem-solving settings?

Along with science, math is considered a crucial gate-keeping discipline (Tananis, 2002). Like in science, the field of mathematics education is seeing a shift in not only what but how content is taught and learned. Today's math is characterized by an emphasis on problem solving, critical thinking through discovery learning, and communication and discussion in groups (NCTM, 2000). There is an increased challenge to all students to not only compute and calculate, but also to be able to talk "the mathematical talk". Related reforms, taking place in science, acknowledge the "talking math"-in-and-for-science connection.

According to the American Association for the Advancement of Science (AAAS, 2003) and the National Science Education Standards (NRC, 1996), there is a push to integrate mathematics with science so that all students will be armed with the mathematical tools with which to carry out scientific investigations. Undoubtedly, knowledge of math and science and the literacies and discourses used within them are highly important for school children in the U.S. On this point, the black political activist-cum-mathematics educator, Robert Moses, has declared that, "In today's world, economic access and full citizenship depend crucially on math and science literacy" (cited in Clements, 2003, p. 13).

The principal aim of this chapter is to understand how "the math initiative", part of a school-wide effort to increase achievement across content areas, worked to promote participation and learning in a middle school science classroom. "The math initiative" was a daily math lesson, in the form of a short warm-up, carried out by small groups of students in their science class. Within that routinized math lesson, students were expected to engage in problem solving to find answers to the problems posed by the teacher. I sought to understand how ELLs, along with bilingual and monolingual speakers of English, used discourse to carry out the daily mathematical routine. Specifically, I aimed to determine whether the math initiative achieved its purpose of promoting participation and the learning of mathematical discourse. In addition, I wanted to know if ELLs had sufficient opportunities for participation in daily routines such as the math initiative, and finally, how the community of practice that developed in the classroom affected student learning. To answer these questions, I selected small group interactions between ELLs and their (bilingual and monolingual English-speaking) group members, and analyzed how the students interacted discursively while carrying out the math initiative or, in the words of the students, "the toothpick activity," so named because toothpicks were always used in initiative-related tasks.

By examining student participation within the math initiative, one aspect of the teacher's role emerged as a critical element in facilitating that participation. As the organizer of grouping structures and facilitator of classroom interaction, the teacher was responsible for appropriately matching students so that they would be successful in the classroom activity. Later in this chapter, I will return to the notion of teacher-imposed grouping and the importance of being aware of power differentials that may emerge in classrooms with culturally- and linguistically-diverse students.

Theoretical Framework

Communities of Practice

The overarching theoretical model I have chosen to employ in this investigation is the Community of Practice (CoP) (Lave, 1991; Wenger, 1998). Within this socioculturally-influenced model, learning is defined as participation in shared practices and is equated with participation in a community (Wenger, 1998). The CoP framework, developed as a learning theory, incorporates the notions of social practice with community, and lays the groundwork for understanding the dynamic role of identity in a learning situation.

Wenger defines a practice as something which includes "language, tools, documents, images, symbols, well-defined roles, specified criteria, codified procedures, regulations, and contracts...." (1998, p. 47). The overall social situation, artifacts, and the activities in which members participate represent a social practice. In a science classroom, for example, the students construct a discursive social practice with the help of the teacher, curriculum, norms of school, as well as each other.

The CoP framework posits that community can be created in the classroom (or wherever a CoP occurs) when three important criteria are met. These criteria include (1) mutual engagement (between the ELL and monolingual English-speaking students in the class, for example), (2) a joint enterprise such as a task or activity (i.e., a math warm-up), and (3) a shared repertoire of negotiable resources (Wenger, 1998, pp. 72–73). The shared repertoire of negotiable resources includes language and specific ways of using language in certain communities, as well as the tools, artifacts, routines, stories and styles that are used in particular practices (Wenger, 1998). In a science class, the shared repertoire includes the academic science and math vocabulary and language functions related to the scientific method or the solving of math problems, as well as the social language necessary for working in groups. Thus, a community in a science class, as defined by the CoP framework, is a cohesive group of learners who engage in shared practices (like talking about and doing science, and understanding what it is to be a 7th grader), but who do not necessarily share homogeneous traits.

Central to the CoP, and vitally important to the conceptualization within this model is the concept of identity. As members progress from "legitimate peripheral participants" (Lave, 1991, p. 68) to full participants in a learning process, their identity shifts and transforms. The progression from the periphery to full participant status reveals the changing identities of students as they move along the path of learning.

The CoP has as its core unit of analysis the community of practice itself (Wenger, 1998), however, examining individual activities within the CoP, such as those which emerge from cooperative classroom lessons—i.e. the math initiative—can reveal aspects of participant interaction, identity, and evidence of learning target concepts and discourse. An examination of patterns of participation in CoP can show how a student moves from peripheral to full participation in the situated community of learners by examining how members interact within group tasks. As a result, the CoP model can be fruitfully employed to provide insight into how individual learners of math participate within groups in the practice reinforced by the reform-influenced science class.

Academic Discourse

Since education reform has brought about changes in how mathematics is taught and learned, understanding how discourse is used in math has gained increased emphasis in both teaching and research. Discourse is defined as both the process and the product of language that is used to communicate in and about a content area; in this case, math. It is critical to point out that the terms that define the notion of academic discourse, including academic literacy or language, subsume academic competence in English. Thus, mathematical discourse in U.S. schools is talk about math *in English*. Academic discourse, then, includes the linguistic [in this case, English] and cultural knowledge that members of specific class-room communities of practice need to know (Gutiérrez, 1995). As a result, "students must have opportunities to develop academic discourse [in English] ... in order to achieve academic competence" (Gutiérrez, 1995, p. 21).

According to O'Connor, (1998) mathematical thinking and the processes involved in conducting mathematical operations are "complex, social practices" and the discourse that learners use to carry these processes out can facilitate math learning. Lemke (1989) suggests that since education can be understood as talk, an analysis of discourse employed by students can be used to understand the various roles, responsibilities, routines, and overall meanings of classroom participants and interactions.

Group Interaction

Particular research studies which have influenced the present investigation are those which highlight advantages to learning in small group interactions. For example, Cobb, Wood, & Yackel (1993) found that within small groups of mathematical discussion, students created a Zone of Proximal Development (ZPD), which allowed them "to engage in more advanced mathematical activity" (p. 91) than they would have been able to alone. In this study, participation and inter-action played an important role in the creation of mathematical discourse and scaffolding. However, their research did not focus on ELLs.

In another study, Cook (1999) discovered that in the Initiation, Participation, Reaction, Evaluation (I-P-Rx-E) type of grouping structure, students learned more than in dyads or other kinds of groups. Cook contrasts the I-P-Rx-E struc-ture with the traditional Initiation, Response, and Evaluation (I-R-E) dyadic structure typical of American classrooms. It should be noted that Cook's research, focusing on attentive listening and reaction turns, was not carried out in the U.S., where teacher student dyadic interchanges are common, but in Japan, where group interaction in school is common.

Another relevant study, also conducted in Japan, examined how kindergarten language learners acquire discourse competence in a second language (Kanagy, 1999). Citing Takahashi and Morimoto (1996), Kanagy describes how student–teacher interaction in Japanese foreign language elementary school "changed from teacher-centered to student-centered discourse patterns" (1999, p. 1469) as students became more comfortable with daily classroom routines. Like Takahashi and Morimoto, Kanagy found that interactional routines affected students' participation in second-language discourse sequences. Verbal modeling and nonverbal teacher demonstrations worked to provide student scaffolding. The combination of routinized activities and appropriately constructed participant interactions successfully promoted student learning.

Research demonstrates that interaction between students, also called cooperative or collaborative learning, benefits ELLs as it provides them with an outlet for practice (see, for example, Echevarria & Graves, 2007; Jameson, 1999; Kessler, 1992). Within collaborative groups, ELLs have the opportunity to process linguistic input that is clear and comprehensible, but they also use the target language, or in this case, mathematical discourse, as output. Opportunities to hear and process input as well as to produce output in the target second language are highly important for ELLs in school. And while small grouping structures can provide these opportunities, what is critical within collaboration is that all students contribute to the group task. An inevitable criticism of cooperative grouping is that some students do not carry their own weight. In cooperative groups with culturally- and linguistically-diverse students, it is important to understand that they may not participate due to reasons related to culture or language. Issues of power, identity, culture, and language all contribute to the effectiveness (or not) of group interactions of students in school.

Methodology

The specific strand of discourse analysis that I employ is that of interactional sociolinguistics. The interactional sociolinguistic approach is grounded in the work of Gumperz and Goffmann, which "provides a description of how language is situated in particular instances of social life and how it reflects and adds meaning" (Schiffrin, 1994, p. 97). Schiffrin (1994) further notes that Gumperz is responsible for stressing the importance of communicative competence in discourse, as integral to the interactional sociolinguistic approach to discourse analysis.

While the specific unit of analysis in discourse analytic-oriented research is the ongoing constructed discourse in the classroom, this research will examine, in particular, participant frameworks and interactional routines that occur in a science class. As used in this research, I will take on Michaels and O'Connor's (1996) conceptualization of participant frameworks as an analytic tool that describes the constellations and "rights and responsibilities" of participants within a discursive situation such as a conversation or negotiation, and also their connection to that discourse. In the words of Michaels and O'Connor, participant frameworks "encompas(s) the ways that ... participants are aligned with or against each other, and ... the ways they are positioned relative to topics and even specific utterances" (1996, p. 69). An examination of developing participant frameworks will shed light on "how ... children come to take on the particular roles and discourse forms that are valued in problem posing and problem solving in school" (Michaels & O'Connor, 1996, p. 63).

Interactional routines are activities mediated through interaction with others, which are carried out on a common, regular basis (Kanagy, 1999). Routines are useful for understanding how participants in a learning situation impart and construct knowledge in that they show how learners develop (linguistic and conceptual) knowledge over time. By examining roles, responsibilities, and alignments between discourse participants in a learning situation, we may begin to understand which participants are taking active (or passive) roles in the classroom CoP. Using participant frameworks and routines to understand how ELLs progress from legitimate peripheral participants to full participants in a classroom CoP can provide critical insights in understanding how learners develop new knowledge.

Setting and Participants

The setting for the present study is a 7th-grade science classroom in a medium-sized (628 students) middle school in an urban area in south Texas. The school, Madera Middle, is one of a few campuses that make up Garner Independent School District (ISD), the smallest district in the large metropolitan area where the district is located. Madera Middle School is situated in a Spanish/English speaking community made up, primarily, of Latino families who have lived there for generations, and also very recent immigrants from Mexico. Many of the students at Madera are considered economically disadvantaged. Although approximately 30% of Garner ISD's elementary school students are designated as Limited English Proficient (LEP), and receive second language or bilingual instruction, all students in the district, including recent immigrants, are transitioned to all-English classes by the time they reach middle school. English as a Second Language (ESL) classes are non-existent at Madera Middle School; instead, all content-area teachers are required to take a minimal amount of ESL training classes (equivalent to several Saturday morning inservices). This counts as certification in Garner ISD.

The specific class selected for this research is a 7th-grade science classroom called group 7B. Its teacher, Ms. Jackson, is also the science coordinator in the school. Ms. Jackson, an English monolingual with nine years teaching experience, was considered one of the best science teachers in the district (Bayley, Hansen-Thomas, & Langman, 2005). At the beginning of the data collection for this project (which was part of an on-going longitudinal investigation lasting the full school year), group 7B consisted of twenty students. Of those twenty students, ten were bilinguals who had attended bilingual classes in elementary school. Five students were monolinguals—two being newcomer Spanish-dominant speakers (ELLs), and three native U.S. born English dominant speakers. With the exception of one African American student, all students were of Mexican descent.

The focal students in this study include two bilinguals, Nellie and Annette; Pedro, an outgoing, but somewhat unmotivated (with respect to academics) ELL who had arrived from Mexico three years previous; two Spanish speakers, Little Manuel and Sandra, who had recently arrived from Mexico; and Marilyn, a bilingual groupmate of Nellie and Little Manuel.

Nellie is a hard-working, enthusiastic student who makes good grades and encourages her group members to do well in science. Annette is a sharp student but is often distracted from school by social pressures including volleyball and gang involvement. To varying degrees, these girls served as mentors in their respective groups. Nellie often helped Little Manuel, and Annette would provide Sandra with translations. Pedro occasionally worked with Little Manuel, but when he did, social discussions in Spanish took precedence over science ones. Little Manuel was named as such by the students since there was a much bigger boy in the class with the same name (Big Manuel). Little Manuel was small in stature, charming, and eager to achieve in school, but very limited in his English proficiency. His size and baby face contributed to his nickname and to his status as the class pet by both the teacher and the students. As a result, his bilingual female lab partners tried to help him do well in science. Sandra was very quiet, challenged by science in English, and had only Annette from which to draw on for help. More often than not, Sandra did not receive the assistance she needed in groups. Annette did not usually want to take the time to translate for Sandra, and Sandra's other lab partners spoke only English.

As is traditional in secondary science classes, the students sat at lab tables. Three to four students were assigned to each table and were often expected to work in small groups of lab table members. During new or particularly challenging activities, Ms. Jackson walked around the class and gave assistance to groups and individuals.

The Data

The data for this project were selected from a corpus[1] of audiotaped data collected as part of a larger research project. All audiotaped data were transcribed using CHILDES (MacWhinney, 1994) transcription program and its conventions.[2] The math initiative was carried out every class day[3] for approximately two months. I chose to analyze discourse selections from October 9 to November 6, examining six classes in total. The days I drew from were the following: October 9, 15, 19, 25, 29, and November 6. As the math initiative is a rather short activity, and was usually presented at the beginning of class, discursive interactions surrounding the math initiative lasted approximately five to ten minutes of the total one and a half-hour class period. For purposes of analysis, I have chunked the first two class days (October 9 and 15) together to be categorized as early interactions because these experiences occurred in the first two weeks of doing the math warm up. The last four days (October 19, 25, 29, and November 6) are classified as later interactions because students had had several weeks of experience in carrying out the task at this point.

I analyzed the data by tracking the focal ELLs' discursive participation longitudinally throughout their interactions with the math initiative. I examined the external, macro-level features of interactions including physical classroom organization as well as the participant frameworks that illustrated how student-to-student and student-to-teacher interactions played out in conjunction with the toothpick task. At the micro-level, and within the participant frameworks that emerged, I examined the language that students were using to further understand their role in the academic task.

The Math Initiative

The math initiative was presented to the students on the overhead projector. The teacher wrote one or two mathematical problems on an overhead slide for the students to solve by creating a solution using toothpicks. The teacher would arrange toothpicks in a certain shape, such as a rectangle, square, or cross, and students would follow her instructions from the overhead. For example, the task on the first day the initiative was implemented (October 9) was as follows:

> Arrange twelve toothpicks.
> A. Remove four toothpicks to form one square.
> B. Remove two toothpicks to form two squares of different sizes.

Later in the month, the tasks became more complex. On October 25 the daily math task was the following:

> Arrange twenty-four toothpicks to form nine squares.
> A. Remove six toothpicks to leave three squares of different sizes.
> B. Remove eight toothpicks to leave two squares of different sizes.

To carry out the day's task, students were expected to first find the solution in small groups by manipulating toothpicks into the correct shape according to the instructions. Enough toothpicks were passed out to each table so that students could do the activity alone, but they were encouraged to do it with lab table members. Students then needed to copy the solution (individually) onto paper to turn in.

Findings

Early Interactions

Early interactions with the math initiative were characterized by traditional Initiation/Response/Evaluation (IRE) and whole class interactional routines, with a focus on teacher-led discourse. On the first day in which Ms. Jackson introduced the math initiative, she spent the majority of the class discussion leading a lecture. The first excerpt reveals a slice of how class was conducted. Much of the discourse was teacher-directed with just a few outgoing students interacting.

October 9, **First Day, Excerpt 1**

MS. J = Ms. Jackson

P = Pedro

MS. J: You need two sheets of paper.

MS. J: One is for the DUST (Daily Understanding of Science Terms), and one is for the math initiative.

P: We are doing math in science?

MS. J: You are doing math in every single class.

MS. J: You will have a math problem every single day in every single class.

MS. J: These math initiatives are being collected by your teachers and then they go to the principal.

MS. J: What you are going to do is you're going to get twelve toothpicks.

MS. J: You are going to get twelve toothpicks and arrange in the diagram.

MS. J: You may work together as a group.

In this first excerpt, the teacher posed all but one turn of talk (turn 3). As she is introducing the activity, students are, for the most part, silent and presumably engaged in her discussion and attentive to her directives. They have not yet been given the opportunity to start work on the task. Ms. Jackson's directive in turn 9 reveals that this activity will be one in which students will be working jointly in groups.

Once students began work on the task, the teacher moved through the room, assisting and scaffolding individual groups. When the majority of students had reached a solution, Ms. Jackson proceeded to initiate a class discussion about the first day's math initiative. In the following example, Ms. Jackson is prompting individuals from the class to provide their answers so that they could discuss the problem and the answer as a whole class. Pedro is chosen to share his solution.

October 9, **First Day, Excerpt 2**

MS. J = Ms. Jackson

P = Pedro

N = Nellie

MS. J: Did anyone figure out the second one?

P: Yeah.

MS. J: How does it go?

MS. J: Tell me.

MS. J: It says you can only remove two ## and you're going to end up with squares of ## two squares.

P: They more look like rectangles.

MS. J: Tell me what to do # How do you do it? # If you remove two what do you do?

P: I forgot.

MS. J: You forgot? # Show me what you did # How did you make two squares of different sizes?

P: Like that.

MS. J: You are so close dude! You get to remove another one # you get to remove
        two # no no two # no, more, you almost had it!

MS. J: Nah that makes it two rectangles # put it back.

MS. J: What did you do Nellie?

N: We didn't do it.

MS. J: You didn't do it?

N: We did these two but they were rectangles.

MS. J: Okay that made it rectangles.

In this primarily dyadic teacher–student participant framework, we see that
Ms. Jackson is first strongly encouraging, then heavily guiding Pedro's actions.
Once Pedro has chosen to engage in dialogue with the teacher (turn 2), the
teacher uses commands such as tell me and show me, and questions (turns 4, 7,
9, and 13) to elicit Pedro's response regarding the manner in which he came to
solve the problem. He is hesitant and tries to avoid answering her (turn 8), but
Ms. Jackson perseveres. When he finally decides to tell the teacher what she
wants to know, he chooses to do so by demonstration (turn 10). Unfortunately,
Pedro does not get the answer Ms. Jackson is looking for. She, nevertheless, continues to encourage and prompt him. Following this interaction, the teacher
engages Nellie in a participant framework (turn 13). Although Nellie is generally
reliable in providing correct answers, she was not in this instance. However, she
proves herself attentive and skillful in rationalizing why she and her group did
not have the answer. In turn 16 Nellie explains that they had come up with an
answer, but as the teacher explained in turn 12, rectangles were an inappropriate
solution.

Although Ms. Jackson created participant frameworks by inviting both Pedro
and Nellie into the discourse, they were somewhat reluctant to interact with her.
Pedro, a rather limited-English-proficient ELL who was generally participatory
in small groups, interacted in a discursive framework with the teacher with the
help of visuals and demonstrative pronouns ("like that"). Nellie's explanation in
line 16 revealed why she could not supply the response the teacher needed, and
effectively cut off the participant framework from further discussion. We see that
although Pedro and Nellie used some mathematical terminology ("rectangle") in
this excerpt, their use of elaborated academic discourse was limited.

This excerpt is particularly representative of early interactions with the math
initiative because it, like many of the other early teacher–student and
student–student participant frameworks which emerge at this early stage of
doing the activity, is characterized by teacher-led discourse and student clarification, and in some cases, (student) reluctance to participate. Some students are
not really clear as to what they are doing and it is likely that they are not yet sure
of the task and how to complete it. With regard to the classroom CoP as a cohesive whole, whole class student interaction is smooth. At this point in the semester, students had been together for over six weeks and were familiar and
generally comfortable with each other and with Ms. Jackson's friendly but direct
teaching style.

On October 15, Little Manuel and his group worked quickly and efficiently on the day's warm-up. Although the activity was still somewhat new, his bilingual group with their shared repertoire created a cooperative environment in which to discursively solve the mathematical task. Ms. Jackson had allotted the students time to carry out the activity and this group took advantage of the time and opportunity to work together.

October 15, Excerpt 3

MS. J = Ms. Jackson
M = Marilyn
N = Nellie
LM = Little Manuel

MS. J: Today it gets a little more complicated, look at the diagram now, now you're going to have seventeen toothpicks # to arrange, I think you guys can do it.

LM: Ah miss.

MS. J: I'm just gonna give you some toothpicks, I don't have time to count.

M: I'll count, I'll count, I'll count.

N: You give us like thirty of them.

LM: Marilyn, qué decía allí? [Marilyn, what does it say there?]

(A few moments later)

LM: ¿Ya le entendiste? # ¿Eh, ya le entendiste, qué dice? [Did you get it, hey, did you get it, what does it say?]

L=M: Oh, pues xxx [Oh, well, xxx].

LM: ¿No lo entiendes? [You don't get it?]

MS. J: It says, remove five toothpicks to form three squares of equal size.

LM: Está (en) inglés ## ¿Tiene que topar hasta mero abajo? [It's in English ## Does it have to hit at the very bottom?]

L: Uhhuh.

LM: Ay (giggling) # guys!

M: Miss like that?

N: Miss, there!

On October 15, the math initiative is still rather new; excerpt three, however, illustrates how Little Manuel's group works productively together. As the excerpt begins, the teacher sets the stage for the activity, but allows the students to do more of the talking. Although she warns the students that the task may be challenging, she provides words of encouragement to all (turn 1). In turn 2, Little Manuel responds to just her, surprisingly in English, and acknowledges the teacher's warning that the task may be hard. Next, Ms. Jackson continues to interact with the class as a whole, indicating that she does not intend to count out toothpicks for each student—they will be responsible for doing so themselves. In turn 4, Marilyn agrees to count for her table. In so doing, her commitment has served not only the purpose of taking responsibility for counting toothpicks, but has also initiated the interaction between the bilingual students at the table. After Nellie jokingly chastises Marilyn for passing out too many toothpicks, Little Manuel engages Marilyn in a participation framework in

Spanish, asking her to translate the instructions for the task that are written in English on the overhead projector (turn 6). Little Manuel clearly sees Marilyn as a trusting partner he can rely on for language assistance. He continues to ask his group if they understand (turn 7), and while Marilyn attempts to respond, her answer is both unclear and unsatisfactory to him. Little Manuel then questions her comprehension of the task (turn 9). Ms. Jackson jumps into the discursive interaction, clarifying the instructions (turn 10). Little Manuel acknowledges his linguistic challenge with the instructions, and then presses his group for a response regarding how to form the toothpicks (turn 11). In turn 12, Marilyn responds, which Little Manuel acknowledges lightheartedly (turn 13), effectively completing both the task and the discursive participation framework. The last two turns of talk show both Marilyn and Nellie confirming the group's solution with the teacher.

Little Manuel plays various roles in a number of participation frameworks throughout this excerpt. In line 6, he begins as a dependent, eliciting assistance. Then, in line 7, he again asks for help, but appears to want to take charge of his participation. In line 9, he begins to make his own choices regarding the solution, and in line 11, he is an equal partner in the mathematical task, but using Spanish to interact with his group. His role as a partner is highlighted in the friendly and playful interaction between the three group partners at the end of the excerpt.

Bilingualism within a cohesive CoP is at play in the success of this particular group. We see the group members joke around with each other and, at the same time, work effectively at finding the solution to the problem. Through his discussions with Marilyn and Nellie, it appears that Little Manuel values his group-mates' knowledge (including English language ability), however, when he discovers that Marilyn cannot provide him clear answers to his questions, he makes his own assumptions and works on his own. Nevertheless, the girls provide support within this small group activity that facilitates Little Manuel's ability to complete the task. At the end of the excerpt, we see Nellie and Marilyn again taking charge, acting as go-betweens for their group (and monolingual Little Manuel) and the teacher, confirming that their small group's solution was correct.

Later Interactions

In general, later interactional routines are characterized by individual work and less group discourse, or more "telepathic" or unspoken discourse. However, the classroom community and individual group members within that community still help each other out. Expert students that Ms. Jackson selects to be the day's "teacher" of the math warm-up, who explain their solution to the class, continue to help others in "teacher"–student (one-on-one dyads) or "teacher"–students ("teacher"–fronted whole-class framework) types of participant frameworks. Students continue to ask very direct questions to find solutions by seeking the assistance of the teacher or student experts.

By October 19, students have become familiar with the math initiative, and know what to expect, so some interactional routines, many of which the groups have started, include beginning the activity as soon as it is presented, working together or independently to finish it, then initiating a whole class discussion to review the answer. Ms. Jackson is less hands-on at this point and generally stays at her desk at the front of the room and uses the warm-up time to take roll and attend to administrative issues. Some students begin to "play teacher" and help others to get the solution to the daily math routine. In addition, because the teacher has started to award the student who is first to come up with and share their answer with the privilege of walking around class and collecting the toothpicks at the end of the activity, there is an incentive to get the answer first. In the following examples from October 19 and 25, Annette and Sandra and their two group members are working on the day's math task. In excerpt 4 and 5, Annette is enthusiastic about the teacher's reward for finishing first. As a result, Sandra does not receive individualized assistance from her bilingual peer.

October 19, Excerpt 4

MS. J. = Ms. Jackson
A = Annette
S = Sandra

A: Miss # Miss, like this?
S: [giggles].
MS. J: We're doing the last one, six toothpicks to form two squares of different sizes.
A: Here Miss we're done! # I'm done.

October 25, Excerpt 5

MS. J. = Ms. Jackson
S = Sandra

MS. J: You're doing B, you did A on Tuesday, today you're doing B.
S: A, B, C, D, E (in English) xx tres cuatro tres cuatro uno dos tres cuatro cinco seis siete ocho, lo voy a doblar, uno dos tres cuatro cinco seis # cinco seis siete ocho, no. [xx three four three four one two three four five six seven eight, I am going to bend it, one two three four five six # five six seven eight, no] —
S: ¿En qué página iba? [What page are we on?]
(no response by anyone in her group)
S: Eh, ¿en qué página iba está, tú, Annette? En la siguiente #? [Hey, what page are we on, you, Annette, on the next?]
S: Eh, Isaac, Isaac, ¿en que página iba? [Hey, Isaac, Isaac, what page are we on?]

In excerpt 4 from October 19, Sandra does not actively attempt to engage in the discourse or the task. Her giggle in line 2 may indicate she is engaging with Annette, who is working on the task, but Sandra does not otherwise participate in the activity. Annette races to complete the task, eliciting the teacher's attention (turn 1) and confirmation (turn 4). Although Sandra appears not to be actively participating in the discursive interaction, it also appears from this excerpt that

Annette, the one bilingual who can easily draw Sandra into the task, effectively facilitating her participation in the framework, is not interested in doing so.

Excerpt 5 paints a different picture. In excerpt 5, Sandra is attempting to engage her groupmates in interaction, but to no avail. As the excerpt begins, the teacher reminds students what they must do, and then leaves students to their own devices, as she commonly does in later interactions. Seemingly inspired by Ms. Jackson's talk of letters, Sandra begins repeating the alphabet in English in line 2. She then switches to numbers, but in Spanish, and then turns to the task at hand. In turn 3, Sandra starts to engage her groupmates in a participation framework, although it is clear that she has not yet gained an understanding of the routine of the math initiative. She asks what page they are on, but no book is involved in the math initiative at all. In the next two turns, 4 and 5, Sandra continues to engage both Annette and Isaac, another (non-Spanish-speaking) lab partner, but they do not respond. Her elicitations of their assistance consistently focus on an incorrect assumption—that the mandated math task involves a book. Despite this, Sandra still does not receive the help she needs in order to do the day's warm-up.

In this situated instance, Sandra repeatedly attempts to engage her lab partners in discursive participation frameworks, but she is unsuccessful. Although she demonstrates limited ability to use English (repeating the alphabet), it is clear that her comprehension level is low. Despite the fact that the class had been working on a consistent routine for several weeks, Sandra's understanding of the task—even after six class days—is lacking. It is likely that her groupmates' reluctance or inability to help her exacerbated her lack of comprehension. Sandra's table had neither the ability nor the interest in aiding her that Little Manuel's table had. In the following excerpts from October 29 and November 6, we see how Little Manuel continues to interact about math in English with his group.

Excerpt 6, October 29
N = Nellie
LM = Little Manuel

N: Move four toothpicks to move, to make five squares.
LM: ¿(Es)tá bien chiquito, no? [That's pretty small, isn't it?]

Excerpt 7, November 6
MS. J = Ms. Jackson
LM = Little Manuel
N = Nellie

LM: Ya lo hice. [I got it].
N: Miss, miss, Little Manuel got it!
MS. J: Who's got it?
N: Little Manuel.
MS. J: Orale [hey] look at you!
(Little Manuel smiles)

In these two excerpts from October 29 and November 6, we see that Little Manuel is an active participant in the math task and in the participant frameworks in which he is involved. Although he does not use English in either of the excerpts, he is a legitimate and active member of his small group's CoP. He plays roles of partner and collaborator on both days. In excerpt 6, Little Manuel works closely with Nellie to begin the task. She reads the day's instructions and then he thinks about and then analyzes the outcome of the task. Their discursive interaction reflects a thoughtful, cooperative partnership comprised of two members who work toward a common goal while using a shared repertoire and language. However, much of this shared language does not occur as standard academic discourse—it is in Spanish, not English.

On November 6, Nellie and Little Manuel continue to work collaboratively, as partners, but on this day, Little Manuel plays the role of expert, while Nellie is another kind of expert as the linguistic go-between. Because of her bilingualism and ability to elicit Ms. Jackson's attention, Nellie obtains credit for Little Manuel for his solution and contribution to the day's math activity. Within this participant framework, Nellie not only translates for him, but revoices (Goffman, 1981; Michaels & O'Connor, 1996) what he has said and his intention to be acknowledged (line 2). In essence, she obtains power and legitimacy for Little Manuel. When Ms. Jackson enters into the discursive framework, she acknowledges Little Manuel's contribution. She attempts to connect with him by praising him in colloquial Spanish. The praise is effective, as Little Manuel is demonstrably proud of his accomplishment. The triadic exchange between Little Manuel, the "principal" (Goffman, 1981) who is responsible for the utterance, the teacher, and the revoicer (Nellie), illustrates how some of the members of the CoP in the classroom work effectively to mediate learning for ELL students.

Discussion

Some of the ELLs and bilinguals in science group 7B had opportunities to interact discursively and learn about math, while others did not. Factors such as high proficiency in English, identity issues such as confidence and outspokenness, and having a background and strong ability in math and science (ergo, an established academic identity) positively contributed to student participation in science class. Academically strong bilinguals Nellie and Annette were able to complete the math warm-ups with success. Pedro, an ELL, participated in whole class discussions, but with attentive direction and scaffolding from the teacher. Although ELLs Little Manuel and Sandra had approximately the same English language proficiency, their participation in the math initiative was very different. Little Manuel had the linguistic and academic support of his group members, including Nellie and Marilyn, but Sandra had only the possibility of help and translation support from Annette, who was often more interested in helping herself than others. Student identity affected participation in that it helped or hindered language learners' ability to interact with others in and about the math initiative.

While a cooperative and cohesive bilingual group facilitated Little Manuel's participation during the math initiative, it did not contribute to his use of academic discourse in English. Since most of the group's interactions were in Spanish, Little Manuel was able to talk about possible solutions to the toothpick activity with his partners. However, the girls were often so willing to speak for him in English, that Little Manuel did not have opportunities to do so himself. Additionally, because the math initiative became increasingly more student centered, and Ms. Jackson played a smaller role as a discourse participant, ELLs like Little Manuel and Sandra had less exposure to her modeling of standard academic discourse.

Sandra was disadvantaged in terms of her use of academic discourse not only because she did not have the linguistic support that Little Manuel's group offered him, but also because of the lessening of teacher scaffolding. As the math initiative became routine and students were expected to carry out the activity without the teacher, Sandra's lab partners quietly ignored her, leaving her without language or academic support. Although she tried to participate, she was unsuccessful in doing so. Neither Sandra's academic language nor conceptual knowledge appeared to increase throughout the math initiative.

In the case of both Sandra and Little Manuel, the group in which each was placed played an important part in their role in their respective participant frameworks, as well as in their participation as productive members of the classroom CoP. Certainly, the teacher's choices with respect to her grouping of students affected the outcome of both Little Manuel and Sandra's interaction with and ultimately, learning of the mathematical concepts and discourse that they needed to know as 7th graders. On the one hand, Little Manuel was advantaged because he was aligned with group members who not only protected and wanted to help him achieve, but they also spoke his language. On the other hand, he did not have many opportunities to use the target language he was responsible for learning—the language of math and science. Although Ms. Jackson made an effort to place Sandra with someone who could help her, external factors precluded her from truly becoming a member of classroom CoP and from appropriating the all-important academic discourse. Clearly language played a role in Sandra's lack of success—in that two of the four group mates did not speak her native language. But why did Annette refuse to help her, when she was a highly proficient bilingual? A definitive answer to that question is unclear, but any number of theories can be explored. Perhaps Annette was uninterested in helping, or she was busy doing other homework, or she was exerting power over Sandra and choosing not to help her. Ultimately, though, Sandra was not getting the scaffolding that the small grouping should have provided her. Finally, what was particularly problematic was that Ms. Jackson was wholly unaware that her grouping strategies were not as effective as she had intended.

Conclusions and Implications

As group 7B students came to understand the math initiative and its purpose, most were motivated to do the daily activity. Different interactional routines and participant frameworks emerged over the course of the month-and-a-half visits to the classroom, and those routines and frameworks became progressively more student-led. When students were first introduced to the math initiative, classroom discourse patterns were characterized by more whole class and IRE sequences. In later interactions, when routines were set, students became more independent and worked in small groups and individually. As well, more students (including ELLs) took on various roles within their small groups, such as that of partner, expert, and helper. Thus, for most of the students, including the ELLs and bilinguals of science group 7B, the math initiative did in fact achieve its purpose of promoting participation and learning of mathematical concepts. For other students, like Sandra, the math initiative did not become routine; thus, she did not have the opportunity to take on roles of increasing responsibility like other learners in the classroom CoP did.

In general, ELLs had sufficient opportunities for participation in the math initiative, but student participation was dependent on the individual, based on their language proficiency, academic identity, and role in the classroom CoP. It was also related to the dynamic of the group in which the students interacted. In Little Manuel's case, his supportive and bilingual tablemates made sure that he both understood the academic task and was appropriately recognized when he was able to successfully carry out the task(s). His status as a teacher's pet, or at least a favored student helped him to integrate into the classroom CoP—despite limited English. In Pedro's case, his more advanced linguistic skill in English allowed him access to whole class and small group participation frameworks. Certainly, his fearless personality helped him to be integrated into, and be accepted by the classroom community. However, not all ELLs in the class had Pedro's skill and personality, or Little Manuel's support. Sandra, who lacked a strong academic identity, English language proficiency, and support by her peers, did not have as many opportunities to participate in small or large discursive interactions regarding the math initiative. While she did in fact try to engage with her peers, Sandra's attempt to participate was unfortunately unsuccessful.

The CoP in the classroom affected student learning of the math initiative. The science classroom in the present study fit the criteria for a true CoP as the three characteristics were all present: The joint goal that group 7B worked on together was the daily math routine; the shared repertoire included the discourse the students used while carrying out the activity, as well as the knowledge and understanding of how to manipulate the toothpicks to reach a solution; the mutual engagement group 7B shared was completing the daily math initiative.

As students in group 7B worked together to carry out their daily task of finding solutions to the math initiative, they interacted and participated in various participant frameworks. They took on different roles and engaged in academic (as well as social) discourse regarding the math activity. As they worked through the math tasks over the course of several months, students took on various identities of learner, expert, teacher, translator, helper, and even friend. When students became familiar with the daily routine, their CoP opened up to differing and various types of discursive student interactions. Interactional routines that emerged began with the students' need for teacher–student interaction in the form of scaffolding and clarification, whereas students later became more independent and motivated to do the activity in student–student interactional sequences.

The teacher herself also worked to create and develop community in the classroom by promoting group interactions and by encouraging students to help and rely on each other for scaffolding and support in academic endeavors such as the math activity. However, there were missed opportunities that Ms. Jackson could have taken to promote both academic discourse development and richer student interactions.

In order to promote academic discourse development, the teacher could extend and elicit discussion, even when students get answers incorrect. In this way, the teacher extends the participation framework and promotes academic talk. She could also highlight the academic language students need to be using through oral means and by writing on the overhead projector or on the blackboard in front of the students. Because ELLs' language development should be continually scaffolded, the teacher should both monitor this more closely and arrange students in appropriate heterogeneous groups to provide both linguistic and academic support for the ELLs. Ms. Jackson could have also brought in home/community knowledge to help Pedro and Little Manuel leverage what they knew about math and science in general. By so doing, she would show students that she valued and recognized their prior knowledge and experiences. Such a tactic would also increase student discourse in and around math and science.

Sandra also needed another mode to both comprehend and demonstrate her knowledge. To do so, Ms. Jackson could have relied more heavily on bilingual resources and translations in order to present the content and academic tasks to her, to ensure that Sandra understood what she needed to do. Because Sandra was a beginning learner of English, she would have benefited from use of her native language to a certain degree. Using it as a tool to teach, as well as to allow her to demonstrate her knowledge of concepts is something that Ms. Jackson could use with Sandra. Explicitly teaching critical vocabulary is another strategy that can be used with beginners like Sandra. Using the newly learned academic vocabulary in English, Sandra could write simple sentences to illustrate knowledge of learned concepts.

In Ms. Jackson's group 7B, one ELL was grouped appropriately, while another was not. Little Manuel benefited from working with supportive, bilingual lab partners who were willing to help him. Conversely, Sandra did not have the consistent support she needed in order to gain access to the mathematical content and language being used in science class. Ms. Jackson would better serve Sandra by making sure that she understood the daily assignments. One way she could do this would be to confirm that her unofficial peer tutor Annette was willing and able to help her. While Ms. Jackson was certainly well-intentioned in her grouping strategies, it is clear that she did not consider all of the potential problems that could have arisen within the small group in which Sandra was placed.

Individual student identities also play an important role in classroom interaction. Creating groups that take into account one's academic identity, as well as monitoring that students are in fact receiving the assistance they need (from peers and from the teacher) are critical teacher responsibilities. By manipulating student groups and highlighting important features of academic language, the teacher will better meet the academic needs of her ELLs. In this way, ELLs will be both better participants and better learners in science.

Targeting Enduring Understandings

1 What are the connections you see between the discussion in Hansen-Thomas' chapter and the four elements of culturally-responsive instruction, as described by Villegas and Lucas?

2 How would you describe the "work" that language is doing in this multi-cultural science context?

Deepening the Reflection

1 Little Manuel worked in a group that helped him gain access to the math activity; however, he did not demonstrate an appreciable use of academic discourse over the several-week period that the math initiative was implemented. What are some reasons why he did not? How could his teacher, Ms. Jackson, have promoted more opportunities for ELLs like Little Manuel to interact with other classroom community members and acquire academic discourse? What do we learn from Little Manuel's case about the relationship between the language of a community and active participation in that community?

2 As Hansen-Thomas notes in this chapter, NCTM's math standards highlight the use of academic discourse and focus on groupwork and interaction. Do the science standards? To what extent do these standards take into account the special needs of ELLs?

3 Schools in Garner ISD do not have specific ESL classes for students, but rather require all content area teachers to be "certified" by taking minimal ESL methods classes. What do you believe are the implications of this educational policy? If you were in a position of leadership, what kind of professional preparation with respect to ELL issues would you require of all science teachers?

Encouraging Engagement

1 Go into a secondary level science classroom and observe culturally- and linguistically-diverse students in cooperative groups. Can you assess who is in charge of the group? What is the role of the other students? Are any ELLs particularly reticent? Why do you think so? After observing the groups and taking notes, interview the students and the teacher. Ask them whom they believe plays the most active participatory role and who is the least participatory. How else do they describe students' science learner identities? Are teacher and student responses the same or different? Try to account for your findings.

2 Follow up your classroom observation with an interview with the science department chair or the school principal. The purpose of the interview is to learn more about how these students' needs are met, particularly with respect to their science learning. What kinds of questions will you need to ask? Use the questions you develop to guide your interview and share your questions (and the answers you get!) with your peers.

Notes

1 Data collection for the present project as well as the larger project was made possible by a Spencer grant awarded to Juliet Langman and Robert Bayley entitled Acquiring Content and English Language Knowledge in the Middle School Science Classroom, 2001–2002.
2 Commonly used include: # = pause of 1 second, — = text removed, xx = unintelligible.
3 Classes were held on alternate days (A/B schedule). Thus students worked on math every other day, or eleven days in total from October 9 to November 6.

References

American Association for the Advancement of Science (AAAS). (2003). Project 2061. Retrieved November 27, 2003, from http://www.project2061.orgt/default_flash.htm.

Bayley, R., Hansen-Thomas, H., & Langman, J. (2005). Language brokering in a middle school science class. In J. Cohen, K. McAlister, K. Rolstad, & J. MacSwan (Eds.), *Proceedings of the International Symposium on Bilingualism 4* (ISB4) 2003. mwp-Publisher: Cascadilla Press.

Clements, D. (2003). Math matters. *Scholastic Parent and Child, 10*(4), 14.

Cobb, P., Wood, T., & Yackel, E. (1993). Discourse, mathematical thinking, and classroom practice. In E. Forman, N, Minick, & C. Stone (Eds.) *Contexts for learning: Sociocultural dynamics in children's development* (pp. 91–119). Oxford: Oxford University Press.

Cook, H. (1999). Language socialization in Japanese elementary schools: Attentive listening and reaction turns. *Journal of Pragmatics, 31*, 1443–1465.

Echevarria, J., & Graves, A. (2007). *Sheltered content instruction: Teaching English language learners with diverse abilities* (3rd ed.). Boston: Allyn & Bacon.

Goffman, E. (1981). *Forms of talk*. Philadelphia, PA: University of Pennsylvania Press.

Gutiérrez, K. (1995). Unpackaging academic discourse. *Discourse Processes, 19*, 21–37.

Jameson, J. (1999). *Enriching content classes for secondary ESOL students: Study guide*. Washington, DC: CAL-Delta Systems.

Kanagy, R. (1999). Interactional routines as a mechanism for L2 acquisition and socialization in an immersion context. *Journal of Pragmatics, 31*, 1467–1492.

Kessler, C. (Ed.). (1992). *Cooperative language learning: A teacher resource book*. Englewood Cliffs, NJ: Prentice Hall Regents.

Lave, J. (1991). Situating learning in communities of practice. In L. Resnick, J. Levine, & S. Teasley (Eds.), *Perspectives on socially shared cognition* (pp. 63–84). Washington, DC: American Psychological Association.

Lemke, J. (1989). *Using language in the classroom*. Oxford, UK: Oxford University Press.

MacWhinney, B. (1994). *The CHILDES Project: Tools for analyzing talk* (2nd ed.). Pittsburgh, PA: Carnegie Mellon University Press.

Michaels, S., & O'Connor, M. (1996). Shifting participant frameworks: orchestrating thinking practices in group discussion. In D. Hicks (Ed.), *Discourse, learning, and schooling* (pp. 63–103). Cambridge, UK: Cambridge University Press.

National Council of Teachers of Mathematics (NCTM). (2000). *Principles and standards for school mathematics*. Reston, VA: Author.

National Research Council (NRC). (1996). *National science education standards*. Washington, DC: National Academy Press.

O'Connor, M. (1998). Language socialization in the mathematics classroom: Discourse practices and mathematical thinking. In M. Lampert & M. Blunk (Eds.), *Talking math-*

ematics in school: *Studies of teaching and learning* (pp. 17–55). Cambridge, UK: Cambridge University Press.

Schiffrin, D. (1994). *Approaches to discourse.* Cambridge, MA: Blackwell.

Takahashi, E., & Morimoto, Y. (1996, November). *Development of young learners' interaction in Japanese classroom and beyond: A longitudinal study of a Japanese FLES program from the sociocultural perspective.* Paper presented at the American Council on the Teaching of Foreign Languages, Philadelphia, PA.

Tananis, C. (2002). What are we learning? Summary report from the 2000–2001 District Profile of Math and Science Indicators. Retrieved July 10, 2003, from http://www.msc.collboratives.org.

Wenger, E. (1998). *Communities of practice: Learning, meaning, and identity.* Cambridge, UK: Cambridge University Press.

11 Science, Culture and Equity in Curriculum

An Ethnographic Approach to the Study of a Highly-Rated Curriculum Unit

Joel Kuipers, Gail Brendel Viechnicki, Lindsey A. Massoud, and Laura J. Wright

Introduction

Considerations of student diversity in K-12 settings often center on within- and between-group comparisons of students of different cultural and linguistic backgrounds. In this chapter, Kuipers and his team use and describe an innovative approach in their consideration of diversity in science classrooms; they examine "diversity in practice rather than diversity in label." Specifically, these authors argue that a simplistic examination and performance categorization of "how the black kids did" or "how the Latino kids did," for example, may not fully capture the differential experience that may be afforded by participation in heterogeneous small group collaborations with innovative curricular materials designed to more fully invite all students into science learning. Instead, the authors consider the opportunities provided by unfolding heterogeneous small group interaction with the curriculum to understand what is meaningful to students as they participate in science learning. They encourage us to focus on the diversity of students' personal experiences in the classroom, rather than the diverse identities of students themselves.

In order to document students' personal experiences with a reform science curriculum, the authors analyzed specific ethnographic and linguistic dimensions of students' talk, writing, and interactions and also examined dimensions of the reform curriculum unit including its narrative structure, organization of lessons, and opportunities for reasoning and reflection. This close examination of the juxtaposition of students' personal experiences and the science materials allows for a deep understanding of the science-learner identity opportunities presented by reform science curricular materials beyond the surface-level characterization of those opportunities in the materials. The authors push us, in this chapter, to see how well-designed curricular materials do indeed interact with who children are, and what they know and bring to their science learning.

As you read through this chapter, think about how you have learned to analyze the benefits of science curricular materials. What were the dimensions along which you examined the unit(s)? In what ways did the materials or the information provided to you encourage consideration of the different kinds of science learner identities, like personal experiences and knowledge, that your students might bring to the classroom? In what ways did the curricular materials or your implementation of the materials awaken you to the possibilities inherent in organizing and supporting diverse experiences in the classroom in support of science learning? After you read this chapter, think about how you might modify or supplement instructions and materials in science curriculum to offer opportunities for students to meaningfully relate to science learning and to usefully leverage the presence of other learners as they engage in science sense-making and science learner identity development.

Philip is an African American boy in an ethnically-diverse middle school science classroom who appears to enjoy the subject. He likes to be the one with the answers. His lab table partners—Sean, a European American boy, and Natalie, a Latin American girl—find him annoying. As they prepare to weigh a mixture of vinegar and baking soda for an experiment, Sean confirms the next step under Philip's watchful eye. Table 11.1 contains a transcript of their exchange.

Sean, often appearing apathetic and unprepared for class, uses a technical-sounding phrase to describe their task: "We have to do the **before weight...** right?" Philip, challenged by Sean's assertive display of this unusual phrase, moves to re-establish control and reaches for the scale. Natalie, watching intently, focuses instead on the task of recording the results. A step ahead of the boys, she notes the weight of the scale—"one oh four"—and proposes what they should write into the workbook as an answer, further challenging Philip's authority. Philip gets control of the procedure again with two unusual expressions: "there might be a **slight difference**" and "there's the **lid weight** that we need now." In the end, he overrules Natalie, and gains interpretive rights over the experiment from Sean. In line 12, Sean asks Natalie how much the mixture weighed—essentially, what they should write down on their worksheets. In line 16, Philip wonders about where to write that number down on their worksheets and Natalie points him in the right direction. Philip then notes that the mixture has continued to lose weight, and reads the number from the scale out loud in line 18 ("point three") and Natalie makes sure that Sean has the right number to write on his worksheet (line 20).

Thousands of such mini dramas occur in schools each day all over the U.S., and, indeed, throughout the world. Most, however, do not occur in a curriculum unit that yielded such impressive results as this one. At the conclusion of this eight-week unit devoted to exploring the idea of conservation of matter, the students did not like Philip any better. But they all participated. Sean used expert expressions, Philip got to handle the equipment, and Natalie did much of the recording. And these acts of participation were not unusual: in the seventeenth

Table 11.1 Transcript

Transcript	Student activity
1. S: *Okay but weigh it all first*- we have to do the before weight, so I'll-pour this in now, right?	Philip – lab procedure *(holds plastic bottle)*
2. P: Yeah. Pour it.	
3. N: Why don't we just put-	*(can't see activity; Sean is standing in front of table, but he seems to be holding objects)*
4. P: Hey- Yeah put that in but don't put it- put the the stuff in.	
5. N: Hey look. Look, look look look.	
6. P: What?	
7. N: Okay it says one O four, that's before, so why don't we just- it's going to be the same thing.	Sean – lab procedure *(pours substance into bottle)*
8. S: No- *it might be different.*	
9. P: It can't- there won't- it might- there might be a slight difference, but we should- so we should be sure.	
10. S: *That and we should put the lid- we should put the lid on here.*	
11. P: Exactly sure. Plus there's- and there's- there's the lid weight that we need now.	
12. S: Okay, how much is it?	
13. N: One sixteen point four.	
14. P: Yes.	Philip – lab procedure *(puts lid on the scale to weigh it)*
15. S: One sixteen point four.	
16. P: Where do we put that? This- side.	
17. S: XX.	
18. P: Point three.	Natalie – clean up *(uses paper towel to wipe water from the table in front of the scale)*
19. S: One sixteen right?	
20. N: Point three.	
21. P: Mhmm.	

largest school district in the country, this curriculum unit, titled *Chemistry That Applies* (*CTA*), not only raised outcome scores overall, but significantly narrowed achievement gaps between traditionally well- and under-served groups. As Figure 11.1 shows, children who were classified by the school district as eligible for Free And Reduced-Price Meal System (FARMS), or who were formerly so classified, showed the greatest improvements in relation to the comparison condition (i.e., children who received a curriculum unit unrated by the American Association for the Advancement of Science (AAAS)).

In this chapter, we take the notion of equity further than disaggregated outcome scores, although these analyses are certainly revealing. In examining *how* the curriculum works for the students, we describe how differences in participation characterize student diversity in terms of what is meaningful to students—what they actually *do* within the classroom. We describe, ethnographically, manifestations of diversity in student practice, and further aim to understand how this diversity, in the context of a highly-rated, hands-on curriculum unit, can be drawn upon to increase student achievement in science.

In examining one particular curriculum unit, however, we are not interested in performing a curriculum evaluation; this has already been done by the AAAS

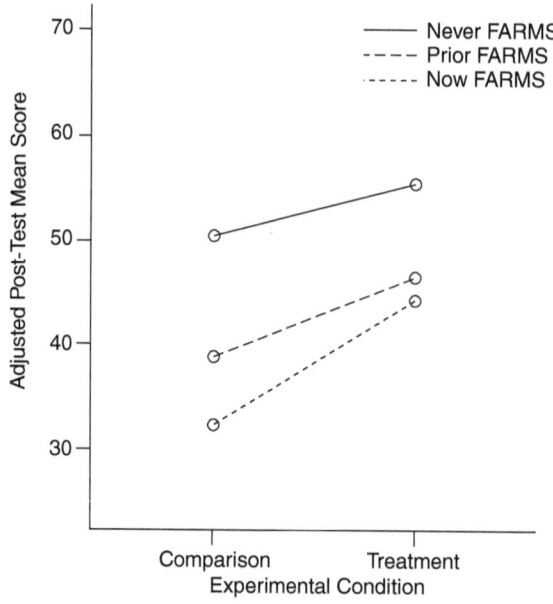

Figure 11.1 Post-test scores for treatment (*CTA*) and comparison students, year 1.

Project 2061 team. Instead, we are providing a linguistic and ethnographic interpretation of how this curriculum unit functioned in its social and cultural context of use. Drawing broadly on the criteria for instructional support formulated by AAAS Project 2061, we examine *CTA* by Michigan State University as it was implemented in a diverse middle school classroom in suburban Washington DC. Adopting the students' perspective, we seek to understand how middle school learners experienced the content, goals, and organization of these curriculum units, as well as the relation of the "official, manifest" curriculum texts to the students' own, informal sense-making practices. Among these latter practices are (a) use of the scientific register, or "talking science"; (b) object manipulation that provides students with direct experience of relevant scientific phenomena; and (c) graphic representations of knowledge through literacy practices, or reading and writing.

American educational theory has a long history of placing importance on student experience in learning. According to the National Research Council (1999b), for children to learn, crucial features of instructional support include: "adults who direct children's attention, structure experiences, support learning attempts, and regulate the complexity and difficulty levels of information." Given the importance placed on instruction as the organization of experience, it is surprising how "little empirical and conceptual work has directly considered student experience" in the analysis of curriculum (Erickson & Schults, 1992). Our paper seeks to address this gap by proposing specific ethnographic and linguistic dimensions for interpreting students' own multimodal sense-making practices. By focusing on student participation, we emphasize diversity in practice rather than diversity in label. In addition to discussing categories that are usually used for disaggregating test scores, such as race, socioeconomic status, and language abilities, we examine how students themselves interact with a curriculum unit in different ways. Diversity can manifest in a number of ways in education and social life more generally and is not merely a collection of categories. We examine diversity here in how students experience it—as differences in participation patterns within the classroom. We then consider the implications of student experience for the analysis of equity.

Background and Rationale

The persistent problem of achievement gaps between under- and well-served groups in the U.S. has defied simple solutions for decades. Billions of dollars and years of effort have done little to improve the dismal and, indeed, worsening, picture. In 1996, the U.S. National Research Center's TIMSS Project (NRC, 1999a) reported that a major barrier to improving science education included the quality of U.S. science curriculum overall. They argued that it was unfocused and fragmented, and the textbooks and curriculum materials used came in for special criticism. Even when improved materials are obtained, teachers have difficulty implementing them and delivering instruction consistent with educational reform goals. Several systemic reform initiatives of the 1990s emphasized the need for a more focused science curriculum and better curriculum materials for teachers to use (aligned with science standards, instructional methods, and assessment/accountability measures). U.S. science textbooks are larger and often more colorful than those of other countries, but they are sometimes less comprehensible, more repetitious, and tend to ask students to do tasks that only require low-level kinds of thinking. As Lynch and colleagues (2003) pointed out in a recent paper, "It is reasonable to assume that improved curriculum materials aligned with science education standards that encourage students to learn with understanding could accelerate the reform process."

In the mid-1990s, Project 2061 carried out an extensive review of the science education research literature and created a list of instructional strategies that the research base suggested ought to be built into curriculum materials. The resultant Project 2061 Curriculum Analysis is arguably the most detailed and thorough curriculum evaluation process of its kind to date (Kesidou & Roseman, 2002, but see also Holliday, 2003). It relies on the research base and the rational judgment of teams of experts trained to use the process. Project 2061 developed a systematic procedure to evaluate written science and mathematics curriculum materials, geared to the teaching of national science standards and benchmarks, with indicators for each criterion to guide the ratings made about the quality of the curriculum materials.

The Project 2061 Curriculum Analysis analyzes curriculum materials to ascertain: first, if there is a focus on a specific benchmark/standard (content analysis) and second, how well twenty-two instructional criteria in seven categories are met (instructional analysis). The seven categories organizing these criteria ask whether the curriculum units (1) convey a sense of purpose; (2) address student ideas and misconceptions; (3) promote engagement with relevant phenomena; (4) promote use of scientific ideas and terminology; (5) encourage student thinking; (6) encourage assessment of progress; and (7) create a learning environment: curiosity and inquiry for all students. The assumption was that curriculum materials with high ratings according to these criteria should result in superior student outcomes on the target benchmark/standard compared to curriculum materials that would not receive a high rating.

However, these 2061 criteria were designed largely for the analysis of curriculum texts and other instructional materials, not for how the students interpreted or used the instructional materials as part of their process of classroom participation. As Erickson & Schultz (1992) observe, there is "virtually no research on how curriculum is actually experienced by children." Moreover, there is little information on how these criteria are interpreted by diverse students. Project 2061, for example, asks if a unit "conveys a sense of purpose," but how can we tell when a sense of purpose has been conveyed to the students enacting a unit that has been highly rated according to this criterion? In other words, what consistent, patterned, student behaviors are observable that would indicate that a curriculum unit has actually achieved this goal? Likewise, what sorts of methods would allow one to study "engagement with relevant phenomena"? What counts as evidence of "engagement" in an actual classroom setting? Even more problematic is "student thinking": What counts as evidence of thinking? Finally, the 2061 project provides little information on how these criteria might be interpreted by diverse students. Does "student thinking" look and sound the same for all students? How about engagement with relevant phenomena? Finally, how can we organize our answers to these questions in a way that reflects a child-centered focus on the curriculum as it is experienced by its intended audience?

In summary, then: How can we devise ethnographically-oriented criteria by which we can evaluate curriculum units *as they are implemented in classrooms with diverse students*, rather than curriculum units on paper only? In what follows, we seek to focus on how curriculum texts are part of an ongoing system of classroom participation. We then attempt to use the results of these ethnographically-oriented criteria for interpretation of variations in outcome scores.

From Sense-Making to Text-Building

In order to understand how students experience a curriculum unit, it is important first to understand what it is that is being experienced. While the term "curriculum" generally refers to a course of study, it also refers to the instructional materials that support those learning activities. In general, those materials take a *textual* form.

While the move toward student-centered curriculum has familiarized educators with the idea of learning as a form of "meaning-making," Becker points out that texts are a crucial framework by which people (not only students) "build" meaning from experience. Whether it is through telling stories, conveying content or communicating intentions, students construct meaning for themselves from curriculum materials by building text-like models of key ideas. If cultural differences consist of differences in systems of meaning, a focus on how students experience and make meaning out of curriculum materials can thus provide a central route for teachers to appreciate, understand, and work with student cultural diversity.

Of course the study of texts is nothing new; centuries of sophisticated scholarship have been devoted to textual exegesis. But while the field of literary criticism has developed elaborate approaches for interpreting written materials, these frameworks are often difficult to implement in the complex, active and richly oral environments of everyday life. Becker (1979) proposes that we view such activities as "text building"—a dynamic, on-going process of using linguistic and social forms to create meanings. The four dimensions he urges us to consider are:

1   *The relation of texts to the outside world.* These are what the text constructs as a "referent" or "content" of the materials. For example, a curriculum unit might be devoted to constructing for students an image of atoms and molecules as the referent or content of the materials.

2   *The relation of the text to the intention and goals of its author.* In a curriculum unit, of course, these are pedagogical in nature. For example, in one curriculum unit, the goal might be to encourage children to learn by memorizing the text and testing them later on their knowledge of the content; another curriculum unit, on the other hand, might set for itself the goal of encouraging children to learn by engaging in and doing relevant activities.

3   *The relation of the parts of the text to one another.* A curriculum unit might be understood as a kind of narrative in that its lessons are organized in a particular way in order to structure students' experiences with the concepts it is trying to convey. Students' experiences are also structured on a daily basis as a unit is implemented. Ethnographically, one can note that some units have students devote considerable time to preparing for and performing group lab activities, while others provide students more time to reflect on experiments. The integration of these various types of classroom activities with one another—its coherence—varies from one curriculum unit to the next.

4   *The relation of the text to other relevant texts outside of it.* Students experience a given text partly in relation to their expectations about how similar texts convey meaning, and according to their own preferences about how to represent the text to themselves and to others. Thus some students expect to write a lot during a curriculum unit, others expect to talk and build verbal representations of the curriculum ideas, while still others expect to build haptic, tactile representations of the text.

Curriculum units as they are implemented and experienced by actual children may thus be thought of as part of a process of "text-building." Somewhat like a script for a play, curriculum units are a set of instructions for action that are interpreted in various ways by individual students and teachers much like actors who each "build" an interpretation of a script in individual ways through various reading, speaking, and acting strategies. The job for the researcher, then, is to interpret the dimensions by which the students make sense of the curriculum—the text itself, the roles intended for them, the phenomena under investigation, and their teacher's actions.

Video Ethnography

In order to apply this model to the classroom data, we conceive of the "text-building" process in ethnographic terms. By "ethnography" we mean a detailed description of the communicative practices of a group, and of the culturally-defined situations in which relevant distinctions are made in that system (Conklin, 1962). Drawing broadly on a sociohistorical framework of analysis, we examine learning as a form of participation in such culturally-defined activity systems.

To carry out the study that we report on in this chapter, students were videotaped throughout the implementation of a highly-rated science curriculum unit, *CTA*. This process resulted in sixty-eight hours of recordings. The videotapes were digitized, transcribed, and analyzed using advanced database software. The resulting keyword-searchable corpus includes approximately 4000 pages of videolinked student discourse, which are coded for class period segments, clarification, object manipulation, scientific term use, and literacy practices.

The Curriculum Unit

The curriculum unit that is examined in this chapter was chosen because it was highly rated by Project 2061, and involves active learning and hands-on manipulation of relevant phenomena. While not explicitly designed for a diverse student body, the theoretical framework guiding the developers appears to have been one that recognizes the social and historical context of learning. The unit is described here briefly.

CTA

This unit is the product of researchers from the School of Education at Michigan State University led by Theron Blakeslee (1993). *CTA* is designed for students in grades 8 to 10, and in this study *CTA* was implemented in an 8th-grade classroom.

The benchmark concept toward which *CTA* is aimed is the conservation of matter. By first observing, and then weighing, four different reactions under different conditions, students are to realize that (1) within a closed environment, mass (operationalized as weight in this unit) is conserved, and (2) that the rearrangement of atoms and molecules can help one to understand the conservation of matter.

The four reactions that are central to this unit—burning, rusting, the decomposition of water, and the reaction of baking soda and vinegar—are revisited in each of the unit's four clusters of lessons. In Cluster One, "Describing Chemical Reactions," students are asked to observe "common, everyday household substances" in various mixtures, to describe the physical and chemical changes they see occurring, and to observe when new substances are formed. Cluster Two instructs students to compare the mass of substances during physical and chemical reactions (by weighing mixtures before and after) in order that they will realize that matter—even invisible matter like gases—is conserved. Students are introduced to atoms and molecules in Cluster Three, and are asked to construct explanations of the weight changes they observed in the four reactions in terms of the rearrangement of atoms and molecules. A fourth cluster, focused on the energy changes in these reactions, was not implemented in this study.

Constructing the "Content" of the Curriculum

The specific content focus or referent of *CTA* is the particulate nature of matter and its conservation. This "content" item is seen as being real (unlike, say, Huck Finn in a literature class) and "natural" (having an external reality independent of the text of the curriculum). This content is then "conveyed" to the student via various means—experiments, illustrative examples, visual aids, workbook questions, etc. (see Project 2061 #1).

But curriculum units may establish the external reality of their ideas in different ways. For *CTA*, one way that the "reality" of the conservation of matter is conveyed is through the transformation of a balloon in its appearance and shape. The balloon thus becomes an icon resembling the process the authors wish to illustrate.

In the first cluster or section of *CTA*, students observe four reactions, like the one described above, that are meant to iconically represent the conservation of matter: a balloon is sucked into a flask, an equal arm balance dips dramatically as the baking soda and vinegar bubble away furiously in a flask atop the balance, and so on. The Law of Conservation of Matter is thus "seen" (in part) by means of these kinds of experiments. Thus, in the first section of *CTA*, the conservation of matter is made visible, so to speak, through these dramatic, iconic experiments which students appreciate without pen or calculator or graph paper in hand.

In the second section of *CTA*, however, the concept of the conservation of matter is made transparent by means of numerical inscriptions: observed "weight changes" are to serve as evidence of the conservation of matter. Students perform a mathematical calculation (subtraction) between what the teacher calls "the before weight" and "the after weight" of the mixture, and it is this inscription, "the weight change," that students are to understand as "evidence" of the Law of Conservation of Matter. The instructional materials and the teacher scaffold this analytic move, albeit in subtle ways. For example, regularly-occurring rhetorical and linguistic forms assist students in appreciating that "the before weight" and "the after weight" are in a relationship, and that a relationship between the two numbers is compelling evidence of the conservation of matter. We have argued elsewhere (Viechnicki, in review) that unusual nominalizations, syntactic structures, and pragmatic forms systematically appear at critical junctures in the text of *CTA* that function to emphasize for students that one set of inscriptions (the numbers written on their worksheets representing "the weight change") is what allows students to "see" the conservation of matter. In *CTA*, students observe several physical and chemical changes without actually weighing anything at first; students are guided through this process over the course of the unit, from the structure of the lessons (discussed in detail below) to the linguistic and rhetorical features of the text.

One can hypothesize that the way that the target idea is constructed is due at least in part to the nature of the curriculum content (or "referent" in our terms). The experiments that the students conduct in *CTA* are not familiar to the students—the bubbling and fizzing and test tubes involved are in fact prototypically "scientific" in nature and thus do not need to be defamiliarized in the same way that they might in a different unit, whose lab activities are more pedestrian, or at least less marked as scientific.

Constructing the Intentions or Goals of the Curriculum

Another way by which the purpose of the lesson (Project 2061 #1) is conveyed is by interpreting the intentions of the authors of the curriculum units. In *CTA*, the purpose of the curriculum unit and, indeed, the individual lessons, is to initiate students into the practice of scientific activity through hands-on experience and group work. *CTA* piques students' interest in the benchmark concept by having them puzzle over four unusual experiments in the first cluster of lessons and, in the second cluster, collect data on these experiments, and puzzle over inscriptions (the weight changes) rather than the actual phenomena.

Another way of thinking about what this unit intends for the students is in terms of models of participation and apprenticeship. The developer of *CTA*, Theron Blakeslee, wants to apprentice students into the world of science by creating opportunities for vivid experiences in which they participate in groups, gather around the apparatus, perform experiments, and discuss the results. While students debate the significance of these experiences, they move from peripheral to legitimate participation (Lave & Wenger, 1991).

Coherence: Narrative Flow

As Becker (1979) points out, another way in which to build meaning through text is through the organization of the parts in relation to one another. In classrooms, this manifests itself in two ways: (1) the sequence of activities across the eight-week unit; and (2) the sequence of activities in any given lesson. As a way of talking about the different kinds of ways in which the curriculum units build stories about the benchmark concepts for children, we call the first "unit flow" and the second "lesson flow."

Unit Flow

In *CTA*, the concept of the conservation of matter is only introduced midway through the curriculum unit. A series of quite different experiences are introduced that all come together to explain a single law of the conservation of matter.

In *CTA*, roughly halfway through the unit, the students are introduced to the Law of Conservation of Matter as a "fundamental law of nature," as well as being introduced to the Periodic Table and the concept of atoms and molecules as authoritative facts. (The authority of these concepts is not only predicated with sentences such as, "Scientists have found," etc. but also graphically marked with capitalization, bold fonts, etc.). Students are basically told that this is what explains the experiments they have performed. The third cluster of the unit requires students to build molecular models of those same four reactions to drive home the molecular explanation of their experimental results. Thus, the narrative coherence of *CTA* hinges on the introduction of an explanation for students' experiments. Students in *CTA* are explicitly told that their observed weight changes are evidence for the law of conservation of matter.

Lesson Flow

Understanding the ways in which this curriculum unit makes use of class time is crucial in constructing a picture of how *CTA* functions within the classroom. To examine this ethnographically, and its implications for students' experience of curriculum coherence, we have focused on class period segments, or the types of activities that the students engaged in at various moments during the course of a school day.

Classroom Period Segments (CPS)

These categories were based on distinctions that appeared relevant to students' and the teacher's participation in activities. Shifts between classroom period segments are often indicated by "contextualization cues," (Gumperz, 1982), such as visual cues from the teacher (e.g., turning lights on or off), students' repositioning at the table in terms of visual and bodily orientation, and changes in students' communication patterns (e.g., if they're talking, to whom, and about what). Bounded by these cues, these segments of activity can last for only a minute or two, or can endure for almost the entire class period. The six class period segments we coded for are: Warm-up, Presentation, Exploration (Group, Class, and Individual), Reflection, Closure, and Transition. The distinctions between these types of classroom activity are detailed in Table 11.2.

The classroom period segments with the highest frequency are Group Exploration, Presentation, and Reflection, occurring in every single lesson of the curriculum unit. Individual Exploration, though it peaks during lesson 16 (Atoms In Equals Atoms Out), makes up a very small percentage of the total segments coded for CPS. Group Exploration and Reflection, on the other hand, occur

Table 11.2 Types of classroom activity

CPS	Characteristics	Contextualization cues	Duration
Warm-up	• Designed to get students thinking or review material presented the previous day	• Indicated by verbal cues like "For your warm-up today, you will" • Students oriented toward the front of the room	• Fifteen to twenty minutes

Presentation	• Teachers familiarize students with what they will be doing, give lesson objectives, and/or introduce a new concept or idea
	     Procedural Presentation: *what* students will be doing, *what* materials they need, and *how* they will conduct the exploration
	     Conceptual Presentation: a demonstration/ lecture, or students answer introductory questions
		• Indicated by verbal cues like "Today, you are/we are…"
		 • Students oriented toward the front of the room
		 • Discourse is focused on the future
			• Usually five to ten minutes
			 • May be longer if it involves a demonstration or lecture/explanation

Exploration	• An experiment, lab, or written activity that provides experiences of the main concepts
	     Individual Exploration: students work alone
	     Group Exploration: students work in groups of two or more
	     Classroom Exploration: teacher leads the class
		• Indicated by verbal cues like "Once you get your materials…"
		 • Indicated by movement of students to gather materials or go to lab stations
		 • Students oriented toward the center of their tables or to their papers where they are recording the activity
			• May be as short as ten minutes or last the entire class period

Reflection	• Students reflect on an activity (Exploration) they've completed • Can occur through teacher-led discussion, but is often a written activity completed individually, in pairs, or in groups	• Indicated by verbal cues like "Now that you've completed the activity, I'd like you to answer…" • Students oriented toward the front of the classroom or toward the table	• Usually ten to twenty minutes but can last the entire class period
Closure	• A comprehensive statement or activity that summarizes the day's activities and/or relates them to a key concept • Sometimes done in the form of exit cards or tickets to leave	• Indicated by verbal cues like "So" or "Okay" • Students oriented toward the front of the room, or the table if they're doing a written closure	• Two to five minutes
Transition	• Movement between activities, natural "downtime," or when students are waiting to leave	• No central focus of student attention • Casual chatter	• A few seconds to ten minutes

frequently and in large chunks. Reflection, in particular, becomes more pronounced toward the end of the unit. Classroom Exploration occurs more frequently at the beginning of *CTA* and diminishes toward the end, which may mirror a trend of greater student autonomy over the course of the unit. Closure peaks in the middle of *CTA* during lessons 6 and 8, which address whether weight changes in chemical reactions and whether gases have weight, respectively. Finally, Warm-up does not appear to be a constant feature of the classroom activity sequence, but does occur periodically during *CTA*. The graph (Figure 11.2) below represents each classroom period segment as a percent of total coded segments for each lesson of the unit.

Intertextuality

As Becker points out, the process of "text building" proceeds according to the expectations that an actor (e.g., a student) has about what a text should be. That is, students build textual representations of their experience by constructing representations of their own that model, or in some way mirror, the ideas being conveyed in the curriculum unit text. In our study, we observe how the students interpret the curriculum by linking it to other sense-making strategies of

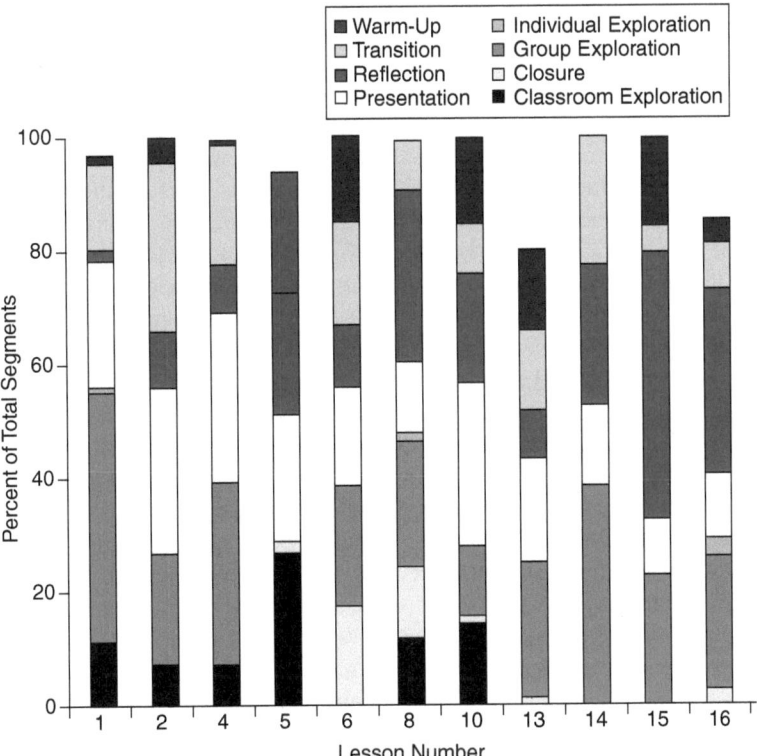

Figure 11.2 Classroom period segments over the course of *CTA*.

representation such as (1) use of the scientific register and scientific terms; (2) haptic representation and object manipulation; and (3) literacy practices. Unlike the first three aspects of text-building (i.e., conveying of content, goals, and narrative flow), this aspect of students' curriculum experience belongs to an informal kind of meaning-making that is often harder to identify, but nonetheless crucial for understanding how the curriculum functions.

Scientific Register: "Weight"

AAAS Project 2061 criterion #4 states that to be effective, curriculum units should "promote use of scientific ideas and terminology." While science educators have long recognized the distinctiveness of a scientific register, ways of dealing with this have differed. Some sought to minimize the distinctiveness of the scientific way of talking by avoiding specialized jargon and terminology; others sought to force students to memorize terms as a way of providing students with access to key concepts. At the forefront of many of the discussions regarding how scientific terminology is taught in the classroom is the issue of diversity. In the U.S., students have a large range of experience generally with the English language—from having only just arrived in the U.S., never having spoken English before, to growing up in a bilingual household, to being raised solely speaking English. Furthermore, students have varied degrees of prior experience with the scientific register, depending on their parents' occupations, or on what books they have read as children. Although students may not be aware of all of these distinctions when they arrive in their science classroom, this prior experience can significantly impact students' experience with the scientific register when they encounter it in class. In our study, we aim to understand more about how students use scientific terminology as part of their sense-making practices in the classroom. We do this by searching the transcripts for evidence of the use of a scientific register appropriate to the curriculum.

Although there seems to be some consensus that teaching science to children involves familiarizing them with how to "talk science," there remains a great divide between the linguistically-oriented researchers who focus on the details of linguistic form in written texts, and the more socially- and pedagogically-oriented scholars who examine verbal interaction in classrooms. There are relatively few scholars who examine the relation between features of linguistic form and how these are used as functional resources by children in classroom learning environments.

There are several reasons for paying close attention to form-function relations. One is methodological. In order to code videotapes (or direct ethnographic observations for that matter) reliably and in a valid manner, a broadly linguistic and ethnographic approach is useful as an analytical framework. When identifying and classifying student activities, unless the categories used correspond to those being used by the students themselves, the categories are unlikely to prove useful in the long term for research purposes. While there are important "limits of awareness" (Silverstein, 1979) among "native" users, the syntactic, semantic, and pragmatic forms of language in use can nonetheless provide reliable and valid categories for analysis.

Scientific Term Use in CTA

After dividing all the videos into Class Period Segments, we focused on the moments of Group Exploration (GE), defined as periods of time when students are exploring the benchmark concepts in the unit in a small group, and this usually involves the performance of hands-on lab activities. As noted above, *CTA* devoted significant amounts of time to group exploration.

We selected the scientific terms for consideration by eliciting them from teachers through a written survey. They responded to the question "What are the key terms in the unit?" We used these terms as the basis for a key word search through the transcripts.

After coding for a set of terms suggested by the teachers, we found the patterns depicted in Figure 11.3.

In *CTA*, students used the appropriate scientific terms with relative frequency, often as part of a strategy of building an argument and making sense of classroom activities.

Example of Scientific Term Use: Nominalization

In *CTA*, the key benchmark ideas are accessed partially through nominalization. By "nominalization" we mean the grammatical process by which verbs or adjectives are transformed into nouns. In general, the trend over the course of the curriculum was an increase in the percent of student term use that was nouns. In other words, students were using more nouns as terms toward the end of the curriculum, rather than verbs, adjectives, or adverbs.

In the first few lessons, students were describing substances and reactions, which prompted more adjectival term use; they were also saying what was

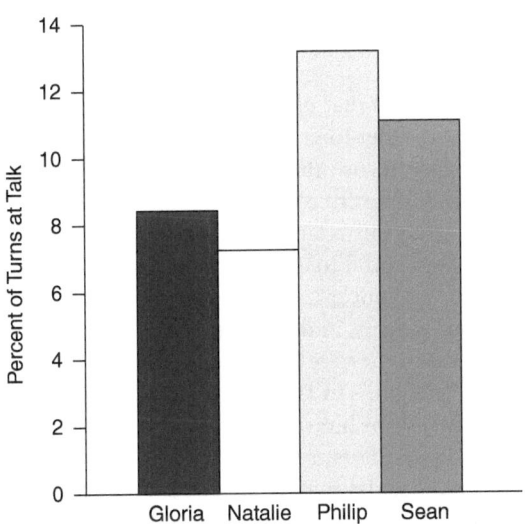

Figure 11.3 Use of scientific terms per turn at talk during group explorations.

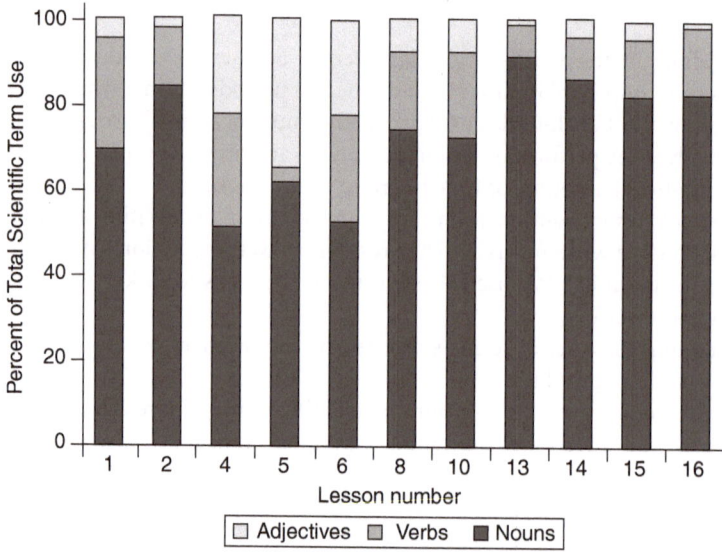

Figure 11.4 Relative proportion of nouns, verbs, and adjectives in students' scientific term use in *CTA*.

happening in the various reactions they were being shown, triggering more verbal term use. By the end of the curriculum, students were talking more about the objects, processes, and concepts behind what they've observed and described—objects like "atoms," "molecules," "oxygen," "hydrogen"; processes like "a weight change"; and concepts such as "the conservation of matter." This objectification, or the turning of processes into object-like entities, signals that students have transformed their scientific talk from descriptions and observations into object-like nominal knowledge—a hallmark of scientific language (Halliday & Martin, 1993).

There is one particular nominalization that plays a critical role in the curriculum's narrative structure. In *CTA*, in order to grasp the central idea of conservation of matter, it is essential to learn to use the verb *to weigh* not only as the nominal "weight," but as the subject of a verb e.g., "the weight changed" or "the weight remained the same." Mastering this relatively unusual locution (unusual in probabilistic terms) is nonetheless central to effective participation in the curriculum and thus "attainment" of the concept. In the example from our Introduction, both Sean and Philip use "weight" in describing the procedures for the lab they are performing. In both cases, they are transforming another statement to include the nominal form of "weight." In the first case, Sean starts to say, "Okay but weigh it all first" and then stops himself, and says, "We have to do the before weight." There is no procedural difference between "weighing it first" and "doing the before weight," but Sean decides mid-sentence to restate the instruction, perhaps invoking more authority by using a nominalized scientific term. In the second use of "weight" in the above example, Philip reconstructs Sean's sentence, but with the addition of the nominal "weight." He takes "lid" from Sean's suggestion about what to do next and creates a novel nominal construction similar to the use of "before weight" and "after weight": the "lid weight." In both cases, the students' rephrasing of previous utterances calls up a nominal compound using the nominalized form "weight."

Insofar as these novel, nominal constructions serve to decontextualize and desubjectivize students' lab experiences, this linguistic detail is critical to how the unit is designed to help students understand the benchmark concept. It brings the focus onto the data that they are to collect and should transparently demonstrate the conservation of matter. In other words, there was a sort of "habit of mind" that this linguistic maneuver was involved in engraining. Otherwise-invisible actions of oxygen atoms and iron atoms (in rusting, for example) can be "seen" so to speak, or made transparent, at least in part, by the appropriate grammatical characterization of one's evidence. If "the weight of a mixture" is in the grammatical subject position of a student's utterance—"the weight increases/decreases/stays the same"—this may indicate that a student understands what counts as data, and may be that much closer to understanding what explanation that data is evidence for.

Moreover, when the children did nominalize in this way, it had consequences for their participation at the table, as again the example in the Introduction illustrates. Philip, for example, generalizes from this linguistic form at key interactional moments, especially when his special status as "table expert" is questioned. This student responds to his lab mates in the interaction by immediately nominalizing what they have just said ("a slight difference" and "lid weight"). As the science expert, this student is responsible for "data collection," and he re-casts what they are doing at the table as the collection of things (first differences and then lid weights), and thereby re-establishes his authority.

In *CTA*, the focus was on observing a single event, describing and explaining its qualitative change (or lack thereof). Thus, nominalizations were provided to students as a resource for explaining the curious phenomena they were observing. For *CTA*, the verbal expression of nominalization was presented as integral with the process of explaining the benchmark idea.

Haptic Representation Through Object Manipulation

The conventional wisdom in science education is that good science curriculum units permit "direct experience of relevant phenomena" (AAAS, 1989), or "hands-on learning." Drawing on Becker's semiotic model, however, we argue that hands-on learning is not a form of "direct experience," but rather a form of *haptic representation*. That is, touching things is but one way—among many—of representing and making sense of one's experience in a science education context. Like other forms of sense-making practices, object manipulation is a way of building a representation—in this case a *haptic* or tactile representation— of a phenomenon so that it can be interpreted, and thus made sense of. Like scientific term use, sketching pictures on paper, or conversation and hands-on activities in science classrooms, are informal ways of "text-building" in Becker's sense because they result in the construction of sensory representations that are in turn linked in the context of practice to the text of the curriculum unit.

A great deal of research on object manipulation in science classrooms focuses on the number of minutes students spend in labs correlated with achievement scores. However, few studies examine the behavior and patterns of individual students and the implications of those patterns on student learning. In their study of autonomous manipulative use in a mathematics classroom, Moyer and Jones (2004) found that students' manipulations differ from the teacher's intended uses. In turn, these innovative manipulations create changes in student participation; some students spontaneously used manipulatives to tutor their peers, whereas others controlled the objects in an effort to direct their group and their classmates. These types of actions would not be possible without the mediating object being introduced into the classroom. Taking this into consideration, we developed a taxonomy of object manipulation types based on framing theory to examine students' use of the objects and how object manipulation affected participation in labs.

The graph in Figure 11.5 provides a picture of the extent to which each child engages in particular kinds of haptic experiences. These students appear to have different patterns of activity related to the objects. While Philip was deeply engaged in completing the scientific procedures as outlined by the text and teacher, Gloria participated in procedural usage three times less than Philip. Gloria often relied on Natalie to retrieve and return goggles for her and frequently issued directives to Natalie about which goggles she wanted. Moreover, of all the students at her table, Gloria positioned the objects away from Philip most. Natalie participated haptically most often through retrieving and returning objects. Sean's object manipulation was by turns playful and procedural; that is, he was as likely to play with the lab objects as to use them for the procedure.

In spite of their differences, however, it is clear that these students spent the majority of their haptic activity in conducting scientific procedure, which is, in general, in keeping with the curriculum's intentions. For three of the students at the focus table, procedure was their most frequent type of object manipulation. As Rosebery, Warren, & Conant (1992) demonstrate, the activity structure of the science classroom has important implications for the development of scientific literacy. Students who are active participants in a scientific investigation are more likely to know how to study something independently and how to build upon their knowledge, rather than treat a text as a set of static facts. Engaging students in active learning becomes important as scientific literacy is not merely about memorizing facts, but about using, constructing, and reflecting on scientific ideas.

In the example given at the beginning of this chapter, all three students were actively participating in the lab procedures. However, this was not the case during the entirety of that particular class period. At the beginning of the class, Sean and Natalie stood watching as Philip handled the objects and completed the procedure. After a while, Sean and Natalie's gaze began to be directed to other areas of the classroom. Sean laid his head down on the table and looked around the classroom while Natalie looked down at the floor. Philip, however, intently

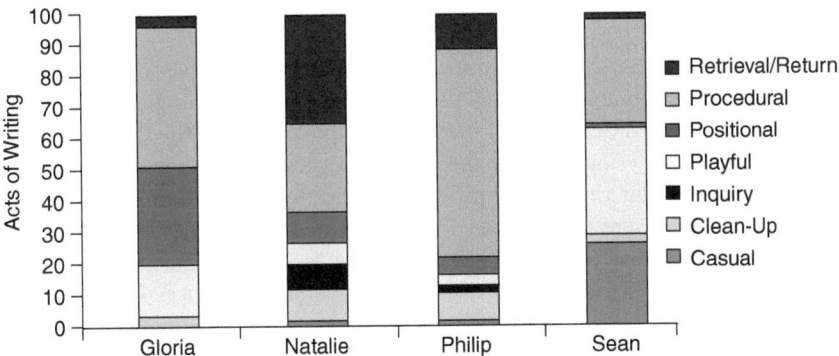

Figure 11.5 Object manipulation types by student.

watched the scale and mused about why the substance was taking so long to dissolve. The students' interest in the lab seemed directly related to the amount of time they spent touching the objects. Prior to the interaction we presented, the teacher had stopped at this table and specifically directed Natalie to measure a substance for the next part of the lab. Sean, too, became more actively involved when Natalie began to participate in the lab. The teacher's intervention and assignment of tasks had an immediate impact on the students' engagement in the lab activity.

Recent research comparing visual and haptic modalities' effect on learning has shown that there are differences in what an individual learns through a particular mode. For example, students given a tactile task will tend to make judgments related to the texture, whereas the visual mode is often better for making spatial judgments (Jones, et al., 2005). However, when comparing effectiveness, the combination of modes is superior to visual or haptic modes alone. Thus, it seems that students who merely watch labs may not develop as rich a judgment of the phenomena as those who engage in active touch and manipulation of the materials. If this is true, it seems especially important to pay attention to the ways in which students participate in labs and the implications that social and behavioral patterns have on lab group dynamics. While conventional wisdom may dictate that "hands-on" does not necessarily mean "minds-on," haptic research shows that deliberate touch has crucial implications on what students learn.

With this particular group, the social dynamics among the students greatly influence the lab procedures. Philip, somewhat of a social outcast in this class, always tries to be heavily involved in conducting the lab procedures. At the beginning of this unit, if he was not given a turn to touch an object, he commented that it was "not fair." Over the course of time, the students at this table expected that Philip would want to be in charge of the equipment—as can be seen in the object manipulation chart. Philip's object manipulation is far greater than any other student in his group. Philip's desire to engage in lab procedures is so great that he and another student (Gloria) got into a heated argument one day

over the use of lab equipment, which led to Gloria's refusal to work cooperatively with Philip.

Evidence of these social dynamics is found in the discourse surrounding the
procedural object manipulation in the example at the beginning of this chapter.
At first, Natalie and Sean looked on while Philip completed the first part of the
lab procedure, offering little in the way of suggestions. When they began to
participate in conducting the lab procedures, they looked to Philip for confirmation of what they were supposed to do. In line 1, Sean states what procedure he is
going to do but then adds a tag question to be certain that his group members
agree (S: *Okay but weigh it all firs- we have to do* **the before weight**, *so I'll- pour
this in now, right?*). Philip responds affirmatively. Then Natalie offers a suggestion of writing the weight they had originally recorded for the first step. Sean
answers her twice, but each time Philip repeats what Sean says almost exactly,
suggesting that Sean's opinion is not enough and that his own is more authoritative. While all students negotiate the procedural aspects of the lab, it is clear in
the discursive moves of this group that Philip is attempting to lead the group. At
another table on the same day, this type of procedural negotiation did not take
place among the students; rather, one student took control of the lab procedures
by issuing directives to his lab mates. Thus, it is not uncommon for one student
at the table to take an authoritative stance among his or her peers.

Hands-on lab work has a number of benefits for students ranging from
increasing students' sense of agency (Katz, 1989), development of scientific literacy (Rosebery et al., 1992), and developing more holistic haptic representations
of the materials and phenomena they are working with (Jones, et al., 2005). Since
these benefits are a result of active touch, it seems important to insure that all
students have access. Furthermore, because access to the objects is directly
affected by social and behavioral patterns in the lab groups, it seems especially
important to pay attention to and understand the diverse ways in which students
use objects in the classroom.

Literacy Practices

Scientific literacy is a major goal of the 2061 project but it is rarely examined in
its social context of use in actual classrooms. In this study, drawing on the work
of Heath (1983), Street (1995), and others, we analyze literacy as not only a skill,
but also as a form of practice. Therefore, we ask not merely what the students are
reading and writing but how they read and write, with whom, when, and in what
contexts. In other words, how does this curriculum unit orient children toward
graphic information? We examine the video data for when and how the students
interact with print, by writing and by reading.

In our study, we coded as a "writing" event whenever any child at the focus
table had a pen in hand, in motion, while in contact with paper. Early efforts are
underway to develop a taxonomic scheme that captures generalizations about
the different kinds of writing that students engage in during the implementation
of the unit. The data will thus be sub-coded for, inter alia, note-taking, recording
data, writing involved in the performance of mathematical operations (averag-

ing, graphing, etc.), and answering worksheet questions. We also are coding for social forms of writing, involving note-passing, and doodling, for example, which appear to play an important role in the ever-shifting and always-important participation structure at the tables, which affects the task structures assigned and assumed by students as they perform laboratory procedures.

By not restricting our attention to only those acts of writing that end up being evaluated by the teacher, we hope to trace the evolution and development of students' reasoning about written answers. In the above example, the students first write down the number 116.4, but as the mixture continues to lose weight, and the scale reads 116.3, there is discussion about whether or not to write down "point four" or "point three." In fact, Philip, Natalie and Sean were unsure about what to write on their worksheets, as the mixture *ought to have weighed* 116.4. Thus in our analysis, we note which students wrote down 116.4, at what point, which students cross that number out or erase it, and then which students write 116.3—in all cases attending simultaneously to the discussion going on around the inscription of this knowledge—what number, in this case, is going to count as the evidence about which to reason. In this class, generally, and in this lab, in particular, Natalie seems concerned about "what to write down." Figuring out an answer to this question means, in this case, reconciling the authority of the instrument in front of her that reads 116.3, with the teacher's directive to ignore the "point one difference" in the weight loss as it is "insignificant."

This part of our coding project thus allows us to describe a sort of natural history of an inscription. In *Laboratory Life*, Latour & Woolgar (1986) watch scientists working and writing in their laboratories, tracing the evolution of the many inscriptions bandied about in a working laboratory to find out which end up in a published research article and which do not. Their ethnographic research indicates that the fate of an inscription depends on many things, including discussions among the researchers in the lab, and shared assumptions about the methodology in use—what is merely an outlier and need not be reported, for example, and what is a "pattern" or "trend" in the data that must be included in a draft of a paper and explained. By attending closely to students' literacy practices, we can see that there are many parallels to Latour & Woolgar's description of a working laboratory, as students here are apprenticed into the cultural assumptions about what "counts" as evidence to be explained, and what can safely be discounted (e.g., a "point one difference"). These are critical distinctions for students whose "job" it is to write things down for evaluation. In other words, what number should be inscribed onto the all-important, *graded* worksheet? Tracing students' literacy practices, and connecting them to a cultural understanding of scientific practice as we do, thus offers us a unique view of students' apprenticeship to the nature of science, their reasoning, and their development of conceptual understanding.

Indeed, as we conceive of learning in terms of apprenticeship and increasingly central participation, coding for student writing seems especially important as the episodes of writing leave definite impressions on the resulting participation structure for the lab tables. As already noted above, of the four students at the focus table, Natalie seems most concerned with "what to write down," and thus directs the attentional resources of the table to this task at various junctures in the labs. The decision of students to read out loud from the text is also critical to the development of participation frameworks important to understanding how a unit functions from the perspective of the students. We note which pieces of the text are read out loud, and by whom, as well as the conversational sequence of those moves. How do these various literacy practices, in other words, affect the participation of the students? Indeed the goal is to see if one can determine different participatory profiles for each of the students at the focus table as measured in terms of their literacy practices.

Coding for literacy events in this way allows us to begin to answer questions about the nature of the curriculum unit that would seem important to understanding how it functions for diverse learners. For example, to what extent are the curriculum materials something to which information must be added (through acts of writing), or is it a source of information, from which students read to each other? Preliminary analyses support a conclusion that, in *CTA*, the worksheets function more as sources of information for solving problems. The text functioned as an aid to participation in group problem solving, rather than summoning them to turn away from the group toward silent reflection. Students were responsible to the text, of course, in writing answers to questions about the labs, but these questions often appeared in the text, *in situ*, rather than in a separate section; that is, as the students performed the lab procedures, they were answering questions about them simultaneously. This structure seemed to facilitate students' working and reasoning together as a group.

In Figure 11.6, the graph depicts the frequencies of acts of writing during Group Exploration, the same class period segment covered in the object manipulation and scientific term use figures above. It is clear that there is an inverse relationship between writing and object manipulation. Natalie writes the most but handles objects the least and Philip handles objects the most but writes the least. In Figure 11.7, it is clear that Natalie and Gloria are not only interested in using

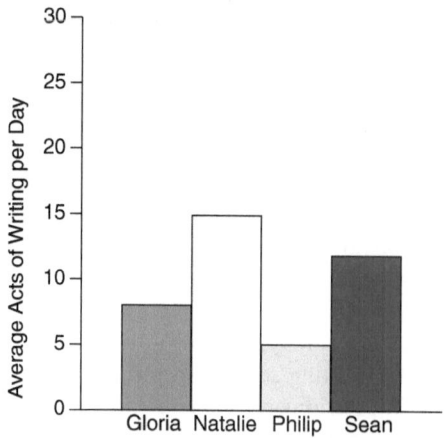

Figure 11.6 Frequency of acts of writing during group exploration.

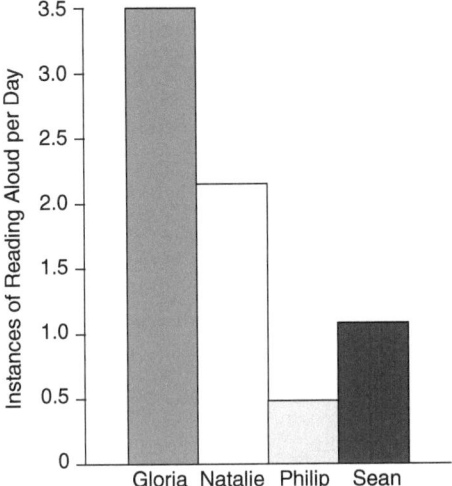

Figure 11.7 Frequency of reading out loud from the curriculum materials.

print to *record* information, but often use it as a source of information. The graph shows that Natalie and Gloria read aloud from the text far more than Philip or Sean. Thus, even if they did not participate in the lab experiment by creating haptic representations of their experiences, they did so by creating graphic representations.

Discussion

Lee and Fradd (1998) point out that the effect of curricular interventions on diverse populations is very poorly understood: "little information is available on [science curriculum] effectiveness in terms of what works and why." This is particularly true when it comes to the analysis of diverse populations: very few effectiveness studies disaggregate the data in terms of ethnicity, gender, and race. While we have presented outcome scores as one picture of what works, what was missing from that interpretation was an analysis of how the curriculum unit was experienced by the children themselves.

"Diversity" is not a word the children themselves often use; rather, it is a word used in institutions—government, education, business—in the context of management questions. How do we cope with it? What shall we do about it? In order to connect the concerns with diversity at an institutional level to the concerns of actual students, it is important to link concepts of diversity to ideas and issues that matter to the target population. One way to do this, we have suggested, is to focus on the varieties in the students' personal experience as they struggle to make sense of their science experience. Viewed from an institutional perspective, the lab table consisting of Sean, Philip, Natalie and Gloria would likely be considered "diverse" in demographic terms. In a broad sense, the diversity of their table resembles the diversity found at other tables in the classroom, and indeed in the school system as a whole. But while we are not claiming that our findings are *statistically representative* (e.g., we do not claim that Philip can stand for all African-Americans), we wish to observe that even if they were, it would not tell the whole story. What would be missing is how to relate the demographic diversity to the subjective experiences of the students in the classroom.

Rather than comparing the students in terms that would sound strange and formal to them (e.g., demographic categories), we have sought a common framework that is closer to their experience. Since students in science classes typically share a common exposure to a curriculum unit, variations among their experiences can be the basis of comparison. Rather than relying on interviews or self-reports to access their experiences, we examine how they use language and other representations to interpret and "build" the curriculum as a meaningful "text." To achieve and experience "science literacy" students must become competent "text-builders" of scientific meanings.

Drawing on Becker's model of "text-building," we examined the ways in which curriculum units provide (1) "content," (2) "goals," (3) "narrative," and (4) "sense-making practices." The variations in text-building practices, in turn, provide an *interpretive framework* (i.e., not a causal or predictive explanation) for the patterns of diversity in the outcome scores. In analyzing how *CTA* builds the content of the Law of the Conservation of Matter, we are primarily interested in the text-building processes that the ideas provoke in actual classrooms; that is, we are interested in the ways in which the curriculum unit requires students to construct and make manifest the content of the curriculum as an experienceable idea. In this case, the curriculum unit employs semiotic media to convey the ideas to students.

We show that *CTA* uses *iconic* means of communication to convey the benchmark ideas. That is, the main content idea of conservation of matter is conveyed by vivid visual transformations (collapsing balloons, rusting steel wool, foaming test tubes) that *resemble* the idea to be conveyed (e.g., the change in the color of the steel wool in a closed flask simulates the idea of the change of matter in a closed environment).

In reading the worksheets and interpreting the directions, the students construct images of what the goals of the curriculum unit are. In *CTA*, when told to act like scientists, students readily donned goggles, and used beakers, flasks, and other items that seemed consistent with that role. Students interpreted the goal as one of participation in problem solving; the distribution and allocation of roles for that participation process became an important concern for them.

The overall narrative structure of the unit and of the individual lessons was organized in ways that reflect the curriculum unit's goals. *CTA* provided time for reflection, closure and debate about their conclusions. The students' own sense-making practices are articulated here as well. While students in *CTA* were given relatively few nominalizations in the textbooks, they invented and used many of their own to describe their experiences: a hallmark of "talking science." Indeed, students in *CTA* increased their use of scientific nominals over time. In addition, students in *CTA* often read aloud from the textbook as a way of solving problems. In *CTA*, students also had hands-on experiences, but the distribution of those experiences was more uneven.

Conclusions

Much has been made, in recent literature, about the value, for learning, of small group activity. Teachers are encouraged to have differentiated ability groupings and to provide all students in their classrooms with challenging and motivating opportunities to learn. In this chapter, we have shown that shifting the focus on classroom learners away from a broad brush characterization of "the black kids," "girls," "Latinos," etc. and refocusing with an emphasis on the transactions that take place between students' personal experiences and their classroom science learning can yield fascinating results. We have attempted to provide a rich characterization of the ways that children use talk, touch, and writing to derive and build meaning in science. Our aim was to both expand the ways teachers and classroom researchers think about what children bring to science learning, and to show how curricular materials can be used to create wonderful meaning-making contexts for students.

There are many implications to be drawn from the examples of Philip, Sean, and Natalie and their interaction with the reform science curricular materials. First, Philip had a strong desire to engage with the curricular materials. We suggest that the design of the materials, a design that allowed students early and regular access to the benchmark lessons of the unit, encouraged this motivation to manipulate the materials and to make meaning in science. Second, Natalie and Sean found roles in their interactions with the curricular materials, and with Philip, that allowed them to carry out the work of science, and interpret and debate the meaning of science outcomes. Third, a close examination of how students make meaning, using science materials and using their multifaceted interactions with each other, can provide teachers with a view into what and how and through what mechanisms (e.g., writing, discussing and debating, and touching/manipulating) students come to know science. Viewing students through this type of lens can potentially enhance instruction, including small group interactions around science learning, and provide teachers with a broad array of means for naturalistic assessment of students' knowledge in the science classroom setting.

Here we aimed to cast a bright light on the organization of student experience during a science activity. We have proposed specific ethnographic and linguistic dimensions for interpreting students' own multimodal sense-making practices (Kress & Van Leeuwen, 2001). It is our hope that a principal outcome of this work will be to provoke increased examinations of, and discussions about, how students can and do experience curricular activities and how their various interactions with and within those activities serve to shape learning.

The approach outlined here creates an interpretive framework within which to understand the outcome data. At a time when educational research is increasingly focused on quantitative results (i.e., "outcome data"), it is especially important to observe that even when the outcome data are disaggregated in a way that reflects cultural and class differences, they still do not speak for themselves, and must be contextualized. Although educators know that the full story is not being told by the numbers, they are often at a loss to provide systematic evidence that provides a fuller picture.

The context proposed here for interpreting "what works" focuses on the nature of student experience of curriculum as text—both oral and written. This multimodal approach to both "talking science" and "writing science" shows how the curriculum functions in actual classrooms. The evaluative framework proposed for examining curriculum implementation in its communicative context permits us to take account of the different content, goals, coherence strategies, and forms of representation by which curriculum units are enacted in diverse classrooms. At the same time, it also allows for the isolation of specific factors that may be related to outcome scores. The "success" of *CTA* in raising outcome scores for all students may be linked to the ways in which it permitted students immediate access to the benchmark concept through multiple channels of representation—visual, verbal, written, and haptic.

Targeting Enduring Understandings

1   What are the connections you see between the discussion in the Kuipers, Viechnicki, Massoud & Wright chapter and the four elements of culturally-responsive instruction, as described by Villegas and Lucas?
2   How would you describe the "work" that language is doing in this multi-cultural context?

Deepening the Reflection

1   This chapter showed that students are diverse in their patterns of participation. What do you think is the relationship between diversity as *participation* and the more traditional definition of diversity as *demographics* (e.g., race, gender, ethnicity, etc.)? What are some of the kinds of diversity that you have noticed in students' ways of participating with science curriculum materials? How may those ways of participating interact or not with demographic diversity?
2   Consider in what ways a student's experience with curriculum materials is different from a teacher's experience with the same materials. What might account for the similarities or differences in experience? How would this affect teaching and learning?
3   What strategies or tools could you as a future science teacher use to better understand how your students are interacting with the curriculum materials?

Encouraging Engagement

1   Identify a curriculum unit that you think has many of the same characteristics as *CTA*. To what extent do you think the unit would appeal to diverse teachers and students (both in terms of participation and demographic diversity)? To what extent does the unit provide multiple routes of access to the benchmark idea?
2   Observe science instruction in a classroom with culturally- and

linguistically-diverse learners on a day in which students will be engaged in hands-on inquiry, like that described in this chapter. Using one or more of the analytic tools introduced in the chapter, account for students' participation in their science learning and their interaction with each other in the inquiry activity.

References

American Association for the Advancement of Science (AAAS). (1989). *Benchmarks for science literacy.* New York: Oxford University Press.

Becker, A. L. (1979). Text-building, epistemology, and aesthetics in Javanese shadow theatre. The imagination of reality: Essays in Southeast Asian coherence systems. In A. L. Becker & A. A. Yengoyan (Eds.), *Imagination and reality* (pp. 211–243). Norwood, NJ: Ablex.

Blakeslee, T., Bronstein, L., Chapin, M., Hesbitt, D., Peek, Y., Thiele, E., & Vellanti, J. (1993). *Chemistry that applies.* Lansing, MI: Michigan Department of Education.

Conklin, H. C. (1962). Lexicographical treatment of folk taxonomies. In F. Householder & S. Saporta (Eds.), *Problems in lexicography: Report of the Conference on Lexicography held at Indiana University November 11–12, 1960* (pp. 119–141). Bloomington, IN: Indiana University Research Center in Anthropology, Folklore, and Linguistics.

Erickson, F., & Schultz, J. (1992). Students' experience of the curriculum. In P. Jackson (Ed.), *Handbook of research on curriculum* (pp. 465–485). New York: Macmillan.

Gumperz, J. J. (1982). *Discourse Strategies.* New York: Cambridge University Press.

Halliday, M. A. K., & Martin, J. R. (eds). (1993). *Writing science: Literacy and discursive power.* Pittsburgh, PA: University of Pittsburgh Press.

Heath, S. B. (1983). *Ways with words: Language, life, and work in communities and classrooms.* Cambridge, UK: Cambridge University Press.

Holliday, W. G. (2003). Comment: Methodological concerns about AAAS's Project 2061 study of science textbooks. *Journal of Research in Science Teaching, 40*(5), 529–534.

Jones, M. G., Minogue, J., Tretter, T. R., Atsuko, N., & Taylor, R. (2005). Haptic augmentation in science instruction: Does touch matter? *Science Education, 90*(1), 111–123.

Katz, D. (1989). *The world of touch.* Hillsdale, NJ: Erlbaum.

Kesidou, S., & Roseman, J. E. (2002). How well do middle school science programs measure up? Findings from Project 2061's curriculum review. *Journal of Research in Science Teaching, 39*(6), 522–549.

Kress, G. R., & Van Leeuwen, T. (2001). *Multimodal discourse: the modes and media of contemporary communication.* London and New York: Oxford University Press.

Latour, B., & Woolgar, S. (1986). *Laboratory life: The construction of scientific facts.* Princeton, NJ: Princeton University Press.

Lave, J., & Wenger, E. (1991). *Situated learning: Legitimate peripheral participation.* Cambridge, UK: Cambridge University Press.

Lee, O., & Fradd, S. H. (1998). Science for all, including students from non-English-language backgrounds. *Educational Researcher, 27*(4), 12–21.

Lynch, S., Pyke, C., & Jansen, J. (2003). Deepening understanding of science and mathematics education reform principles: Novice teachers design web-based units using project 2061's curriculum analysis. *Journal of Science Teacher Education, 14,* 193–216.

Moyer, P. S., & Jones, G. M. (2004). Controlling choice: Teachers, students, and manipulatives in mathematics classrooms. *School Science and Mathematics, 104,* 16–31.

National Research Council (NRC). (1999a). *Global perspectives for local action: Using*

TIMSS to improve U.S. mathematics and science education. Washington, DC: National Academy Press.

National Research Council (NRC). (1999b). How people learn: Bridging research and practice. Washington, DC: National Academy Press.

Rosebery, A. S., Warren, B., & Conant, F. (1992). Appropriating scientific discourse: Findings from language minority classrooms. Journal of Learning Science, 2, 61–94.

Silverstein, M. (1979). Language structure and linguistic ideology. In R. Cline, W. Hanks, & C. Hofbauer (Eds.), The elements: A parasession on linguistic units and levels (pp. 193–247). Chicago, IL: Chicago Linguistic Society.

Street, B. V. (1995). Social literacies: Critical approaches to literacy in development, ethnography, and education. New York: Longman.

Viechnicki, G. B. (In review). "The evidence from your experiment is a weight loss": Grammatical processes of objectification in a middle school science classroom. Linguistic and Education.

12 The Importance of Objects in Talking Science

The Special Case of English Language Learners

Doris Ash, Kip Tellez, and Rhiannon Crain

Introduction

When you think about learning science, what kind of environment comes to mind? Do you think of a classroom, a teacher, a whiteboard, and some laboratory equipment? These are undoubtedly central elements of the school science setting. But are there other places where students learn science? If so, how do the science understandings gained in one setting transfer to another? These are the questions Ash and her colleagues would like us to reflect on as we read their chapter on informal science learning in an aquarium and marine biology discovery center. Many of us have been to such centers. They are often on our list of "things to see" when we visit new cities. On rainy days in our home communities, we may consider them a dependable place to fight off the doldrums. We frequently have fond feelings for these centers because they remind us, when we were younger, of the excitement of the very occasional special field trip. But have we thought seriously about the important role such centers play in science learning? Here the authors ask that we consider how essential such centers are because of the way they provide access to a multitude of objects that no single science classroom could ever hold and because of the way, even further, they make science learning a family matter. Both of these unique characteristics of discovery centers, these authors argue, make them useful tools in enhancing science learning for a growing segment of the U.S. school population—English Language Learners (ELLs).

As you read this chapter, think about how the interactions it documents between members of Latino families visiting an aquarium and marine discovery center challenges the often deficit-oriented notions that schools attach to this population. As you read about the Rios father holding out a piece of kelp for his daughter to touch, about Eva surprising her mother with all she knew about whales, about the Honda father and son becoming engrossed with the Leopard Shark, think about how these vignettes provide us alternative ways of knowing about science learning among ELL Latino students that take us beyond their frequent "limited English proficient," or "under-performing" classifications. Pay attention to how the objects available in the discovery center facilitated focused interaction among family members, advanced instructional conversation related to such interaction, and promoted acquisition of both the content of science and proficiency in English. After reading Ash's and her colleagues' account, you'll have an expanded understanding of what the science learning environment could be and, indeed, was for these ELL students outside the classroom walls. You'll also be intrigued by the questions this chapter raises about the ways these students take what they learn inside those walls to make sense of their science-learning experiences outside of them, and the work that language does in connecting that learning.

Success in science is a powerful gatekeeper to future academic success, yet Latinos, and other non-white populations, are disproportionately underrepresented in the most prestigious science professions (NSF, 2005). Like other chapters in this text, we are concerned with issues of equity and equal access to talking and thinking about science. Unlike the other chapters, however, we explore issues of access to the "talk" and "objects" of science for culturally and linguistically diverse learners, both in and out of classroom settings, and, in our case, aquaria and marine science centers. The term objects, in this chapter, refers to the "things" of science, such as live and stuffed animals, plants, pictures, models, text and other similar tools for teaching and learning.

Almost one quarter of the U.S.' forty-five million school-aged children live in households in which languages other than English are spoken (National Center for Education Statistics, 1997). In California, alone, Latinos and Asians will comprise half of the state's population by 2020 (Hajnal & Baldassare, 2001). Barton & Osborne (2001) have suggested, "an education for marginalized children involves rethinking foundational assumptions" (p. iv). In this chapter we argue that one such fundamental assumption is access to the "objects of science."

We know from classroom-based research that objects (often called *realia* in studies of classroom teaching and learning) play an important role in providing opportunities for English learners to practice the language and ideas of science (Heaton, 1979). Access to the "objects of science," like other teaching materials, has been uneven in classrooms; such uneven distribution often reflects culture, class, gender, or other differences. Science museums and other informal places of learning, conversely, are object-rich, but these institutions have only recently begun to explore how objects (typically called *artifacts*) enhance science understanding for those new to English. Native English speakers, rather than those new to English, often frequent informal places of learning museums in America. In both classrooms and informal places of learning, therefore, it appears that culturally- and linguistically-diverse learners are less likely to have contextualized language experiences with science objects.

We are interested in how learners come to talk about the objects (and ideas) of science (in any language) in science classrooms and aquaria, and how such ideas, and the words used to express them, progress over time, and migrate from aquarium to classroom and vice versa. We rely on theoretical and practical foundations taken from both classroom and informal setting research, theories of second language learning, theories reflecting the social nature of scientific knowledge, highlighting particularly the role of science discourse.

We draw particularly on sociocultural theories of learning and teaching (Vygotsky, 1987) and the foundations of second language learning and teaching. We are interested in the complex interrelationship between the objects of science (such as animals and plants in a seawater tank in an aquarium), and how students and their families come to appropriate the language and ideas of science. The phrase "talking science" has come to reflect the view that "language is a culturally transmitted resource for making meaning socially" (Lemke, 1990, p. 4); such activity can be practiced in either classroom or aquarium. How learners come to "talk science" in either English or Spanish is currently the subject of considerable debate (Yerrick & Roth, 2005).

A related goal is to use such new understandings with teachers who routinely use science objects when working with culturally and linguistically diverse students. It has been suggested that in intercultural education effective teachers are able "to reflect on the relation between their teaching and the wider social, economic, cultural and political context in which they function" (Angelides, 2006, p. 3). Many researchers (Villegas & Lucas, 2002; Nieto, 1999) have explored ways to promote reflective teaching within "difference"; reflective practices based on these ideas are central to the suggested activities following the chapter.

Research Method

We illustrate our points with data from a longitudinal study conducted on family dialogue in an aquarium and a marine biology discovery center in northern California. Participants included both Spanish- and English-speaking Head Start[1] families, but the majority was Spanish-speaking. Data were collected over three years with twenty family groups. Visits lasted approximately two hours per family. Family members were invited to explore the aquarium at their leisure, but were asked to visit a specific "focus area" where intensive data collection took place. To capture dialogue, family members wore remote microphones and were videotaped throughout the visit. Visits ranged from twenty to eighty minutes. Data collection was naturalistic (Ash, 2004; Moschkovich & Brenner, 2000); a Spanish-speaking biologist served as a bilingual mediator. Conversations were subsequently translated and transcribed from the video data by a professional bilingual transcriber. Ethnographic notes augmented the data.

A team of researchers, including biologists and education graduate students, participated in several layers of iterative data analysis with the transcribed and translated data. We then coded the dialogue by systematically looking for biological topics and themes such as habitat, reproduction, or classification. We prepared narrative accounts of each family's "significant events" (Ash, 2002) during the visit, and members of the research team conducted moment-by-moment analyses of object-use throughout the visits.

An Exemplar and the Tensions It Reveals

As we are interested in how learners make use of objects, we start with a vignette from an aquarium visit by a Spanish-speaking immigrant family in central California. The Rios family episode highlights how one family, new to science and to English, coped with novel content, materials and setting. We have excerpted below a section from a bilingual specialist's ethnographic notes on the family's twenty-five-minute visit.

The Rios family stood at the seawater "touch tank" at the aquarium. The father gently placed a piece of kelp in his five-year-old daughter's hand. They both had a look of wonder on their faces. The docent described the particular type of kelp, how it got its name and why it was important to the ocean ecosystem. The mother, holding her baby and three-year-old daughter, watched the interaction intently from only a few feet away.

Such interactions occur often during family visits to this exhibit. This particular event, though, was somewhat unique; the parents spoke very little English, the docent spoke only English, the signs were in English, and the family had never before visited this type of learning environment. The Rios parents did not know many of the Spanish or English names for the living objects; they did, however, use the resources at hand to have a memorable experience.

This vignette underscores essential tensions concerning how we view teaching and learning science in places like classrooms or aquariums. Such tensions include: the role of the teacher, in this case a museum educator; the value of authentic objects, in this case kelp; the language choice in scaffolding learners' scientific sense-making, in this case English; and the family's collective ability to use prior knowledge and other resources, in this case gestures, pictures and Spanish, to understand the science in the exhibit. In the research reported here, we start with science talk in informal science learning spaces, within which family members are free to use home-based, school-based languages to make sense of science objects.

We know that family participation in this particular research project stemmed from parents' concern for their children's academic advancement, taking part so that "they [their children] can learn science." Most parents in this research study did not speak English; their children, however, spoke Spanish and English to differing degrees. The parents viewed the aquarium as a place where family members could participate in dialogs about the objects they saw. They recognized, as we are suggesting, that engaging objects offer speakers of any language, of any science background, a chance to participate by discussing what they *see*. More specifically, the Rios family, just like immigrant parents across the country, wanted their children to be academically literate in the dominant language (Espinosa, 1997; Bayley & Schecter, 2003).

In the past, public schools have been seen as "the place to gain academic competence in the dominant societal language" (Bayley & Schecter, 2003, p. 188). Public schools, thus, have carried the main burden for helping immigrants learn English (Crawford, 1991). The home has been viewed as an important place to maintain native language practices. One major question raised by research with families such as the Rioses is, what is the role of informal learning contexts, such as an aquarium, in fostering the language and ideas of science?

A growing body of research now focuses on learning in non-school, educational organizations (Leinhardt & Crowley, 2002; Paris, 2002), such as aquariums, which often reflect both home- and school-based ways of talking. Such non-school, educational settings are ideal test beds for new ways to think about language and science in tandem. Such hybrid (Bhabha, 1990) locations also allow both everyday and academic language, as well as both Spanish and English to emerge. Ideally learners, surrounded by compelling and authentic objects of science, might use prior experiences, words, ways of thinking and social practices to access the ideas and words of science. Like other real world settings, such as stores and the home, museums and aquaria can provide opportunities for interacting in multiple languages.

Episodes such as the Rios' also highlight the "inequitable distribution of linguistic resources" (Bayley & Schecter, 2003, p. 192) and the related unequal access to authentic objects of science for English learners, by highlighting the "newness" of such an event for this family, where it is more common for a middle class, European-American family. The Rios family visit also underscores the challenge for both classrooms and aquaria to develop strategies for helping those new to English. Just as with classroom teaching, the difficulty for informal places of learning is in making decisions regarding sign, human, video, or other mediation, using languages other than English, hiring bilingual museum educators, and deciding when, where, and how such scaffolds are appropriate.

We have learned much from classroom-based research centered on underrepresented populations and scientific sense making (Lee, 2005; Moll, Amanti, Neff, & Gonzalez, 1992). Museums are just now beginning to benefit from research focusing on the learning resources of culturally and linguistically diverse populations (Ash, 2004; Crowley & Callahan, 1998; Rogoff, 1998). The research described in this chapter takes advantage of available expertise from both classroom and non-classroom contexts. We argue that educators must criss-cross educational learning institutions if they are to effectively manage difference and to insure that all students enjoy the potential for equal opportunity in talking about science objects and ideas.

This chapter draws upon the socioculturally-based theories of object learning and the foundations of second language learning and teaching to reappraise the view that objects are crucial for language and science learning for teachers wishing to criss-cross between classroom and informal settings. We focus on three areas in the following sections: lessons about English learning gained in classroom research and practice, the role of objects in talking science, and how, together, these factors impact issues of equity for those new to English.

We hope to engage classroom teachers and museum educators in partnering with each other and with families like the Rioses to enhance both English language and science content understanding. We believe this goal can be reached when researchers and teachers alike have a better understanding of the intertwined role of language and objects in learning science. Last, we share several teaching activities that we believe will help classroom teachers and museum educators use objects for science and English language learning.

English Learners in the Classroom

Talking about novel objects, such as kelp, takes effort in any language; it is more difficult still for those who are new to science or to English. Recent research has demonstrated that immigrant families learning English can achieve a complex and layered bilingualism that allows both for school success and a strong cultural identity (Pease-Alvarez, 2002). Children raised in bilingual homes, speaking Spanish at home and English in the classroom, do equally well in language and mathematics (CLiT, 2005) especially when they have active engagement with texts, "deciphering signs, … translating official materials, … and elaborating in talk about concepts encountered in written text" (Bayley & Schecter, 2003, p. 192). It seems clear, therefore, that immigrant families wishing to acquire English and learn science need not give up their native language.

While we cannot provide a detailed account of language learning, we do provide a brief overview of past views on language acquisition. For many years, social scientists believed that language was learned largely through the environment. Babies listened to their caregivers, imitated parts of what they heard, and were reinforced (i.e., rewarded with verbal praise) when they used words in the appropriate context (Skinner, 1957). Chomsky (1959) argued that, because children make use of language so quickly and with such ease, the environment could not possibly be the only source of knowledge. Chomsky suggested that humans have the structure for language already in place. Chomsky asserted that humans need to acquire language in the presence of a social community of speakers, thus comparing language acquisition to walking, maintaining that both are nearly unstoppable biological processes of development which are not "taught" in the traditional sense.

Another line of research centered on how we learn our first words focused on connecting words to objects. The linguist and developmental psychologist Eve Clark (1973, 1993) hypothesized that early word learning was based on the physical features of the object itself. She discovered that the more dramatic and circumscribed (or "foregrounded") the object, the more likely it will be for a child to acquire a reliable term for it. Her studies determined that a child's first words almost always referred to (a) animated objects, (b) objects unique in the local environment, (c) objects that fit into child's field of vision, and (d) those with features that interacted with the child's sensory systems (i.e., they could touch it, smell it, hear it).

Clark (1993) asked parents to keep journals tracking their child's first word; "doggy," "kitty," or the name of a family pet was commonly reported. Sometimes the first word was the name of a valued stuffed animal; "mommy" or "daddy" were often acquired after the pets' names have been mastered. Educators and researchers who worked with second language learners began to wonder whether the propensities for early word learning in the first language might also apply to second language learning. Clark's research suggests that, by combining naturalistic experiences, complex social milieu and compelling science objects, we might glimpse second language as it is used to make sense of science.

We use an example from the Jeronda family at the same aquarium the Rios family visited, using the same data collection and analysis methods, to develop this latter point in more detail. This next vignette highlights a ten-year-old native Spanish speaker, Eva (along with her eight- and five-year-old brothers and mother and father) who saw a life-size fiberglass blue whale hanging from the ceiling of an aquarium. The sheer size captured her interest. Eva excitedly asked in Spanish, "*¿Es esa ballena, verdad?*".

At this point in the visit, a teacher/guide/video/sign might provide the scientific and common name for blue whale in English. In this case, however, we know that Eva knew the words for whale in both English and Spanish; she also knew the English name blue whale. She facilely switched from English to Spanish, using *ballena* when talking with her family, and whale when talking with English speakers. The other family members spoke with the bilingual mediator primarily in Spanish. Eva explained, to both the English-speaking researcher and her parents, that she had studied whales in school. Eva then turned to a whale video located below the whale model, both watching it and simultaneously translating phrases for her mother. She explained, in Spanish, how whales move through water. The mother expressed interest and wonder at the extent of her daughter's knowledge.

The mother heard the word whale spoken several times in relation to *ballena*; she heard about their swimming, communication, and other abilities. More importantly, she experienced her daughter's excitement and interest in whales. Eva's father and two brothers witnessed these events as active listeners. Eva's family "learned" the term "whale" in English, but because the interchange of information between Eva and her mother was public and social, they also learned about whale behavior.

While linking word to object seems obvious in first language development, it may be less apparent in second language acquisition. Second language learners such as the Jerondas faced the challenge of reconciling their developing word sense and word meaning in English with the equivalents in the native language, as well as with their personal meanings. We see in Eva's naming episode a preliminary linking of name to object, as well as a beginning context for her family for understanding whale behaviors.

We see in the Jeronda vignette that social interaction fosters beginning competencies beyond identification. The inclusion of a variety of dialogic and physical interactions, such as talking to others, pointing, touching, looking at videos, and looking at models, can enlarge the context within which deeper understandings can begin. In this case the combination of a video, bilingual mediator, several languages, and a well-placed conversation with others, opened new areas of understanding concerning the word whale. Clearly the mother learned as much or even more than Eva through this interaction. Museums are built to do just this, to encourage dialogic, visual, and gestural forms of activity, in order to help visitors to make meaning beyond naming. Such events remind us that language learning is a social phenomenon, which relies on multiple participants, objects, and their interactions.

We also saw that objects can provide language learners with multi-sensory impressions of the language which, Rivers (1983) notes is "learned partly at least through seeing, hearing, touching, and manipulating" (p. 47) items. Second language educators and researchers have long argued that such "real" objects help make new language comprehensible, building "an associative bridge between the classroom and the world" (Heaton, 1979, p. 79). Such associative bridges are created within activity, as family members with unequal competencies (in English, Spanish, or science) use objects as the focus of science talk.

We have seen from the above example that science talk about objects is not dependent on one particular language. We see in the next example (from a Marine Science Center in central California) how the Honda family father and son expanded on a naming event. The English translations are in bold text.

FATHER: Leopard shark (in English reading from sign). *Oh, ¿Sabes cómo se llama? Andy!* [**Leopard shark. Oh, do you know what it's called? Andy!**]

ANDY: *¿Eh?*

FATHER: *¿Sabes cómo se llama?* [**Do you know what it's called?**]

ANDY: *Tiburón.* [**Shark.**]

FATHER: *No, pero aquí está su nombre, mira.* [**No, but here's its name, look.**]

ANDY: *¿Dónde?* [**Where?**]

FATHER: *Este.* "Leopard shark." (reading the sign again in English) *Leopardo. ¿Sí ves?* (pointing) *¿Los colores?* [**This one. "Leopard shark." Leopard. Do you see? The colors?**]

ANDY: *Sí.*

FATHER: *Por eso se llama leopardo.* [**That's why it's called leopard.**]

After looking into the tank and the nearby sign, the father matched the striped colors with the name Leopard, saying the word leopard excitedly in both English and Spanish. Matching English and Spanish cognates is an often-used strategy in second language classrooms. This particular episode included words and gestures, the living shark, the sign, and two people moving toward a common goal of identifying the shark. This seemingly simple episode, like the two preceding vignettes, illustrate how museum environments, with their attractive and multi-layered artifacts, provide the rich environments for beginning speakers to talk safely about new things and to enter into the ideas of science, and perhaps to acquire new words and their meaning in either English or Spanish. The Honda father, later in the same visit, asked what the leopard shark eats, how it grows, and how dangerous it is to humans.

FATHER: *¿Y crecen grandes esos, los tiburones?* [**And do these, the sharks, do they grow big?**]

(Several turns)

FATHER: (To daughter) *Es un tiburón.* (To bilingual museum educator) *¿Y qué les dan de comer? ¿Pescado?* [**It's a shark. And what do they feed them? Fish?**]

(Several turns)

FATHER: *¿Pero todos comen diferente comida, no?* [**But they all eat different foods, right?**]

The father also told the five-year-old daughter the name of the animal, and queried the bilingual museum educator further. We know that this family has remained interested in sharks because, six months later, the father again asked about sharks:

FATHER: (Pointing to the same shark) *Bueno, ¿Estos sí ven?* [**Well, do those see?**]
(...thirteen turns)
FATHER: *Porque he oído por ejemplo que ... por, por la sangre, cosas así.* [**Because I have heard, for example, that ... because, because of the blood, things like that.**]
(...one turn)
FATHER: *¿Por—es el olor, o sea el sabor?* [**Because—is it the smell, in other words the taste?**]
(...one turn)
BILINGUAL MUSEUM EDUCATOR: *El color, y el sabor porque el olfato y el sentido de, de, de sabor están muy relacionados.* [**The smell, and the taste because smell and the sense of, of, of taste are very related.**]
(...three turns)
BILINGUAL MUSEUM EDUCATOR: *—ven como nosotros también cuando vamos por la calle, estamos oyendo para que no pase un carro—* [**—they see like we do also when we're going along the street, we're hearing so that no car comes—**]
FATHER: Oh.
BILINGUAL MUSEUM EDUCATOR: *—o, pero también estamos viendo así en frente, entonces, usamos todos los—* [**—or, but we're also looking forward like this, so, we use all the—**]
FATHER: *Sentidos.* [**Senses.**]

Such naming and classifying activities in Spanish allow people to relate to unfamiliar organisms. During their activity at the rocky reef tank, the Honda father and son constructed meaning using words and ideas more complex than the original naming activity might have indicated (Ash, forthcoming). The other three family members watched, listened and sometimes interrupted. Such basic, though not simplistic, ways of thinking and talking provided this family, as well as the Rioses and the Jerondas, with the raw material for establishing a common language. Just as with the blue whale example, after the name Leopard Shark was collectively established, the ideas behind the words gained additional meaning.

This family (and others in our research) remembered such names and ideas over time (Ash, 2004). Families taking part in such research generally talked about personally relevant topics at such marine science exhibits, as they made linguistic connections with each other and with the science objects. Family members named organisms, characterized and then placed them within their family's larger social understanding of, and relationship with, the natural world

(Ash, 2004, forthcoming). Such episodes confirm Lemke's (2001) view that "What matters to learning and doing science is primarily the socially learned cultural traditions of what kinds of discourses and representations are useful and how to use them" (p. 68).

Relating the results of this research to classroom-based studies, we find that science learning research in second language classrooms has shown that students need to be engaged in authentic science-learning activities, in which students ask and answer questions that have relevance in their daily lives, and which allow for discourse in each learner's main language to be linked to formal science concepts in the language of learning and teaching (Warren, Ballenger, Ogonowski, Rosebery, & Hudicourt-Barnes, 2001). The implications of such research provide clues for exhibit and research design. Such linkages suggest that exhibits, books, signs, and teachers must scaffold new material using an object or concept they may already recognize, such as a shark or a whale.

Objects

Objects, such as kelp, shark, and whale models, played a crucial role in learning science-related words and ideas for the Jeronda and Honda families. Such an object-based epistemology (Conn, 1998) has been frequently discussed in museums. Second language classroom teachers have routinely used objects in teaching English, arguing that students achieve greater communicative competence when objects are used in their science teaching (Enright & McCloskey, 1988; Heaton, 1979). Enright & McCloskey (1988) have argued that: "If language learning truly involves meaningful communication and reflection, often about concrete and mutual referents, then it is crucial that you have all kinds of different enticing objects for display and use in your classroom, and ones that appeal to all the senses" (p. 111). We note that objects, when used explicitly in language-learning activities concerning words with multiple meanings, facilitate a better performance on word-defining tests than do pictorial and verbal lessons alone, for both English learners and English speakers in classrooms (Rule & Barrera, 2003a). Classroom instructional practices linking words and objects have included graphic organizers, juxtaposed text and images, multi- and hypermedia, film, and objects (Tang, 1992). When paired with oral descriptions in the target language, learners can mimic how we learn our first language. This argument has had great impact on second language learning classroom practices; such pedagogy has been regularly adopted in classrooms. Such research, however, seems confined to pictures and images, a consequence of the constraints of working in classrooms rather than in object-rich museums.

Much less research has focused on second language acquisition with objects in informal learning settings, where collecting, managing, and interpreting real, man-made, living or non-living objects have dominated museum activity for the past century. Such an emphasis on objects in learning is not surprising. Scholars from a variety of theoretical traditions have long claimed that interaction with objects is important to learning (Dewey, 1991), acquiring new languages (Thorne, 2002), developing concepts of self (Csikszentmihalyi & Rochberg-

Halton, 1981), building vocabulary (Rule & Barrera, 2003b), and making connections with the real phenomena of science (Oppenheimer, 1990).

Many theoretical frameworks touch on objects, language, and learning. Belief in the power of real objects reflects Piagetian views of active participation with the natural world in constructing knowledge. Such reliance on direct interaction with phenomena is at the heart of inquiry-based teaching. Objects (typically called artifacts or tools) are also central to socioculturally-based accounts of learning (Cole, 1996). Such socioculturally-based views have noted that thinking and speech are intimately intertwined and that language itself is a powerful mediating agent in learning (Vygotsky, 1987). Such a stance focuses on the relationships between person, language, object, and intended outcome, and also views objects as bound by the meanings they carry in socially, historically, and culturally situated contexts.

Both the view that objects (realia/artifacts) benefit language learners and take their meaning from human activity are rooted in a theoretical stance that recognizes that objects and their meanings are always products of human activity. In the sections below we discuss several characteristics of language and artifacts that reflect social and cultural aspects of such activity. The first aspect is the aesthetic and personal connections between people, their language, culture and the object/artifact. Paris & Mercer (2002) hypothesized that "visitors search for features of their personal lives, both actual and imagined selves, during their exploration of artifacts and museums, and their searches may lead to confirming, disconfirming, or elaborating understanding of their own identities" (p. 402). Creating such identities with science objects touches directly on issues of equity and access.

We know from such research that there are strong connections between artifacts and people's age, family status, and gender. We do not, however, deeply understand these connections between personal experience and object. It has been noted, that perhaps, the object itself is less important than story about the object. Such views suggest continuing research into the many meanings with which objects are imbued (Gurian, 1999). We know, for example, that our research families viewed sharks in many different ways: an animal that can be fished from the pier; an animal that is not exactly fish, but probably is related to them; an imposing dangerous animal, which is the subject of movies and the popular press; and, part of the ocean food chain. At stake in any learning context is, who gets to write the story of objects? Do we rely on exhibit developers, teachers, learners, texts, or a combination of sources? A pluralistic view of social interaction as learning reflects "a more complex notion of learning-in-community, often among unequal participants, with a significant role assigned to power relationships and differences of age, class, gender and sexuality, language and cultural background" (Lemke, 2001, p. 74).

Equity

We often think of museums as institutions dedicated to serving all income and ethnic groups, and as places where obvious power relationships might be mitigated. Recent research has called this view into question. Milligan & Bray-field (2004) have found that, in spite of targeted efforts by a university-sponsored art gallery to recruit low-income schools for cost-free visits (even transportation was paid), the wealthier schools were the only consistent patrons. Milligan & Brayfield do not speculate why the lower-income schools chose not to participate, but current social, political, and cultural conditions suggest several reasons. Other research has indicated more sanguine findings. A recent Center for Informal Learning and Schools (CILS) survey indicated that 40% of Informal Science Institutions (ISIs) serve schools with low-income and minority students. We have, however, little data on the outcomes of such programs.

While field trips for underrepresented students seem to be increasing, attendance by culturally- and linguistically-diverse families seems to be decreasing. A recent report by the Ralph and Goldy Lewis Center for Regional Policy Studies indicates that Latino attendance decreased (for art and cultural museums) between 1984 and 2005, while white attendance increased. Differential attendance patterns seem to be growing. Informal learning institutions want to serve non-English speaking and culturally diverse visitors, but they remain uncertain of issues of language, culture, and access. In the following paragraphs, relying partly on past classroom-based research, we explore how museums might begin to accomplish such aims.

Berwald (1987) has noted that objects "are not only a series of artifacts that describe the customs and traditions of a culture, but they are also a set of teaching aids that facilitate the simulation of experience" (p, 147). Objects are unique to life in a specific culture and are key tools for assisting immigrant students in understanding a new culture. A fast food hamburger, for instance, has no peer as a cultural object when immigrant learners try to help make sense of life in the U.S. Second language theorists (Clark & Hernandez, 1999; Short, 1991) have suggested that objects help second language learners to understand a new culture. Roney (1994) noted the work of a Spanish language teacher who used an ornate dollhouse in the classroom to teach family names, relations, and adjectives such as outside and inside; other second language educators have had similar results using objects (Hess & Sklarew, 1994; Zukowski-Faust, 1997) in the target culture.

Such teaching by "cultural metaphor" requires that the museum (and schools) understand the culture of the learner and know how its artifacts can be connected to that culture in ways that enhance understanding. Imagine a museum displays a totem pole (more properly, a Gyáa'aang). Most visitors are not Native Americans of the northwest coast cultures, and therefore are not familiar with this cultural artifact. If we wish to explain this object to our U.S.-born, white, middle-class visitor, we might tell them that the totem served the purposes of an elegy and a coat of arms, two cultural features that may be familiar to them. We recognize that this cultural link fails to explain fully, and may somewhat distort the true purpose of totems, but such a link makes use of native culture to understand a new artifact. With this initial understanding, the educator may be able to nuance further understanding among learners.

If we can agree that cultural connections are valid means to assist museum or classroom learners in learning about artifacts, we can shift our attention to U.S. ELLs, whose culture may be different from those of their museum and classroom teachers. Demographic studies have shown that the vast majority of ELLs are native Spanish speakers, whose cultural heritage is based in rural Mexico (Kindler, 2002). Although we are not suggesting that museums and classrooms should link their artifacts or pedagogy only to immigrant families whose heritage is based in rural Mexico, educators need to make their best effort to foster the learning of native Spanish speaking, Mexican-descended families. Our fundamental point here is that educators may correctly argue that their programs cannot possibly be oriented toward all language learners from all cultural groups, and yet when we look more closely at who most needs English language assistance in both in and out of school settings, it is the population we have just described.

We know that learners comprehend a new language better when the discussion relates to their own culture (Au, 2002); the challenge for educators is finding an optimal balance between endorsing ELLs' native cultural understandings and helping them to learn about new artifacts and a new language. In a search for this balance, the educator can never be certain of the proper mix, but those who continually consider and reappraise the role of native culture in artifact learning are likely to be most effective (Tellez, 1999).

Discussion

The purpose of this chapter has been to explore commonalities concerning the role of objects in dialogue in learning and teaching across classroom and informal learning settings. We have noted in past research how objects in marine science centers enhance dialogue and encourage people to link words to objects, thus, creating opportunities to "talk science" (Ash, 2004, forthcoming). We have suggested that the objects of science do not merely enhance learning; instead, they are essential mediational means for scientific sense making in any language. Participating in science talk is essential to student access to academic discourses (Valdes, Bunch, Snow, & Lee, 2005) especially for those new to science. The capacity to "talk science" relies in turn on learners' access to the objects of science.

Objects are not all alike; some are common in a language classroom or at home, and some are not. While the role of objects across learning settings may seem elusive at first, understanding their role through empirical research is a key way to ensure that instruction in any setting will be effective. Classrooms cannot contain the many artifacts language represents. Museums and aquaria are full of such objects. Practitioners and researchers in both contexts have claimed that object-rich settings can best serve the underrepresented, yet, for both museum and school practice, such claims are undertheorized.

Such insights underscore the critical role of objects in museums and classrooms for learning content in any language. Second language learners may gain most when science objects play a prominent role in their instruction. Ironically, due to institutionalized inequities within school systems, English learners often must rely on outside-of-school educational institutions (e.g., museums, field trips, parks) for access to the science objects that middle-class, European-American families take for granted. Lack of familiarity with a new culture, access to reliable transportation, and a host of other socioeconomic factors can prevent immigrant families from engaging with the objects of science in their community. We have also seen that such informal settings do not have a long history of serving such populations. English learners in the U.S., therefore, are less likely to have contextualized experiences with such science words, objects, and ideas either in school or out.

The life-size fiberglass blue whale, kelp, and a shark represent objects that can provide both language and science content opportunities. These are not something a teacher can purchase for personal use, or even something a school district is likely to own, but they may be at a local aquarium. The question we pose is how to maximize English learners' access to multiple learning environments, each with their own sets of science objects, while designing linked and coordinated learning goals and teaching strategies. Such interactions can maximize language learning by offering greater potentials for understanding. Science talk stems from spaces in which a sense of relationship can be theorized from the connections between personal and shared use of words. Museums and schools each have their strengths and weaknesses; yet, together, they represent the various contexts in which language learners need access if they are to interact with a wide-range of objects, and thus, participate in a wider range of related activities.

Learners lacking opportunities to engage with objects in the development of their science knowledge are selectively disadvantaged. We have argued that language, whether first or second, can be acquired through complex activity with objects, both alive and inanimate, and that object-laden interactions provide contextualized learning experiences, during which learners come to understand the meaning of a new word, phrase, idea, or explanation. We saw that such discourse often starts as a naming activity. We know, however, that such a naming activity is often a prelude to greater participation in science activity, because names of objects include subtextual information about participants, tools, purposes, and contexts of activity.

One important research goal, then, must be to investigate the role of "talking science" across learning settings, from classroom to aquarium and back. As yet, there is but a small body of literature contributing to our understanding of the generative capacity of objects across learning situations. Researchers have only just begun to explore how English learners appropriate scientific ideas and language through contextualized events involving objects. We can only speculate, at this point, how such appropriation may create the basis for academic science in either English or Spanish. Reconciling what is currently understood about object use with the pedagogy we already use in classrooms or aquarium is an area ripe for future research.

Implications for Teaching

Because science objects offer learners entry points to the complex ideas and talk of science, assessment of such non-traditional activity allows teachers (either in museum or classroom) to "see" the content in the talk. Viewed from this perspective then, objects are not just "nice" additions to science learning environments; they are central and essential features. The task, for teachers working with learners who are new to English, is to intentionally orchestrate such opportunities, observe students as they participate over time (see the activities at the conclusion of the chapter), assess current levels of understanding, and guide toward deeper understanding. This cycle of activity could occur at a local bird feeder, which fosters student talk about (and teacher observation of) scientific content, such as birds' structures and their functions, species identification, or typical behaviors. Imagine students then explain to their families (in either English or Spanish) how species differ. Imagine further such content being guided toward understanding biological adaptation, a fundamental principle of biology. Such content is consistent with local, state and national science content standards.

In the aquarium examples, students used both Spanish and English to make sense of science, often clarifying their thoughts and explaining ideas to their parents in Spanish. We saw that alternating languages (what linguists call code-switching) while talking science, didn't get in the way of understanding. Instead such code switching allowed learners (both adults and children) to access their own "funds of knowledge," while speaking about what they knew. Such insights into students' culture are important. Paying attention to existing language and ideas directly impacts how students come to "talk and write science." Teachers who choose to access their students' cultures and language in this way may find themselves and their students fundamentally transformed.

Targeting Enduring Understandings

1   What are the connections you see between the discussion in the Ash team's chapter and the four elements of culturally responsive instruction, as described by Villegas and Lucas?
2   How would you describe the "work" that language is doing in this multicultural science context?

Deepening the Reflection

1   Reflect on the science classrooms you have been in as a learner or a teacher. Would you describe them as object-rich or object-poor? What do you think the number and kinds of objects in these classrooms communicates to students about science?
2   What is the relationship between the emphasis that the Ash team places on objects in this chapter and current understandings of best practice in science education? What theories could you use to take a stand, with the Ash team, that object-rich environments are essential for science learning?

3   How often as part of your formal science education did you visit aquariums, marine discovery centers or other community science-based institutions? How did you understand the relationship between those visits and your formal science learning? How does your answer inform your understanding of science schooling?

Encouraging Engagement

1   Observing Science Talk. Teacher research tells us that looking closely at learners' activities transforms the way we teach. In this activity the goal is to observe closely one or more learners as they interact with living objects in an informal learning setting such as an aquarium, backyard, park puddle, or playground.

   a   Pick a reasonable site where you can watch unobtrusively, without talking, while taking notes.

   b   Schedule a fifteen-minute segment. You should pay attention to the learner(s) and to the nature of the relationship.

      1   What kinds of things do they say?

      2   Did they speak in English and/or another language? Write down exact words when possible.

      3   What gestures do they make?

      4   Naming: Do they name objects? Scientific or everyday? What kind of naming? (Butterfly vs. Insect vs. Monarch Butterfly vs. Bug)?

      5   Properties: Do they talk about its properties (e.g., Look, it's eating, it's moving, it looks like it's having babies)?

      6   Do they seek clarification from an external source? A sign, a book, a docent, each other?

   c   Organize and review your notes.

   d   Share what you found with colleagues who are working toward similar understandings of such interactions. What ideas emerged from your first observation? Are there similar language issues? Are names important? Does more than one person mention the same action, words or gesture?

   e   Return to make another fifteen-minute observation. It does not matter if the learners are the same, but it should be the same or a similar object. Focus on an idea that emerged from your first observation. For instance, you might focus your second observation on how/if learners seek expert scientific information from external sources and how they use it. In this case you would be interested in how scientific terms are appropriated and how they propagate?

   f   After the second observation, you should have a better idea about your ongoing questions. Was the (living) object rich enough to keep their attention? Were several learners focused on the same object? Did an expert interrupt a dialogue and "explain" the object? Was there science and how was it expressed? Did students bring everyday names and ideas into the science talk?

   g   This experience should give you some insight into how living things can foment ideas and talk, and about some ways to use them in the classroom. For instance, after engaging in close study with living objects, we find that teachers often find ways to engage students in journal writing or picture-drawing about these living objects. Such activities also inform learning station design in classrooms.

2 Creating a learning station: An opportunity for further research on objects and language in the classroom. The aim of this activity is to put in practice the ideas and insights gained from the first activity. The emphasis is reflection on the interplay between learners, language and objects within the classroom.

    a   Collect appropriate live objects for your sample learning station. Snails are quite useful for illustrating behaviors; they are slow, very common, and can be returned to the wild. Animals that "behave" in interesting ways and easy to collect and keep are preferable (snails, earthworms, fish, crickets, slugs).

    b   Create a simple learning station: cages with covers, several animals, sample food, drawing paper, etc.

    c   Rely on student questions and comments from Activity 1 to inform the design of activities/questioning in your station. Prepare two to three activities/questions related to the animal, e.g., Do they move fast?; Do they move toward the light?; What do they eat?

    d   Identify a local science classroom in which to place your learning station.

    e   Observe and record students' conversation.

    f   Take detailed notes.

    g   Listen to any taped science talk; Is it reminiscent of Activity 1? What is different?

    h   Does students' activity suggest further inquiry?

    i   If you were to use that station in your own classroom, how would you change it to reflect your new insights?

This work has been supported in part by NSF REC Grant # 0133662 to Doris Ash and NSF ESI 0119787 Grant to CILS.

We wish to thank Joanna Sherman Gardiner, Lewis Ingham, and Stephanie Van for their editorial, technical, and organizational expertise.

Note

1 Head Start families were invited to the aquarium and marine science center, and aided with transportation and food. Admission was waived by the museum for each visit. Each family included one child between the ages of four and six, one child between the ages of eight and eleven, and at least one parent. Families were recruited from a local Head Start program.

References

Angelides, P. (2006). *Building collaborative networks for improving student teachers practice.* Paper presented at the 19th International Congress for School Effectiveness and Improvement, Fort Lauderdale, FL.

Ash, D. (2002). Negotiation of biological thematic conversations in informal learning settings. In G. Leinhardt, K. Crowley, & K. Knutson (Eds.), *Learning conversations in museums* (pp. 357–400). Mahwah, NJ: Lawrence Erlbaum Associates.

Ash, D. (2004). Reflective scientific sense-making dialogue in two languages: The science in the dialogue and the dialogue in the science. *Science Education, 88,* 855–884.

Ash, D. (forthcoming). Making sense of living things: The need for both essentialism and activity theory. *Cognition and Instruction.*

Au, K. H. (2002). Multicultural factors and the effective instruction of students of diverse backgrounds. In A. Farstrup & S. J. Samuels (Eds.), *What research says about reading instruction* (pp. 392–413). Newark, DE: International Reading Association.

Barton, A. C., & Osborne, M. D. (2001). *Teaching science in diverse settings: Marginalized discourses and classroom practice.* New York: Peter Lang.

Bayley, R., & Schecter, S. (2003). *Language socialization in bilingual and multilingual societies.* London: Clevedon.

Bhabha, H. (1990). Interview with Homi Bhabha: The third space. In J. Rutherford (Ed.), *Identity: Community, culture, difference.* London: Routledge.

Berwald, J. (1987). *Teaching foreign languages with Realia and other authentic materials. Q & As.* Washington, DC: ERIC Clearinghouse on Languages 1–6.

Chomsky, N. (1959). A review of B. F. Skinner's *Verbal Behavior. Language, 35,* 26–58.

Clark, E. (1973). What's in a word? In T. Moore (Ed.), *Cognitive development and the acquisition of language.* New York: Academic Press.

Clark, E. (1993). *The lexicon in acquisition.* Cambridge, UK: Cambridge University Press.

Clark, G. B., & Hernandez, C. (1999). The bus pass is cheaper: Learning and acculturation. *Radical Teacher, 56,* 7–12.

CliT. (2005). *Positively plurilingual.* London: National Centre for Languages.

Cole, M. (1996). *Cultural psychology: A once and future discipline.* Cambridge, MA: The Bellknap Press of Harvard University Press.

Conn, S. (1998). *Museums and American intellectual life 1876–1926.* Chicago, IL: University of Chicago Press.

Crawford, J. (1991). *Bilingual education: History, politics, theory, and practice* (2nd ed.). Los Angeles, CA: Bilingual Educational Services.

Crowley, K., & Callanan, M. (1998). Describing and supporting collaborative scientific thinking in parent–child interactions. *Journal of Museum Education* (Special issue on Understanding the museum experience: Theory and practice, Scott Paris, Ed.), *23,* 12–17.

Csikszentmihalyi, M., & Rochberg-Halton. (1981). *The meaning of things.* London: Cambridge University Press.

Dewey, J. (1991). Common sense and scientific inquiry. In A. Boydston (Ed.), *The later works, 1925–1935, John Dewey, volume 12: 1938, Logic: The theory of inquiry* (pp. 66–85). Carbondale, IL: Southern Illinois University Press.

Enright, D. S., & McCloskey, M. L. (1988). *Integrating English: Developing English language and literacy in the multilingual classroom.* Reading, MA: Addison-Wesley.

Espinosa, K. E. (1997). Determinants of English proficiency among Mexican migrants to the United States. *International Migration Review, 31,* 28–50.

Gurian, E. (1999). What is the object of this exercise? A meandering exploration of the many meanings of objects in museums. *Daedalus, 128,* 163–183.

Hajnal, Z., & Baldassare, M. (2001). *Common ground: Racial and ethnic attitudes in California.* San Francisco: Public Policy Institute of California.

Heaton, J. (1979). An audiovisual method for ESL. In M. Celce-Murcia & L. McIntosh (Eds.), *Teaching English as a second or foreign language.* Rowley, MA: Newbury House.

Hess, M., & Sklarew, S. (1994, Fall). Realia and American culture. *WATESOL Journal*, 10–12.

Kindler, A. (2002). *Survey of the states' limited English proficient students and available educational programs and services: 2000–2001 summary report*. Washington, DC: National Clearinghouse for English Language Acquisition.

Lee, O. (2005). Science education and English language learners: Synthesis and research agenda. *Review of Educational Research, 75*, 491–530.

Leinhardt, G., & Crowley, K. (2002). Objects of learning, objects of talk: Changing minds in museums. In S. Paris (Ed.), *Perspectives on object-centered learning in museums*. Mahwah, NJ: Lawrence Erlbaum Associates.

Lemke, J. L. (1990). *Talking science: Language, learning and values*. Norwood, NJ: Ablex.

Lemke, J. L. (2001). Articulating communities: Sociocultural perspectives on science education. *Journal of Research in Science Teaching, 38*, 296–316.

Milligan, M. J., & Brayfield, A. (2004). Museums and childhood: Negotiating organizational lessons. *Childhood, 11*, 275–301.

Moll, L. C., Amanti, C., Neff, D., & Gonzales, N. (1992). Funds of knowledge for teaching: Using a qualitative approach to connect homes and classrooms. *Theory into Practice, 31*, 132–141.

Moschkovich, J., & Brenner, M. E. (2000). Integrating a naturalistic paradigm into research on mathematics and science cognition and learning. In A. E. Kelly & R. A. Lesh (Eds.), *Handbook of research design in mathematics and science education*. Mahwah, NJ: Lawrence Erlbaum Associates.

National Center for Education Statistics. (1997). *Confronting the odds: Students at risk and the pipeline to higher education*. L. Horn & C. D. Carroll (Eds.). Washington, DC: U.S. Department of Education, Office of Educational Research and Improvement.

National Science Foundation (NSF). (2005). *US doctorates in the 20th century: Special report*. L. Thurgood, M.Golladay, & S. Hill (Eds.). Washington, DC: NSF, Division of Science Resource Statistics.

Nieto, S. (1999). *The light in their eyes*. New York: Teachers College Press.

Oppenheimer, F. (1990). A rationale for a science museum. In H. Hein (Ed.), *The exploratorium: The museum as laboratory* (pp. 217–221). Washington, DC: Smithsonian Institution Press.

Paris, S., & Mercer, M. (2002). Finding self in objects: Identity exploration in museums. In S. Paris (Ed.), *Perspectives on object-centered learning in museums*. Mahwah, NJ: Lawrence Erlbaum Associates.

Paris, S. (Ed.). (2002). *Perspectives on object-centered learning in museums*. Mahwah, NJ: Lawerence Erlbaum Associates.

Pease-Alvarez, L. (2002). Moving beyond linear trajectories of language shift and bilingual language socialization. *Hispanic Journal of Behavioral Sciences, 24*, 114–137.

Rivers, W. (1983). *Speaking in many tongues* (3rd ed.). New York: Cambridge University Press.

Rogoff, B. (1998). Cognition as a collaborative process. In D. Kuhn & R. S. Siegler (Eds.), *Handbook of child psychology: Cognition, perception and language* (pp. 679–744). New York: Wiley.

Roney, M. W. (1994). Moving beyond the tricks of the trade, or using common, everyday objects as realia. *Hispania, 77*, 298–300.

Rule, A. C., & Barrera, M. (2003a, Summer). Improvement of third graders' science vocabulary through use of descriptive adjective object boxes. *Montessori Life*.

Rule, A. C., & Barrera, M. T., III. (2003b). Using objects to teach vocabulary words with multiple meanings. *Montessori Life* 15(3), 14–17.

Short, D. (1991). *Integrating language and content instruction: Strategies and techniques.* NCBE Program Information Guide Series, 7 (Fall). Retrieved June 7, 1997, from http://www.ncbe.gwu.edu/ncbepubs/pigs/pig7.html.

Skinner, B. F. (1957). *Verbal behavior.* Englewood Cliffs, NJ: Prentice-Hall.

Tang, G. (1992). The effect of graphic representation of knowledge structures on ESL reading comprehension. *Studies in Second Language Acquisition, 14,* 177–195.

Téllez, K. (1999). Mexican-American preservice teachers and the intransigency of the elementary school curriculum. *Teaching and Teacher Education, 15,* 555–570.

Thorne, S. (2002). *Activity theory and communicative practice: Participatory genres, strategies and tactics, and inter-discursivity.* Paper presented at the 8th Annual Meeting of the Sociocultural and Second Language Learning Research Working Group. Retrieved December 12, 2005, from http://www.oise.utoronto.ca/MLC/SCT-L2L/longabstracts.htm.

Valdes, G., Bunch, G., Snow, C., & Lee, C. (2005). Enhancing the development of students' language(s). In L. Darling-Hammond, J. D. Bransford, P. LePage, & K. Hammerness (Eds.), *Preparing teachers for a changing world: What teachers should learn and be able to do.* San Francisco, CA: Jossey-Bass.

Villegas, A. M., & Lucas, T. (2002). Preparing culturally responsive teachers. *Journal of Teacher Education, 53*(1), 20–32.

Vygotsky, L. S. ([1934] 1987). Thinking and speech. In R. W. Rieber & A. S. Carton (Eds.), *The collected works of L. S. Vygotsky, Volume 1: Problems of general psychology.* New York: Plenum.

Warren, B., Ballenger, C., Ogonowski, M., Rosebery, A., & Hudicourt-Barnes, J. (2001). Rethinking diversity in learning science: The logic of everyday sense-making. *Journal of Research in Science Teaching, 38,* 529–552.

Yerrick, R. K., & Roth, W. M. (2005). *Establishing scientific classroom discourse communities.* Mahwah, NJ: Lawrence Erlbaum Associates.

Zukowski-Faust, J. (1997). What is meant by realia? *AZ-TESOL Newsletter, 18,* 9.

Part IV

Assessing the State of Science Education

Towards Promising Practices for Responsive Instruction

13 Linguistics and Science Learning for Diverse Populations

An Agenda for Teacher Education

Lisa Patel Stevens, Julian Jefferies, María Estela Brisk, and Stacy Kaczmarek

Introduction

As the title of this chapter implies, creating culturally-responsive approaches to science instruction means changing business as usual in teacher education. In providing the curriculum through which prospective science teachers gain their professional knowledge and skills, teacher education programs, if they are committed to preparing culturally-responsive science teachers, must alter their curriculum to make such responsiveness a realized, not idealized, outcome. This means infusing into science teacher education itself information about and experiences with culturally- and linguistically-diverse populations. This infusion must take account of the role of language in science learning.

As you go into this chapter, think about how much information you received in your teacher education program about the work that language does in science learning. Did you get any such information? If so, who provided it to you? Was it in your science methods courses or did you pick it up elsewhere, like your literacy or multicultural education courses? How does your own language background contribute to this understanding? What difference do you think it makes whether you get such information as part of your science methods courses or not? If you were to design a science teacher education program committed to culturally-responsive teaching, what do you think, based on these reflections, would be the best way to provide that information to students?

Reading Stevens and her colleagues' chapter will help you understand the need for science education professionals to be asking themselves these very questions. The chapter will get you thinking about language in science teaching and learning, and, more generally, about the different perspectives that exist, when it comes to science, on what "cultural responsiveness" means to begin with.

You will see that this chapter is calling for a paradigm shift in the way we think about our specialized "expert" identities. No longer can we afford to carve up our collective professional expertise into rigidly-bounded areas of "science" or "language" or "culture." Instead, we need to take on interdisciplinary identities, develop interdisciplinary knowledge, and work, when our institutional identities try to box us back in, in cross-disciplinary ways. Science, *and* language, *and* culture intersect in the everyday classroom lives of students. To best serve these students, we must recognize this intersection and mirror it in the professional preparation of their teachers.

Recent immigration patterns into the U.S. have raised logistical questions across many social fields. These patterns have impacted educational settings in predictable and unforeseen ways. For example, recent immigration trends have resulted in bifurcated success trends among children of immigrants, with split streams of high-achievement and low-achievement (Suarez-Orozco & Suarez-Orozco, 2001). Adding to the complexity of the linguistic and cultural diversity found within today's schools are powerful policy initiatives, such as the abolishment of bilingual education in several states, accompanied by suites of pedagogical and curricular implications. Embedded within these shifting and multifaceted arenas, linguistically- and culturally-diverse populations are navigating their way through science pedagogy and curriculum. Teachers face the challenges of making their curricular objectives accessible and achievable to students in increasingly high stakes assessment contexts. In these situations, language can and does act as both a potential mediator for access and exclusion.

In science teaching, the relationship between the knowledge frames and the linguistic structures and patterns used to linguistically represent these frames are often left implicit, fluently leveraged by science teachers but often experienced as impermeable and confusing structures by students from nonmainstream linguistic backgrounds (Lee, 2004). In this chapter, we take up the work of other scholars who have argued that the linguistic structures used in science learning require particular attention for efficacious content learning to occur (e.g., Lemke, 1990; Schleppergrell, 1998). To inform this type of efficacious content learning, we first need to know what pertinent research knowledge exists and what current requirements there are that address this area of second language pedagogy. Working through a review of both professional certification requirements in the U.S. and a review of the literature relevant to teaching science to linguistically- and culturally-diverse students, we propose an agenda for science teacher education to address a current gap in pedagogical content knowledge through the use of educational linguistics (Halliday & Martin, 1993).

Theoretical Framework and Modes of Inquiry

Working from a situated view of language and meaning making (Vygotsky, 1986), we regard language as inextricably linked to the contexts in which it is used. In other words, what counts as expertise and fluency depends on the situation, the participants and the task at hand. And, we extend this view to acknowledge the tight connections between language and power (Foucault, 1979). In this sense, we regard language and discuss it within this chapter as a constitutive force, one that creates possibilities and constraints among participants in social, cultural, political, and economic contexts. For students of any language background, the ability to access the dominant register in a situation is a key gate-keeper, and classrooms represent no less of a potentially inclusive or exclusion-ary discourse community (Cazden, 2001). In particular, for children from nonmainstream linguistic backgrounds, the ability to both enact and talk about linguistic and cultural difference acts as a key conduit to positive experiences and academic achievement (e.g., Cummins, 1994; Gutierrez, Asato, Santos, & Gotanda, 2002; Rymes & Anderson, 2004).

These views of language as situated, mutually implicated with power and pos-sibility, frame the inquiries explored in this chapter: what does the research liter-ature say about teaching science to students from diverse cultures and language, and what are the requirements that science teachers currently undergo to address these challenges? To explore these questions, this chapter provides both a research literature review and review of teacher licensure and education require-ments in the U.S., followed by a detailed discussion of the placement of educa-tional linguistics in teacher education.

Review of Research Literature

In conducting a review of the research literature addressing science and students from diverse linguistic and cultural backgrounds, key terms, including "science pedagogy," "science," "ESL," "limited English proficient," "language minority students," "minorities," and "English language learners (ELLs)" yielded a few notable trends. In particular, the review of the literature revealed either macro socio-cultural studies concerned with issues of culture and scientific inquiry, or microanalyses of linguistic structures found in the academic language required of science.

Within the first trend, research has been conducted by research teams such as Science for All (e.g., Lee, 2002) and the Cheche Konnen Center (e.g., Warren, Ballenger, Ogonowski, Rosebery, & Hudicourt-Barnes, 2001) and has looked to include students' cultural and linguistic funds of knowledge (Moll, Amanti, Neff, & Gonzalez, 1992), while examining how these intersect with scientific prin-ciples. Within the second research trend, smaller scale studies have attempted to describe the linguistic features found within science, in order to facilitate the acquisition of language that students need in order to complete science assign-ments. In the following sections, we detail examples and studies found within both trends and then discuss the implications and limitations in informing the existing research on second language learning in science.

Socio-Cultural Studies Measuring Instructional Interventions

The "Science for All" research group, which started in the 1990s in the southeast-ern U.S., and the Cheche Konnen Center, which started in the 1980s in the northeast, have both measured the effectiveness of educational interventions concerning science learning, instruction, curriculum, and teacher education. This literature review will first focus on student learning, analyzing both research groups' views on science, language, and learning; second, the instructional practices encouraged to facilitate the learning of science in the language minority student population will be analyzed.

Views on Science, Language and Learning

Both research groups have the common goal of promoting science learning and scientific inquiry for students from various linguistic and socio-economic backgrounds, and agree that:

a   Teachers need to incorporate cultural and linguistic funds of knowledge that students of diverse backgrounds bring to science,
b   Teachers need to examine how students' everyday knowledge and language intersect with scientific principles, and
c   Students of diverse backgrounds are capable of learning science and engaging in scientific inquiry

(Lee, 2002, p. 56)

Additionally, both groups recognize that science teachers usually have little understanding of the "intellectual resources that these students bring to science" (p. 56). However, these two programs have different views of science, language, and learning.

For the "Science for All" research group, the way students learn science and language is heavily tied to their culture, and is influenced by their community and home practices. The assumption is that ethnically- and linguistically-diverse students bring values and practices that differ from the knowledge, values, and practices of the Western science tradition. Thus, the knowledge that the students bring is not compatible with the knowledge of Western science. The aim of teaching, then, is to realize where this incompatibility lies, and address it in an explicit way, drawing from a sociocultural consciousness of linguistic and cultural differences.

Learning and teaching in this group, then, is viewed from a cross-cultural perspective that draws from the multicultural education literature, where students are able to access the high-status knowledge of mainstream science while at the same time acknowledging and valuing alternative ways of knowing (Lee, 2002, p. 27). The goal is to lead "students to acquire the language of science as well as their home languages, to understand the culture of science as well as their own cultures, and to behave competently across social contexts" (Lee, 2003, p. 467).

More specifically about science and learning, the researchers of "Science for All" highlight the concepts of "scientific inquiry" and instructional congruence. They take the term "scientific inquiry" from science standards documents, and go beyond the definition of "a process of starting with a question, making observations and reaching a conclusion based on evidence," (Lee, 2003). They take it a step further to talk about an investigation into authentic questions generated from students' experiences.

Additionally, they take the notion of cultural congruence, which occurs "when teachers engage in culturally appropriate discourse patterns in communication and interaction with students, consider students' cultural values and beliefs related to science, and use students' cultural artifacts, examples, analogies and community resources" (Lee, 2002, p. 45), and integrate this idea with Ladson-Billings' (1995) notion of culturally relevant pedagogy. Thus, the notion of instructional congruence is introduced. This mediates the nature of academic disciplines with students' linguistic and cultural experiences to make academic content in science meaningful, accessible, and relevant to all students (Lee & Fradd, 1996, 1998, 2001). To highlight the effect of instructional congruence, instructional interventions in this research program pair students and teachers according to their cultural background.

An example of effective instruction for the "Science for All" group is illustrated when a Hispanic teacher addresses the topic of measuring temperature in the class: first, he recognizes that some students are more familiar with Celsius than Fahrenheit, and compares knowing both systems of measuring temperature to being bilingual. Then, he relates this topic to cooking, something that is named as being familiar to Hispanic students. This research group concludes that Hispanic students can relate the scientific concepts to their prior experience, and have something to build on to start learning science.

Based on the cognitive science perspective and sociology of science literature, the Cheche Konnen ("Search for Knowledge" in Haitian Creole) Center began its research focusing on how students learn to talk, think, and act like scientists in collaborative learning communities. Contrary to the "Science for All" project, this group does not see an incompatibility between students' ways of knowing and those of the scientific community. In fact, they do not frame this difference as a "compatibility" and "incompatibility" of students' knowledge, values, and ways of knowing, but frame it instead as this knowledge being "continuous" or "discontinuous" (Warren et al., 2001). In their view, what diverse students bring can be seen as a continuous, and aligns well with the views of the scientific world. Thus, they are interested in the ways in which poor and minority children's ideas and ways of talking and knowing are related to those of the scientific community.

The Cheche Konnen Center also talks about "scientific inquiry," but focuses on "understanding the productive conceptual, meta-representational, linguistic, experiential, and epistemological resources students have for advancing their understanding of scientific ideas" (Warren et al., 2001). In this sense, they are influenced by research in social studies in science and history of science, where the definition of scientific practice is expanded. Thus, they do not see everyday experience as a form of misconception or informal language as inadequate to the task of precise description, but see it as an asset that students can use to approach the scientific phenomena.

Although the Cheche Konnen Center draws from cognitive development literature, it sees students' experimental reasoning in a different way than the cognitive development tradition. Instead of only focusing on tasks that reward logical inference or hypothetical deductive reasoning, they use open-ended tasks where experimentation is approached as an exploratory process of constructing meanings for emerging variables (Warren et al., 2001). In this sense they move away from the characterization of science of the "Science for All" group, which aligns it closely with the science standards documents. Indeed, they have a more broad definition of science. Among the questions asked in the research are:

1 What do children do as they engage in experimental tasks?
2 What resources—linguistic, conceptual, material, and imaginative—do they draw on as they develop and evaluate experimental tasks?
3 How does children's scientific reasoning correspond to the nature of experimentation as practiced by scientists?

As we can see, the Cheche Konnen Center is interested in scientific inquiry and argumentation, but also on language. For them language is seen in two ways. First, the first language of the speaker is considered a resource for their learning and should be taken advantage of. Also, they pay attention to how language is used to learn patterns of school-based practices and patterns, providing familiarity with the discourse of the school. Thus, they look at the discourse of the classroom in order to provide teachers with examples of how they can use the students' language resources to apprentice them into the language of science.

Instructional Approaches

In terms of instructional approaches, research in the "Science for All" program has suggested that students need explicit instruction in order to conduct inquiry on their own. This is especially true for students who come from home environments where questioning beliefs is not encouraged. However, the goal is for the classroom to become gradually more student-centered, and so they have introduced the notion of "teacher-explicit to student-exploratory continuum," where teachers gradually reduce assistance and encourage students to take initiative and assume responsibility for their own learning, as explained by Lee in 2002:

> At the beginning of the year, realizing that most students were unfamiliar with using basic measurement instruments or doing investigations, the teacher gave explicit guidance about how to use equipment, do measurement, represent data in multiple formats, recognize patterns, and draw conclusions.... As students gained experience in basic skills and concepts, the teacher moved to more exploratory types of inquiry. In measuring temperature differences at different levels in the classroom, students recognized a pattern: Temperatures were highest near the ceiling, lower at desk level, and lowest near the floor level.... Applying the theory that hot air rising and cool air falling, students discussed why they would lie on the floor during fire drills and where they would place a heater or an air conditioner in a room.

(p. 48)

Explicit instruction then refers to the practice of not assuming that students know what teachers assume mainstream students already know (in this case, how to make measurements), and to starting the unit at the point which allows them to gain this background knowledge.

Fitting their view of science and learning, the Cheche Konnen Center, on the other hand, encourages teachers to provide an environment where collaborative scientific inquiry emerges as students learn how to think and act as members of a science learning community. Although practicing science in schools is not exactly like scientific practice in the world, "this relationship will help clarify what it means to teach and learn science" (Warren & Rosebery, 1995). There is not a pre-determined course of inquiry; it grows directly out of students' beliefs and questions. This enables teachers to use the cultural capital that is brought by the students and engage them in the work. The role of the teacher is to facilitate and guide students' investigations of these questions.

An example of this kind of instruction is reported by Warren et al. (2001) when a student in a science class that had been observing mealworms move through the stages of metamorphosis, asks the question: "Why, if the people eat and eat, don't they change their skin, don't they transform, the way insects do?" The teacher immediately asked the students to focus on this question and comment. Some students commented that the humans actually shed their skin, another commented that human beings do not "transform," while another said: "God did not create us like insects." The students that first commented on human beings shedding their skin gave an example: "If you play basketball, you get dirty; when you bathe, your skin comes off with the dirt" (p. 535).

These examples give us a window into how students negotiate the meaning of the concept "change" within the context of insects. In this talk, they brought their every day reasoning as a resource to make sense of the issue in science that they were exploring. Although they lacked a conceptual understanding of the difference between "growing" and "developing," their scientific reasoning served as a good starting point for the teacher to explain those concepts and build on what the students had already begun theorizing (Warren et al., 2001).

Research in this group has included the role of everyday knowledge and informal language as an asset that students can bring to the learning process; thus, teachers are encouraged to let students use it. However, they stress the fact that children must not remain at this stage, but have to be apprenticed into using more formal language as it is used in the scientific communities. For this goal, explicit instruction can be a way to help students construct meanings through inquiry and formal scientific language (Ballenger, 1997; Warren & Rosebery, 2001).

Discussion

Within the first trend found in the research literature, both of these research groups have the aim of "incorporating cultural and linguistic funds of knowledge that students of diverse backgrounds bring to science" (Lee, 2002, p. 56), they both do so to varying degrees. For the "Science for All" group this incorporation is realized when teachers are explicit about the kinds of requirements that practicing science in the Western tradition entails. In other words, although the students' existing knowledges is taken into account, the students' ways of knowing, how they have engaged in processes and inquiry to build their world-views, are not. Students are not encouraged to practice science inquiry according to the different tradition that they bring to school, but the knowledge that they bring is used as a stepping stone to build knowledge in the Western tradition. This knowledge, moreover, is assumed to be coming from the cultural back-grounds of the students, leaving little space for individual variation within cultures. In this way, the "Science for All" group uses a definition of culture that assumes a unitary match between students' ethnic backgrounds and their ways of knowing. The kind of orientation runs the significant risk of reifying stereotypi-cal views of students from various backgrounds, instead of using contextual information about specific students to inform instruction.

The Cheche Konnen Center, rather, defines culture as the everyday know-ledge, practices, and beliefs that students bring into the classroom. This research group has a stronger role for the cultural and linguistic funds of knowledge of the students in that it uses their everyday experience and commonsense know-ledge to build proficiency in science. Furthermore, it does not understand science only as it is being practiced in the Western tradition, but has a broader definition for it. By expanding the notion of experimental reasoning as not only concerning logical inferences and deductive reasoning, it embraces the different ways of knowing that diverse students can bring. It realizes this theory in the classroom by allowing students to bring up questions and guide discussions, which in turn reveals a window into their assumptions of the subject and the steps they are taking to access higher knowledge of science.

This difference is reflected in the fact that the "Science for All" group is more closely aligned with science standards document, while the Cheche Konnen Center re-defines the notion of scientific inquiry. The first group's alignment with these standards, therefore, obscures the participation of the funds of know-ledge that students bring to school. The second group is less superficial in its attempt to include students' backgrounds, but this presents a challenge in the way that it re-defines science instruction in a major way, which has strong implications for teacher training. It also poses important questions as to how prepared students will be to perform well on standardized tests where Western funds of knowledge are only those rewarded.

Linguistic Research in Science

Within the second branch of research, the focus has not been as much on the role of cultural differences that may affect the learning of science, but has focused on the language that students need in order to learn science. This branch of research looks at what kind of linguistic demands students face when interact-ing with the academic language of science. Drawing from socio-cultural theories on learning, second-language acquisition theories, as well as systemic functional linguistics[1] and the sociology of science, this research attempts to make explicit the relationship between language and the learning of science concepts.

These studies' findings concur that some students struggle because they do not master the academic register in science. Schools require that students express knowledge in staged, purposeful uses of language and, therefore, teachers need to show explicitly the linguistic features embedded in these uses. This is regarded as especially helpful for non-mainstream students who are not necessarily exposed to these genres outside of school. From a functional linguistics perspective, scholars have detailed the grammatical features of these genres as they appear in scientific texts, school textbooks, science classroom discourse and student writing.

In terms of looking at scientific texts, scholars working early in the systemic functional linguistics tradition sought to describe the evolution of canonical scientific discourse and linguistic patterns by looking at the writings of Newton and Darwin (Halliday & Martin, 1993). These writings attempt to describe how scientists unconsciously used grammatical resources to build a particular form of reasoned argument. They describe how this process came to be represented in the grammar of the English language through a set of grammatical features that are sometimes opposite to those used in everyday language (Halliday & Martin, 1993). More recent work has looked at scientific discourse across a range of social contexts, such as the genres of popular science texts (Fuller in Martin & Veel, 1998) and science fiction texts (Cranny-Francis in Martin & Veel, 1998).

Most of the work of the systemic functional linguists, however, has described the science discourse in educational contexts. One focus has been to look at the linguistic features that build the genres required in science school textbooks. Unsworth (1991, 1997a, 1997b, 1998, 1999a, 1999b, 2001) analyzes the language used in both elementary, middle and high school textbooks, focusing almost exclusively on the genre of scientific explanations. Martin (Halliday & Martin, 1993) does the same with high school science texts, while Veel (in Martin & Veel, 1998) analyzes the language of high school ecology texts.

An example of a specific linguistic feature that is common in scientific academic language is nominalization. Our everyday language and experiences are construed around categories of "process," where clauses usually contain active verbs: "The girl <u>swam</u> very fast across the pool..." In academic language, on the other hand, those verbs are usually turned in to nouns so that processes are turned into things: "The fast swim of the girl across the pool..." This allows for a greater condensation of information in a clause, and is a defining characteristic of academic language.

Another focus has been to look at the classroom discourse of science as it occurs in school science. In a study of science in secondary education, Lemke (1990) argues that the instructional dialogues that construct meaning in science are grounded in grammatical structures and rhetorical forms that are not familiar to students. He provides a description of the major instructional strategies used by science teachers and how they make up themes and principles in science. Most importantly, he criticizes science teachers for not making explicit the relationship between the meaning of science and the linguistic patterns and genres used to represent the meaning. He argues that students usually are exposed to these patterns of language in teacher talk, but this is not enough for students to appropriate it. In order for students to appropriate scientific language, teachers have to explain how language and other meaning-making tools, such as nominalization, function in science so that students understand the instructional dialogues in a better way.

Gibbons (1998, 2003), in a similar manner, examines a successful strategy of students' progressing from oral and informal language to a more formal written language in an elementary school science class. By showing examples of how two teachers mediate between the students' current linguistic levels, the discourse of science, the educational discourse, and specialist understandings of the subject, she gives examples of how teachers can apprentice students into the specialist discourse of the school curriculum.

Some of the focus within scholarship has been to look at student writing within the systemic functional perspective, something that Schleppegrell (1998, 2002) analyzes at the high school and undergraduate level. By focusing on writing tasks such as description and lab reports, she distinguished the features of successful academic writing by analyzing the features of these two writing genres. Huang & Morgan (2003) looked at high school students' drafts of scientific classifications over a period of time, and the linguistic features that improved the expression of these scientific concepts. Within the implications for both studies is that teachers should pay explicit attention to the linguistic features required in these genres.

Other studies have looked at instruction combined with student writing, such as Duran, Dugan & Weffer's (1998) study of high school sophomores. They analyzed the students' writing in the context of an instructional activity to engage them in constructing meanings in science through mediational means such as science language, signs and symbols, and technology. Highlighted as particularly effective is the use of diagrams and visual representations, while giving the students' the opportunity to appropriate the scientific discourse moving from "ventriloquating" teacher talk toward expressing concepts independently.

Discussion

The research that has looked at linguistic analyses of the language of science does not discuss assumptions about what is good science practice, or the role of culture in students' learning. Although some of the research has linked language learning to a particular approach to science (Duran et al., 1998), most studies focus solely on the language of the classroom. Furthermore, there is no discussion of including students' cultural practices, values, and ways of knowing in the science classroom.

This focus on the micro, textual, level has benefits but leaves gaps in implications. As far as benefits, it provides students with potential explicit tools to access academic language in English. This does not happen with the "Science for All" and the Cheche Konnen Center's work, where explicit teaching refers to literacy in the broad sense, and where they talk about the students' language but do not provide specific ways to facilitate the students' acquisition of the scientific academic language.

Research in this vein either assumes that teachers have a working knowledge of linguistics, or proposes that teachers should have it. On the other hand, no systematic study has looked at science teacher's opinions or beliefs about teaching explicit features of language in their subject, or has tried to instruct teachers and measured the effectiveness of such instruction on science classroom practices.

Implications of the Literature Review

One of the first implications from this synthesis of the literature is a lack of studies, from either research branch, that address the role of language in science learning in situated contexts. For the most part, the studies reviewed here do not explicitly address how language can be used in specific contexts, with specific participants, texts, and objectives. In terms of suggestions or implications for the purposeful teaching of the language of science, neither the "Science for All" nor the Cheche Konnen Center address explicit approaches to language in science. From the research published from the "Science for All" group, the emphasis, rather, is on students' background knowledge and accessing their prior knowledge in order to move them toward ways of thinking and speaking in science inquiry. From the Cheche Konnen Center's research, students' language practices were tapped as a source to build content knowledge without explicit teaching of the language found within content texts. Inclusion of explicit instruction of linguistic features that realize the genres required in science could help teachers instruct students and aid their language acquisition at the same time.

Research that has dealt with the relationship of language and learning in science addresses specific linguistic features of discourse and text but leaves silent the contexts in which these texts are being used. Without being explicit about the role of these analyzed texts within specific contexts, including the learning task, the linguistic and cultural background of the learners, the pedagogical objective, the applicability of these types of analyses is left unclear from the studies.

The inclusion of explicit instruction of linguistic features of science genres must be done while engaging in a discussion of the effectiveness and validity of the scientific academic register itself. Within this frame, more research is needed on the linguistic demands experienced by diverse learners as they engage with various genres of writing and speaking, academic, and otherwise. Without this focus on various linguistic resources, a misinterpretation of existing research might be that students should assimilate to mainstream culture and standard English, leaving their own cultures and registers behind. Amid the English-only atmosphere in some states, the role of the students' first languages in learning remains unclear. More research is also needed on how teachers can find ways to use the students' first language to foster development in content while still teaching in English.

In essence, then, both macro foci upon cultural ways of knowing and micro foci upon linguistic analyses of the language of science reveal gaps in the research literature that, if filled, could inform policy, practice, and theory of linguistic diversity and science learning. Alongside each other, these gaps mutually constitute a lack of a coherent and pervasive message and indicate inertia when it comes to systematically informing teacher education.

Another finding from the literature review is that while individual studies have respectively addressed pedagogical features such as the cultural context of the classroom, backgrounds of students, content demands of the lesson, pedagogical structures, and language demands of the lesson, no studies have researched the integration of these factors in students' learning. Without investigations into these dynamics of situated learning, the implications of existing research have limited potential influence on classroom pedagogy. In fact, this finding is particularly salient for practicing teachers, who are best poised to contextualize the various demands experienced by linguistically culturally diverse students learning scientific content.

While existing research on science education can act as one possible source of influence on classroom practices, other influences include educational policy, including licensure requirements. In the next section, we turn to an analysis of states' certification requirements relevant to language in the content areas.

Review of Certification and Programmatic Requirements

In education, daily classroom practices are the result of myriad influences, pressures, innovations that have some bearing on teaching and learning. Many of these influences are idiosyncratic, ebbing from the professional skills, knowledges, and attitudes of particular teachers. For example, a teacher who is an immigrant and bilingual may take it upon herself to help her students navigate the particular semantics and syntactical structures unique to science learning. In addition to personal influences, however, there are also systemic or policy sources that can act as influences upon science pedagogy, including the requirements that preservice teachers must complete along the road to certification. To build an agenda for educational linguistics in teacher education, one must consider not only what existing research demonstrates as areas of need but also what policies are in place that might address this bank of skills, knowledges, and attitudes. Teacher certification requirements offer one such pressure point—a site where current areas of priority will be apparent and also a site that can be adjusted to better meet the needs of linguistically diverse students learning academic English.

State licensure requirements draw upon three fields of language: linguistics, content area literacy, and bilingualism/teaching English as a second language. While the first area marks a clear potential arena for addressing educational linguistics, more states, might be likely to require a content area literacy component and/or an ESL component for science teachers. While it is by no means probable that educational linguistics would be addressed within a content area literacy or ESL course, the likelihood is higher than without this requirement, and these types of course might, again, serve as potential sites for the infusion of educational linguistics into existing pedagogical content knowledge.

In their 1996 survey of reading coursework requirements for middle and high school content area teachers, Romine, McKenna & Robinson found that a total of thirty-seven states, plus the District of Columbia, reported at least one course for middle and/or high school certification in one or more content subjects. This survey, however, did not discern if any requirements existed regarding linguistics. Furthermore, this study did not distinguish between the requirements for science teaching and pedagogical knowledge for other academic disciplines. However, the 1996 survey provides an historical reference point, showing that, while linguistics may not have been a priority or explicit agenda item for teacher education, some conversations had taken place about the particular nature of texts in the content areas, perhaps potentially informed by the work of micro analyses of science texts. The survey informing this chapter's inquiry into the place of educational linguistics in teacher education was designed to specifically address this topic, with respect to the science content area, while also accounting for other course requirements where the concepts might be addressed.

Because states vary in their specificity about both competencies and course requirements, we asked questions via email and phone conversations regarding the state's requirements for teacher competencies and if these competencies were specifically required through prescribed coursework. We compiled results about each kind of requirement as an indication of current priorities for teacher competencies. Also, because certification requirements for teachers differ from state to state, all 50 state departments of education and that of the District of Columbia were surveyed. In reviewing the certification requirements, each state department's website was consulted for an initial review of the requirements of linguistic knowledge. This electronic review was then followed up with phone interviews with certification specialists in each state. The information obtained via the website was first confirmed, and all specialists were then interviewed using a standard protocol of the following questions:

1 What are the exact requirements for teacher licensure; do they include an educational linguistics component or a course in content area literacy?
2 What standards, specific to linguistics or content area literacy, must be met in order for a teacher education program to become a state approved program?
3 Once a teacher has been practicing for several years, must he or she complete any professional development workshops/courses in educational linguistics or content area literacy?
4 Are there any state-wide initiatives that address bilingualism for teachers and/or students?
5 Are there any conversations of changing these requirements for educational linguistics and/or bilingual education for teacher candidates?

This review included both elementary and secondary certification requirements, even though most of the studies analyzed and represented in the literature search stem from secondary contexts. State teacher licensure requirements, through both traditional, four-year programs and guidelines for alternative licensure were reviewed, particularly for requirements that may lead to knowledge about linguistics, including any requirements for content area literacy and/or the linguistic needs of a diverse school population. For the most part, the departments' websites and the certification specialists addressed the competencies that

typically fell into university coursework. In some instances, certification specialists indicated that linguistics were expected to be addressed under content area
literacy courses; in these cases, this requirement was counted under the content
area literacy requirement but not as a specific requirement of educational linguistics.

Results

In general, the results from the review of teacher certification requirements in
the fifty states reveal a mixed view of requirements, recommendations and priorities toward educational linguistics (see Table 13.1). Very few states, six, specifically, address educational linguistics as a source of competency for some of its
teachers. Furthermore, only one state, California, requires all K-12 teachers to
meet a "Developing English Language Skills" requirement by completing a comprehensive course in reading instruction. The course includes: systematic study
of phonemic awareness, phonics, decoding, literature, language, comprehension,
diagnostic and early intervention techniques," (California Department of Education, 2005). However, as detailed as this list of requirements is, none of these
areas is synonymous with educational linguistics, which would encompass the
specific study and use of the linguistics structures found in genres, writing such
science lab reports. Much more prevalent was the requirement that teacher
candidates be able to address reading and writing in the content areas, a macro
approach language, and texts that do not address linguistic patterns (Lemke,
1990).

Educational Linguistic Requirements

Of the fifty states surveyed, six states (California, Louisiana, Mississippi, Missouri, Pennsylvania, and Vermont) had specific requirements for teacher candidates to take a course in linguistics in pursuit of their certification. Of these five
states, Louisiana, Pennsylvania, and Vermont maintained this requirement only
for teacher candidates pursuing a specialization to work with ELLs. Mississippi
required all English Language Arts teachers to complete a course in linguistics,
while Missouri maintained this requirement only for its teachers certified to
teach English at the secondary level.

Content Area Literacy

Thirty-one states, and the District of Columbia, required some or all of its
teacher candidates to either complete specific coursework in content area literacy, or addressed content area literacy as a competency expected of teachers. Of
these states, twelve maintained this requirement for secondary education majors,
fifteen states mandated this knowledge for all of its teacher candidates, and three
only required content area literacy knowledge of its elementary teachers.

The results from this survey show a decrease of six states from the previous
study of content area literacy requirements (Romine, McKenna, & Robinson,

Table 13.1 Literacy requirements of secondary education teacher preparation programs by state

State	Course in Linguistics	Content Area Literacy/Reading	Working with ELLs
Alabama			
Alaska			
Arizona			
Arkansas		x	x
California			
Colorado			
Connecticut		x	
Delaware		x	
Florida			
Georgia			
Hawaii			
Idaho		x	
Illinois		x	
Indiana			
Iowa		x	
Kansas		x	x
Kentucky		x	x
Louisana		x	
Maine			x
Maryland		x	
Massachusetts			
Michigan		x	
Minnesota		x	x
Mississippi		x	x
Missouri		x	
Montana		x	
Nebraska			
Nevada		x	
New Hampshire			
New Jersey		x	
New Mexico		x	x
New York		x	x
North Carolina		x	x
North Dakota			
Ohio		x	
Oklahoma			
Oregon		x	
Pennsylvania			
Rhode Island			
South Carolina		x	
South Dakota			
Tennessee			
Texas			
Utah		x	
Vermont			
Virginia			
Washington			
West Virginia		x	
Wisconsin		x	
Wyoming		x	x

1996). In speaking with the certification specialists in these states, the competencies were shifted due to changes in state demographics and other requirements for teacher candidates. In one case, the content area literacy requirement was eliminated to make room for specific coursework in multicultural education. In three other cases, the certification specialist expressed a belief that all teachers would be expected to learn about literacy pedagogy throughout all of their methods' coursework. While this may or may not be the case in practice within various teacher education programs, our documentation of these requirements provides a partial potential picture of the role of policy in influencing the place of educational linguistics within teacher education.

ELL Component

In addition to specific requirements in linguistics and content area literacy, teacher education programs might also address the linguistic needs of ELLs within courses addressing linguistic diversity or bilingualism. Of the states surveyed, all had certification requirements for teachers wishing to become specialists deemed highly qualified to work with ELLs. However, four states (Arizona, California, Minnesota, and New York) also required knowledge of the needs of linguistically diverse students for all of their teachers. Additionally, certification specialists in seven states (Kansas, Kentucky, Maine, Mississippi, New Mexico, Wisconsin, and Wyoming) cited their states' requirements within multicultural or diversity education as a potential area where teacher candidates might learn about the particular linguistic needs of students learning academic English.

Discussion

At best, the results of this survey of state departments of education show an uneven status of educational linguistics within teacher education, and more specifically, science teacher education. The most direct and explicit item reviewed, the requirement of a course or specific competency in educational linguistics, yielded only a handful of states with this requirement. Further complicating this specific inquiry, though, is what type of linguistics might be incorporated through coursework or professional development addressing the requirement. The prospect of addressing linguistics is daunting enough for most teachers who might not even see themselves as particularly oriented to macro features of text, such as content-specific vocabulary. Adding in the nuances of linguistic structures, semantics, and features found in scientific text and science discourse greatly complicates the prospect of addressing the role of language in science learning.

Beyond the specific requirements for competencies in applied linguistics, mandates for courses or skills in content area literacy is a potential, albeit delimited, site for addressing educational linguistics. Content area literacy or reading has not historically addressed educational linguistics. More common to this area of scholarship, practice, and policy are macro textual reading strategies such as pre-reading activities, approaches to teaching content vocabulary, and comprehension strategies (e.g., Readence, Bean, & Baldwin, 2004). Geneologically connected to the field of reading comprehension, these courses have not been specifically required to address linguistic structures found within different content areas, including science. While specific linguistic analyses and pedagogical explorations of explicit teaching of linguistic structures may be taken up within the context of a content area literacy course, this is up to the discretion of the program and instructor and is not compelled by certification requirements.

A similar de facto limitation is found with requirements for working with linguistically diverse students. Often couched under course requirements or competencies associated with multiculturalism, diversity, and/or inclusion, these arenas again have the potential to address educational linguistics as a specific pedagogical and curricular feature to be considered and incorporated into planning. However, the chances are much more likely that such courses address more sociocultural elements of classroom instruction, attending to such macro features as creating an inclusive classroom community, tapping students' funds of knowledges, and including references to multicultural sources of information and texts. As such, these topics found within content area literacy and ESL courses could provide a possible entry point for considering where educational linguistics might be taken up more purposefully within teacher education.

The dearth of requirements in educational linguistics poses several questions worth exploring. Does this lack of attention communicate an inattention to ELLs and their linguistic needs? Does education consider this to be the domain of a specialist rather than a mainstream classroom teacher? Is the prospect of addressing educational linguistics too daunting for teacher education programs that are pushing at their seams with the existing requirements for methods and survey courses? All of these are likely reasons for the current list of priorities that does not adequately include educational linguistics as a necessity for teachers' pedagogical content knowledge and skills. However, the antidote to the results of this survey is not as simple as the injection of a number of linguistics courses into existing teacher education programs. Considered alongside the growing cultural mismatch between today's linguistically and culturally diverse students and the largely white, monolingual, and female teacher teaching force (Sleeter, 2001), addressing educational linguistics cannot be conceptualized as a simple add-on. Incorporating module-like formats onto existing frames of knowledge found within teacher education would not result in language being considered alongside other factors in the classroom. In the next section, we turn to a detailed discussion of the ways in which educational linguistics can be taken up in teacher education and the other factors that must accompany it.

An Agenda for Teacher Education: Pedagogical Content Knowledge of Language

Language is not a variable of learning than can be separated from other integrated factors, such as context, culture, topic, and other participants. Therefore, supporting the content and language needs of linguistically diverse learners is a charge that must consider multiple classroom factors. While teachers could feasibly add language objectives to accompany their science learning objectives, there is a significant danger of this being an idiosyncratic exercise, not dissimilar from the ways in which an overdeveloped fascination on teaching methods can quickly lose sight of the needs of actual learners (Bartolome, 1994). To avoid a "fetish" orientation to academic language, and teaching it explicitly, for students to grow in their strategic uses of various linguistic registers, including academic language, teachers need to have several kinds of knowledges at work in their planning and teaching. In this section, we map out the essential knowledges for the effective scaffolding of students' learning in the science classroom. These knowledges are put forth as individually necessary but also incomplete; that is to say that the knowledges must be used concurrently and approached in integrated fashion through a teacher preparation program.

The first knowledge that teachers should have is that of their students. While this seems patently obvious, it is more complicated in application. For example, the socio-cultural studies from the Science for All research group used a definition of culture that may or may not have intersected with students' actual background knowledge and experiences. The knowledge of the students must occur within the classroom setting and should include knowledge of the students as learners, including their ways of knowing; that is to say, their ways of interacting with texts, questions, discussions, new information, authority, etc. Additionally, this knowledge should include knowledge of the students' language abilities, including their first and second languages, and commands of different registers within each language.

Teachers also need to consider the pedagogical structures they use for classroom learning. These structures might include a whole-class environment which is teacher directed to the whole class, a small group environment where students talk among themselves, a student addressing the entire class, students working in pairs, and students working individually. Determinations for how to use which structure and grouping include knowledge of the students' content and language abilities, what approach best matches the learning task, and classroom management considerations.

Another kind of knowledge teachers need to have is the content of science, including the conceptual schema, or how the concepts of science relate to each other organizationally. Also embedded within these knowledges are the advances in the field of science that teachers need to be aware of in order to inform their classroom practice. Within this knowledge is the teachers' definition of science; for example, science inquiry, and how this term is defined and put into pedagogical practice. As we saw in the literature review above, inquiry can be defined as starting from the students' questions about the world or it can be defined by the state standards. However key concepts like inquiry are being defined, the teachers need to be conscious about these determinations and need to be explicit in articulating these definitions and frames to their students.

In every classroom lesson, additionally, there will be a variety of textual resources, including print, oral and multimodal. Therefore, teachers have to know about the genres and linguistic structures found in these resources, including difficulties the academic language of the discipline presents. Within the genres of science reading and writing, teachers have to be explicit about the characteristics of the genres and know the linguistic features that construct them. For example, if the students are writing a procedure, the imperative mood should be used along with impersonal language, avoiding references to themselves. In order to address these types of difficulties, the teacher must have a way of talking about the language of science that is both helpful to learning the content and the students' acquisition of academic language.

These preceding knowledges, of student, content, pedagogical structures, and language, are all situated within social, political, and cultural contexts that surround classroom practices. This final realm of knowledge that teachers need is of these various environments. Teachers and students are impacted by the spheres of the school, district, nation and the world, determining educational policies regarding curriculum, assessment, funding and certification, among others. Beyond educational contexts, though, teachers' decision making should reflect knowledge of other national and global factors and forces. Teachers need knowledges of the economic and political contexts that shape new and growing waves of immigration, bringing new combinations of linguistic and cultural diversity into their classrooms. In working with native-born and foreign-born students, teachers also need knowledges of information age global economies, to better prepare their students linguistically and conceptually for these marketplaces.

While these knowledges can be discussed as conceptually different from each other, the ways that they are enacted are dynamic, working alongside each other within the daily teaching and learning moments in the classroom. In the next section, we provide an example of how these knowledges are considered together when looking at a particular classroom activity.

In the context of a high school biology classroom in an urban context, the content lesson objective was to use information about genetics to predict the genetic coding of offspring.[2] Along with this content objective, though, was an embedded language objective that students need to achieve in order to reach the content objective. Within this lesson, students needed to understand and use conditional sentences to express possibilities of genetic coding. For example, "if a man with no freckles has the dominant gene and he marries a woman with freckles, then their children will carry the dominant gene." In determining that this language objective was appropriate, the teacher considered her knowledge of the students and their linguistic abilities in academic English. In other words, objectives could only be determined by considering the content and language in relation to the existing knowledge and abilities of the students in the class.

To structure the pedagogical activities in the classroom, the teacher then needed to consider what kind of grouping to use and what sources of information to use. In this example, a direct lecture, with the aid of key terms and diagrams on the white board, was given to provide students with an example of how to read and parse out a word problem dealing with genetics, fill in a visual diagram, a punnet (a graphic used in biology to map genetic coding), and then write a short explanation. In particular, this lecture drew students' attention to the use of the conditional sentence in the explanation. Following this large group lecture, then students worked in small groups to follow this sequence with other similar word problems about genetics. The way that these small groups were organized was according to the teachers' knowledge of the students content and language abilities. For example, students who had high content knowledge but more limited abilities to express that knowledge in academic English were grouped with students who have higher linguistic abilities.

Finally, framing this lesson and its determination to be taught was the teachers' knowledge of the local and state context. In this example, students were to take an assessment required for high school graduation, and genetics would be part of the section on science. Therefore, even though genetics was a complex and challenging area of biology that some of the students might not have been quite ready to approach, the teacher decided to focus on this content objective to prepare the students for the upcoming high stakes assessment.

The Place of Language Learning in Teacher Education

To a visitor dropping into this classroom lesson, it may have looked, superficially, like this lesson was simply about genetic coding and punnet squares. While this was the central focus, many other kinds of knowledges were at play in determining the pedagogical structure, the attention to language, the dynamics of the lesson, and the sequence to use. The teacher who planned and executed this lesson, in those processes, brought forward many sources of knowledge about the content, the language, teaching approaches, her students, and social contexts. Where and when teachers can acquire these knowledges about supporting language acquisition in science is far from a simple determination. The most logical place to initially conceptualize language learning as informed by multiple factors is within their initial teacher education programs. Within these programs, preservice teachers typically take a suite of courses meant to give them pedagogical methods and approaches to working with learners. Because the framework of knowledges suggested here is not only about language but also about teaching approaches and school contexts, we argue that the teaching of these knowledges should begin intently and intensely within a science methods course, by teacher education faculty, who are best positioned to situate the teaching of language in the science content area alongside other pedagogical content knowledges.

The prospect of science teacher educators taking on the work of supporting the acquisition of academic English is a daunting one. While it would be far simpler, organizationally, to add courses in linguistics to teacher education programs, these courses would likely be taught outside of a pedagogical context, in an applied linguistics department. While these courses and departments offer appropriate content for those seeking to deeply focus on the nuances of linguistics, the curriculum of these courses is rarely situated within the context of education. Because of this, students would need to make the applied connections between the linguistics content and the implications for their future teaching—a daunting, and arguably, unrealistic task. Rather, a better solution is to infuse the knowledge of linguistics within science teaching methods courses, so that the knowledge is situated with other pedagogical fields. However, this solution has immediate implications for the science teacher educators who teach these courses.

This approach would require science teacher educators to be prepared and trained in three kinds of knowledges: linguistic diversity, applied linguistics, and how to analyze the linguistics found in science texts and discourse. The first area of learning is an orientation to linguistic diversity. This involves the assumptions toward registers, standard English, nonstandard varieties of English, the politicization of language, and understanding language dominance as political and sociologically framed, not as a matter of inherent quality (McWhorter, 2000). In addition, in the second set of knowledges, that of applied linguistics, there are the components of language to be studied, such as phonology, morphology, lexis, semantics, and syntax.[3] Also, there is the topic of research around second language acquisition; that is to say, how languages are learned, including the stages of language acquisition, effective pedagogies for language acquisition, and classroom approaches to supporting these processes. Finally, the third set of knowledge includes how genres of writing are learned and performed in ascending order, such as beginning with personal accounts and then moving to factual. Developing support around this agenda faces many challenges and obstacles. Existing pressures of time, content, and learning might lead many teacher educators to resist this agenda, along with a potential view that issues of language do not fall within their charge. Be that as it may, we, pursuant to a charge of social scientists, including teacher educators, as public intellectuals (Said, 1996), ascribe the first locus of responsibility with teacher educators for it is us who must prepare a teaching profession who has the knowledges, skills, and attitudes necessary to support all students in navigating the conceptual and linguistic complexities of science.

In the case of teacher education programs, what must take place is the transformation of faculty capacity. At one such program, the linguistic education of faculty started with a seminar that emerged from faculty interest in the education of bilingual learners. In this seminar, faculty discussed broad social and political themes connected with the education of language minority students. The bimonthly sessions further addressed effective school and classroom practices in relation to various disciplines, science included. Most enlightening to the faculty was the analysis of text drawn from content area textbooks and tests. Participants in the seminar felt that they had general knowledge and understanding of bilingual learners and of culture but, as one participant stated, "What was new for me was the issue of language" (Costa, McPhail, Smith, & Brisk, 2005, p. 107). Following the seminar, collaborative and focused activities in the context of science methods courses continued between science faculty and graduate students and faculty and students with strong background on language and the education of bilingual learners (Brisk, 2007). In addition, Nevárez La-Torre, Sanford-DeShields, Soundy, Leonard, and Woyshner (2007) consider faculty researching questions related to addressing language and content area teaching and sharing the results among themselves, an essential component of faculty development. A review of the research literature led Lucas and Grinberg (forthcoming) to conclude that including language in the preparation of content area teaching requires collaboration among faculty with different expertise.

Such work with teacher education faculty can greatly impact the ways in which preservice teachers conceptualize not only pedagogical content knowledge but how language plays intricate roles in the construction of that knowledge. However, this orientation is one that requires maintenance and support while teachers are interacting with students in their classrooms. As teachers work alongside their students in their classrooms to convey content, get to know their students, and explore the complex relationships between concepts and the language used to convey them, they are best poised to inform what knowledges are needed throughout preservice and inservice teacher education to benefit linguistically- and culturally-diverse students.

Targeting Enduring Understandings

1   What are the connections you see between the discussion in the Stevens team's chapter and the four elements of culturally-responsive instruction, as described by Villegas and Lucas?
2   How would you describe the "work" that language is doing in this multicultural science context?

Deepening the Reflection

1   Reflect upon the messages and knowledge that you received about language in your science teacher education program? What did you learn about educational linguistics and meeting the needs of ELLs? What was done well and where do you see room for improvement?
2   In this chapter, the Stevens team outlined an agenda for teacher education and for science teacher educators. What do you see as the advantages and drawbacks to extending science teacher education to include this approach?
3   In the discussion of the SFA and CK approaches, which felt most familiar to you? If you had to situate yourself within one of the approaches, which would it be and why?

Encouraging Engagement

1   Gather information about linguistic diversity in your state. How many linguistically-diverse students comprise your state's student population? Then, research your state's statistics on the number of science teachers with endorsements in bilingualism or teaching English as a second language. Be prepared to discuss the implications of what you find.
2   Observe a science classroom with ELL students. In what ways does the teacher accommodate his or her instruction to meet their language needs? Now examine the science textbook this teacher is using for its linguistic demands. In addition to content-specific vocabulary, what types of sentence and linguistic structures might prove challenging for an ELL? How did this teacher, or how might you, provide support for textbook reading in the

course of science instruction? What kind of professional development do science teachers need to best provide this kind of language-cognizant instruction?

Notes

1 Systemic functional linguistics is a field of applied linguistics that draws upon the work of Michael Halliday and works from the context-based view of language. The approach emphasizes that the structure of language, as well as semantics, communicates and carries meaning.
2 The scenario described here is taken from a longitudinal research study conducted by the first two authors into academic English support with immigrant students in a secondary school (Field notes, 2006).
3 For a review of the linguistic structures found in academic texts, see Halliday and Martin, 1993 and Schleppegrell, 1998.

References

Ballenger, C. (1997). Social identities, moral narratives, scientific argumentation: Science talk in a bilingual classroom. *Language and Education, 11*, 1–14.

Bartolome, L. (1994). Beyond the methods fetish: Toward a humanizing pedagogy. *Harvard Educational Review, 64*(2), 173–194.

Brisk, M. E. (2007). Program and faculty transformation: Enhancing teacher preparation. In M. E. Brisk (Ed.), *Language, culture, and community in teacher education*. Mahwah, NJ: Lawrence Erlbaum Associates.

California Department of Education. (2005). Renewal and initial credential information. Retrieved May 15, 2006, from http://www.ctc.ca.gov/credentials/default.html.

Cazden, C. (2001). *Classroom discourse: The language of teaching and learning*. Portsmouth, ME: Heinemann.

Costa, J., McPhail, G., Smith, J., & Brisk, M. E. (2005). Faculty first: The challenge of infusing the teacher education curriculum with scholarship on English language learners. *Journal of Teacher Education, 56*, 104–118.

Cummins, J. (1994). The acquisition of English as a second language. In R. Pritchard & K. Spangenberg-Urbschat (Eds.), *Kids come in all languages* (pp. 36–62). Newark, DE: International Reading Association.

Duran, B. J., Dugan, T., & Weffer, R. (1998). Language minority students in high-school: The role of language in learning biology concepts. *Science Education, 82*, 311–341.

Foucault, M. (1979). *Discipline and punish*. New York: Heinemann.

Gibbons, P. (1998). Classroom talk and the learning of new registers in a second language. *Language and Education, 12*, 99–118.

Gibbons, P. (2003). Mediating language learning: Teacher interactions with ESL students in a content-based classroom. *TESOL, 37*, 247–273.

Gutiérrez, K., Asato, J., Santos, M., & Gotanda, N. (2002). Backlash pedagogy: Language and culture and the politics of reform. *Review of Education, Pedagogy, and Cultural Studies, 24*(4), 335–351.

Halliday, M. A. K., & Martin, J. R. (1993). *Writing science: Literacy and discursive power*. London: Falmer Press.

Huang, J., & Morgan, G. (2003). A functional approach to evaluating content knowledge and language development in ESL students' science classification texts. *International Journal of Applied Linguistics, 13*, 234–262.

Ladson-Billings, G. (1995). Toward a theory of culturally relevant pedagogy. *American Educational Research Journal, 32*, 465–491.

Lee, O. (2002). Promoting scientific enquiry with elementary students from diverse cultures and languages. In W. G. Secada (Ed.), *Review of research in education* (vol. 26, pp. 23–69). Washington, DC: American Educational Research Association.

Lee, O. (2003). Equity for linguistically and culturally diverse students in science education: A research agenda. *Teachers College Record, 105*, 465–489.

Lee, O. (2004). Teacher change in beliefs and practices in science and literacy instruction with English language learners. *Journal of Research in Science Teaching, 41*, 65–93.

Lee, O., & Fradd, S. H. (1996). Interactional patterns of linguistically diverse students and teachers: Insights for promoting science learning. *Linguistics and Education: An International Research Journal, 8*, 269–297.

Lee, O., & Fradd, S. H. (1998) Science for all, including students from non-English language backgrounds. *Educational Researcher, 27*(3), 12–21.

Lee, O., & Fradd, S. H. (2001). Instructional congruence to promote science learning and literacy development for linguistically diverse students. In D. R. Lavoie & M.-W. Roth (Eds.), *Models for science teacher preparation: Bridging the gap between research and practice* (pp. 109–126). Dordrecht, Netherlands: Kluwer Academic.

Lemke, J. (1990). *Talking science: Language, learning and values.* Norwood, NJ: Ablex.

Lucas, T., & Grinberg, J. (forthcoming). Responding to the linguistic reality of mainstream classrooms: Preparing all teachers to teach English language learners. In M. Cochran-Smith, S. Feiman-Nemser, & J. McIntyre (Eds.), *Handbook of research on teacher education: Enduring issues in changing contexts.* Mahwah, NJ: Lawrence Erlbaum Associates.

Martin, J. R., & Veel, R. (1998). *Reading science: Critical and functional perspectives on discourses of science.* New York: Routledge.

McWhorter, J. (2000). *Spreading the word: Language and dialect in America.* New York: Heinemann.

Moll, L., Amanti, C., Neff, D., & Gonzalez, N. (1992). Funds of knowledge for teaching: Using a qualitative approach to connect homes and classrooms. *Theory into Practice, 31*, 132–141.

Nevárez La-Torre, A. A., Sanford-DeShields, J. S., Soundy, C., Leonard, J., & Woyshner, C. (2007). Faculty perspectives on integrating linguistic diversity issues into an urban teacher education program. In M. E. Brisk (Ed.), *Language, culture, and community in teacher education.* Mahwah, NJ: Lawrence Erlbaum Associates.

Readence, J. E., Bean, T. W., & Baldwin, R. S. (2004). *Content area literacy: An Integrated Approach* (8th ed.). Dubuque, IA: Kendall Hunt.

Romine, B. G., McKenna, M. C., & Robinson, R. D. (1996). Reading coursework requirements for middle and high school content area teachers: A U.S. survey. *Journal of Adolescent and Adult Literacy, 40*, 194–198.

Rymes, B., & Anderson, K. (2004). Second language acquisition for all: Understanding the interactional dynamics of classrooms in which Spanish and AAE are spoken. *Research in the Teaching of English, 29*(2), 107–135.

Said, E. W. (1996). *Representation of the intellectual: The 1993 Reith Lectures.* New York: Vintage.

Schleppegrell, M. J. (1998). Grammar as resource: Writing a description. *Research in the Teaching of English, 32*, 182–211.

Schleppegrell, M. J. (2002). Challenges of the science register for ESL students: Errors and meaning-making. In M. J. Schleppegrell, & M. C. Colombi (Eds.), *Developing advanced literacy in first and second langauge: Meaning with power* (p. 119). Mahwah, NJ: Lawrence Erlbaum Associates.

Sleeter, C. E. (2001). Epistemological diversity in research on preservice teacher preparation for historically underserved children. *Review of Research in Education, 25*, 209–250.

Suarez-Orozco, C., & Suarez-Orozco, M. (2001). *Children of immigration.* Cambridge, MA: Harvard University Press.

Unsworth, L. (1991). Linguistic form and the construction of knowledge in factual texts for primary school children. *Educational Review, 43*, 201–212.

Unsworth, L. (1997a). Explaining explanations: Enhancing science learning and literacy development. *Australian Science Teacher's Journal, 43*, 34–49.

Unsworth, L. (1997b). Scaffolding reading of science explanations: Accessing the grammatical and visual forms of specialized knowledge. *Reading, 31*, 30–42.

Unsworth, L. (1998). "Sound" explanations in school science: A functional linguistic perspective on effective apprenticing texts. *Linguistics and Education, 9*, 199–226.

Unsworth, L. (1999a). Developing critical understanding of the specialised language of school science and history texts: A functional grammar perspective. *Journal of Adolescent and Adult Literacy, 42*, 508–521.

Unsworth, L. (1999b). Explaining school science in book and CD ROM formats: Using semiotic analyses to compare the textual construction of knowledge. *International Journal of Instructional Media, 26*(2), 159–179.

Unsworth, L. (2001). Evaluating the language of different types of explanations in junior high school science texts. *International Journal of Science Education, 23*, 585–609.

Vygotsky, L. (1986). *Thought and language.* Cambridge: MA: MIT Press.

Warren, B., & Rosebery, A. S. (1995). Equity in the future tense: Redefining relationships among teachers, students, and science in linguistic minority classrooms. In W. G. Secada, E. Fennema, & L. B. Adajian (Eds.), *New directions for equity in mathematics education* (pp. 298–328). New York: Cambridge University Press.

Warren, B., & Rosebery, A. S. (2001). *Teaching science to at-risk students: Teacher research communities as a context for professional development and school reform* (Final Report, Project 4.1). Cambridge, MA: TERC.

Warren, B., Ballenger, C., Ogonowski, M., Rosebery, A. S., & Hudicourt-Barnes, J. (2001). Rethinking diversity in learning science: The logic of everyday sense-making. *Journal of Research in Science Teaching, 38*, 529–552.

14 Experimenting in Teams and Tongues

Team Teaching a Bilingual Science Education Course

María E. Torres-Guzmán and Elaine V. Howes

Introduction

In this final chapter, Torres-Guzmán and Howes share with us their experiences trying to infuse and integrate ideas of science, language, and culture into a particular teacher preparation course. As they make clear, the experiences they share are not meant to be exemplary; their story, as they describe it, "is not one of unbridled success, but one of stumbling and occasional revelation as we attempted to help our students (and ourselves) use the complex networks made visible as we focused on the inextricable connections among language, culture, and science." For this reason, this chapter is a fitting one to conclude this volume. It takes up the call to develop interdisciplinary, language- and culture-cognizant, approaches to science education, while modeling the importance of reflective practice and, as part of that, being honest about the challenges such a paradigm shift entails. But with the sense of honesty about their process that Torres-Guzmán and Howes develop comes, as well, a powerful sense of promise. Hearing what these scholars, a bilingual teacher educator (Torres-Guzmán) and a science teacher educator (Howes), learned from their collaboration, and, further, what their students learned from the collaboration, helps us envision what opportunities moving beyond business-as-usual could afford us. These scholars' "unusual business" helps chart a different course for the future of science teacher education, a course in which the kind of culturally-responsive pedagogy they attempted to model is dead center.

As you read this chapter, consider what elements of the science education course that these authors describe feel unfamiliar given your own teacher preparation experience. As you identify these, reflect on how, had those elements been present in your own experience, they might have enriched your learning. Then think about how creating similar opportunities for students of your own in a science classroom might be beneficial. What would it take to translate some of Torres-Guzmán's and Howes' practices into your own teaching? What would you leave behind? Why? This exercise will help you clarify the values you hold with respect to science teaching, values which are the foundation of your professional identity.

An important goal of this volume has been to expose you to a different way of thinking about what "teaching science" means, a way that acknowledges the work that language does in the multicultural science classroom. As you read this last chapter, use your reaction to it to gauge the changes you may have experienced in your thinking since being exposed to the first chapter. Do you feel more aware now of the work of language in multicultural science classrooms? Have you begun to consider what this awareness means for your science teaching? Are you eager to learn more about culturally-responsive science instruction? We hope that this chapter puts into play for you some of the insights developed in previous chapters and, lastly, makes clear that at times your best resources for continued learning will be your colleagues and, even more so, your students themselves.

In this chapter, we report on a teacher-research study of a course for bilingual teacher education students in which we—María, as the bilingual teacher educator, and Elaine, the science teacher educator—team taught. Our aim in developing the course was to weave together issues of science, pedagogy, language, and culture. We believed that this approach would engage our teacher education students in considering pedagogies for children who are learning science in two languages. Being that team teaching was new for us, we engaged in inquiry about our teaching. The inquiry focused on how our use of concepts and methods might help teacher education students think about teaching language and science together in bilingual education classrooms.

We grounded our teaching and research in the theoretical understanding that meaning is created, in its major part, through language. For instance, the meanings that we name "scientific knowledge" are produced through human social activities that depend upon language, as are all other forms of shared knowledge. This reality applies to the classroom as well as the communities of professional scientists. Scientists, teachers, and students all depend upon language to comprehend and create meanings about the world. If we are to assist children from all language backgrounds in developing scientific concepts, it is vital that we recognize and draw upon their particular language and cultural resources in order to help them add science to their cultural "tool kit" (Wertsch, 1991).

The concept of language as a resource system (Halliday & Martin, 1993) and as a mediator in learning (Vygotsky, 1962) necessitates that we, as teachers, be aware of how we provide access to multiple language, including that marked by the discursive rules specific to science (Gallas, 1995; Lee, 2001; Lee & Fradd, 1998; Lemke, 1990). We hoped that "both social interaction and classroom discourse would fruitfully interact" (Lemke, 2001, 302) as we focused on the convergence of issues of language and science when thinking about teaching science to language minority students or to students learning in two languages. In this chapter, we discuss how issues of language, culture, and scientific knowledge were used in our planning and played out in our teaching. Specifically, we will examine how everyday and science discursive rules can complement and conflict as this occurred in one of our assignments, the *Ethnomedical Community Study,* and how multiple languages can assist rather than hinder the development of scientific concepts.

In our teaching and our inquiry, we consciously blurred the boundaries of both our individual areas of study and those of teaching and research (Cochran-Smith & Lytle, 1993; Cochran-Smith, 2003). We were both subjects and objects of research. Our inquiry did not focus solely on what we did in the classroom but started from the moment we began our planning through to the analysis and write-up of our new understandings. Our conceptualization of teaching encompassed all that went into the doing and learning—it was what we thought, what we said, what we wrote about, and what we did. Our collaboration was mediated by networks of linguistic and cognitive systems that presupposed the social nature of thinking. The most valuable conscious learning came as a result of our analysis as it represents layers of recursive moves in our thinking and doing. In other words, we construed what happened in the classroom and what happened in the classroom construed us (Lemke, 1990; Halliday & Martin, 1993). Our story is not one of unbridled success, but one of stumbling and occasional revelation as we attempted to help our students (and ourselves) *use* the complex networks made visible as we focused on the inextricable connections among language, culture, and science.

While we did not attempt to correlate what we learned with our students' learning, as we believe they are two distinct albeit interrelated processes, we use students' work to illustrate how what we did may have influenced them, but more importantly, how their ideas facilitated our thinking and learning about language in teaching science to diverse student populations.

Theoretical Perspective

In this chapter we bring together various theories to conceptualize our work. We bring a sociocultural theory of science education, multiliteracy theories, and multilingualism/multiculturalism within the context of teacher education.

Sociocultural theory proposes that language is part of the resource system humans have created to communicate with each other and through which we create meaning about what occurs in our natural as well as social worlds. Language and sign systems are mediating forces for creating scientific concepts as well as for the development of language itself. Particularly relevant to science, language can be viewed as a semantic resource system, that is, it gives life to different kinds of meaning by matching them with actions that are then themselves embodied in words.

Halliday and Martin (1993) propose that the historical roots of the English scientific language, or what they call the English scientific register, and the root of formal English grammar and, thus, literacy, are found in the social construction of scientific knowledge. English grammar, which is quintessentially embodied in the English scientific register, is a material form of the differentiation between the standard language and the commonly spoken language. It is how the powerful differentiate themselves from the common crowd. We would say there is a similar history in other languages as well.

Rationale for the Course

The pedagogical problem is how to bridge these historically opposing worlds, represented by the scientific linguistic register and everyday language. This challenge is intensified in our present context as we must deal with the tension arising from the call for higher academic standards, greater accountability, and testing, on the one hand, and social justice, language equity, and the transformation of the socio-economic hierarchies they represent, on the other.

Within science education, current recommendations include the grounding of teaching in language usage of all types—reading, writing, listening, speaking—and in the development of varied modes such as textual, graphical, and pictorial representations of scientific processes and knowledge (AAAS, 1989; Gallard, 2003; Lemke, 1990; NRC, 2000; Wallace, 2004; Yore, Hand & Florence, 2004). The call is more a description of what language ought to look like in the context of learning; it does not necessarily give guidance on the role of language and other sign systems in the development of scientific concepts and in facilitating the development of systematic understanding of how abstract and/or technical scientific language might be in the service of our everyday lives, locally, or that of humankind, in a more global or temporal sense. It does not address the passageways for bridging the everyday language nor does it provide for the understanding of scientific concepts as connected with everyday life.

Providing strategies was not the central goal of this course. Instead, we intended to assist students in demystifying science for themselves by learning to see it as created and embedded within language and culture. We believe that if teachers can develop this perspective—viewing science itself as shaped by culture and language—they might be better able to value what cultural meanings about the natural world children bring to the classroom, and use this knowledge to help children connect to science as well as maintain and develop their home knowledge and language.

Language Registers and Codes

All teachers must contend with the difficulties of the different English language registers. Registers are variations in language use "in the sense that the speaker has a range of varieties and chooses between them at different times" (Halliday, 1964). Teachers faced with English language learners have the added dimension of distinct language codes. This alone calls for teachers' understanding the relationship between learning and language as qualitatively different. They must facilitate the learning of multiple languages and literacies and the linguistic scientific registers for children that not only attends to their particular developmental stage but that is at the same time engaged in trying to decipher and experiment with a range of language functions that differ from that of the monolingual child (Grosjean, 1989). This is true whether the child is in an English-only or a bilingual setting. In the bilingual setting, however, space for the use of the range of language functions as a resource exists for the learner, and teachers are better equipped in assisting children with a diversity of linguistic and cultural needs. It is an environment that is created consciously to support students in learning to be multiliterate—in their home language, in the English language, and in the language of science (Crawford, Kelly, & Brown, 2000; Warren & Rosebery, 2002). Focusing on bilingual language development and language use through a lens of learning to talk and write science is an important approach to learning to teach science with English language learners (Stoddart, Pinal, Latzke, & Canaday, 2002). As bilingual environments are not always possible, the issues about how to create environments that provide access to the language of science while respecting the cultural and linguistic ways of the students and their communities is always at the heart of teaching English Language Learners (ELLs).

Even teachers who are expert in teaching in or with multiple languages do not necessarily include science in their curriculum; if they do, it is often a traditional vocabulary exercise structured as rote learning. The conceptualization of learning science is thought of as a mental accomplishment of the child rather than as a discursive social performance in the act of scientific inquiry (Kelly & Breton, 2000; Rodriguez, 1997). Thus, it is fair to say that most teachers are not well prepared to teach science with ELLs nor are they prepared to create the kinds of classroom language environments that would facilitate the development of scientific understandings. In our initial thinking about teaching this bilingual science education course jointly, we agreed that helping teacher education students learn to view science itself as a cultural construct, along with developing strategies for bridging everyday language and the traditional scientific discourse, would provide an entry point for discussion of the role language plays in creating knowledge and how to value that which children bring to the classroom as a resource for furthering their linguistic and cognitive development. We aimed to demonstrate to our teacher education students that we valued their language and cultural strengths and that language was central in creating scientific knowledge. Thus, we wanted to model for them the approach to science pedagogy that we hoped they would embrace in their own teaching (Howes, 2002; Rodriguez, 1998). We also thought of our modeling as a vehicle for intentionally exposing our students to the benefits as well as the struggles that come through team teaching across subject areas (Kluth & Straut, 2003).

Study and Course Context

The course, entitled *Curriculum and Methods for Bilingual Teachers: Science*, was the context of our teaching and inquiry. It was required for the preservice (initial certification) and in-service (professional certification) students in the Program in Bilingual/Bicultural Education at Teachers College, Columbia University but open to others. The students met with us six times in a compressed period of four weeks during the summer in order to fulfill the requirement of thirty hours of in-class instruction. In-class meetings were four hours long, with exception of one Saturday when we met for eight hours. The curriculum was divided into two parts. The first part was organized around the physical science topic of buoyancy and associated language concerns; the second segment focused on the life science

topic of ethnomedicine and issues of language and culture. Pedagogical constructs and issues, i.e., constructivism, cooperative grouping, language grouping
and lab experiments, scientific inquiry, and technology, among others, were
incorporated and discussed throughout. Our concept of the course was that we
would be engaged in a dialogue about science, language, bilingualism, and
instruction, structured in such a way that students could participate.

This was the first time we taught this course jointly. Maria had taught the
course several times on her own, and had developed a clear curriculum, associated activities, and assignments. Elaine had never taught this course, but had
several years of experience in teaching science methods courses for elementary
and secondary preservice teachers. Elaine had also been conducting research
with a local elementary school with an active dual language program and a population of largely bilingual (Spanish/English) students with whom Maria worked.

Analysis of Data

There were four data sources in the form of audio-taped discussion and email
communication, reflective journals, field notes, and student writing. The following is a description of each of these data sources.

Audio-Taped Discussions/Email Communication: As a beginning analysis, as
well as a way to generate data, we audio taped our planning and reflective conversations. These conversations commenced well before the beginning of the
course, and continued regularly (twice a week) for the five weeks of the course's
running. Eight audiotapes were audible and have been transcribed. Our audio-
taped discussions are supplemented by email communications.

Reflective Journals: We each wrote reflective journals during the duration of
the course. At course's completion, we shared our journals, and responded to
each other in a written conversational format. This interactive journal writing
served as a second phase in our analysis.

Field Notes: We each created field notes of our classroom observations. One of
us would record classroom talk and interactions as the other led the class; during
group work, we both often took notes. These writings became central to our
journal conversations.

Students' Writing: We documented our evaluation of students' work by saving
copies of their writing. In addition, all of the students' writing—classroom
observation assignment, ethnomedicine assignment, biweekly interactive journal
writing, and science unit plans—were submitted on-line, and thus are an accessible data source.

Three distinct phases characterized our recursive analysis of the data. First, we
identified the events or activities that were salient in shifting our thinking. We
analyzed our journal and reflective writing, discussions, and emails to note these
shifts. From conversation and collaborative writing around these points, we
created conceptual propositions (van Dijk, 1980) that described what it was in
our learning that seemed powerful to each of us. In the second phase, we examined the conceptual propositions as they played out in the classroom observations and reflections of our work with our students and their written work

Finally, in the third phase, we analyzed the propositions from the conceptual framework established.

Findings on the Conceptual Propositions

We organize this section around grounded conceptual propositions about teaching. While we believe that every context is unique in actors, time, and situations, we also think that our academic exercises must bridge to everyday life in schools. Within each proposition, we include explanatory evidence from our planning discussions and our teaching, and present specific representations and development of the propositions evidenced in our learning. We also believe that they embody insights that could help teachers of science who are working with ELLs in English-only and bilingual settings.

Proposition A: Carefully Created Assignments Can Give Students Opportunities to Explore Relationships Among Language, Learning, Culture, Politics, and Science that are Important to the Students and Their Communities

In the Ethnomedical Community Study assignment design, we brought interests that were distinct and yet seemed, for us, to have potential for exciting connections: Maria's interest and experience with a variety of healing approaches, and her knowledge concerning the maintenance of indigenous languages and cultures, linked with Elaine's interests in the role of culture and language in the creation of scientific knowledge outside as well as inside the classroom, and her knowledge of and interest in biodiversity. Inspired by our reading and discussion of literature concerning the relationships among language, cultural, and biological diversity (Balée, 2001; Maffi, 2001; Nazarea, 1998; Rothschild, 1997; Skutnabb-Kangas, 2000), we created a "community study" concerning healing to support students in comparing and making connections between Western scientific knowledge and traditional folk knowledge. We believed that in this assignment we could bring our interests and concerns to bear to help students observe how different world views create different perspectives on wellness and healing, and in the process, reconsider what counts as science.

We were initially calling our ethnomedical assignment "the ways of healing or different healing ways." We knew that we would get into ethnobotany but it was not until Elaine came across the word "ethnomedicine" that we really began to explore it as such. This word opened up a world and became the center of our world for the second half of the course and made us more conscious of language as symbolic representations of the world.

We asked students to choose a malady and focus on treatments from different perspectives: "new age," "scientific" (meaning pharmaceutical), and folk remedies based on plants. The latter would lead into the life science portion of the class, in which we would introduce some basic botany (parts of plants and their functions), and enhance this with a visit to a botanical garden. The assignment itself was the result of our language and our thinking together, mediated through our planning, where our different expertise merged. Our openness to creating a new science activity that would bring together the two areas of knowledge also served to create a space for considering new ways of introducing science within the context of multiple cultures and languages.

In the following examples, we explore two networks of connections around the relationship between social justice and Western science. In one instance, the connections center on how students' competing conceptualizations of relationships between pharmaceutical companies, environmentalists, and locals problematized the role of "white progressives" in minority or developing world communities. The second illustrates students' attempts to negotiate relationships (or the lack thereof) between traditional knowledge and scientific knowledge concerning healing. These connections were constructed through students' responses to the Ethnomedical Community Study.

ETHNOMEDICINE AND ENVIRONMENTALISM: THE INSIDERS
AND THE OUTSIDERS

When students were asked to interview community and traditional medicine experts, issues of social justice within the contexts of ethnomedicine and ethnobotany came to the forefront. Class discussion, in this area, centered on the exploitation of indigenous or local peoples by pharmaceutical companies and the exhortations by "well-meaning outsiders" to the people of the third world to protect their environment. For example, a Dominican student, Marielis (a pseudonym, as are all of the student names used in this chapter) voiced very strongly how in the Dominican Republic, "progressive whites" did not understand the reality of survival and what occupies the lives of people in the Dominican Republic. Thus, the progressive outsiders were less successful in their attempts to attend to issues of the environment and, in addition, erroneously labeled Dominicans as nonprogressive on the issues of the environment. Marielis described a more complex situation within the Dominican Republic, and other developing countries, wherein pharmaceutical companies send scientists to find out about plants and healing. She noted that the riches gained from the marketing of traditional ethnomedical knowledge from developing countries were never returned or shared. She felt that rather than focusing on the "car batteries in people's backyards," environmentalists should focus on the waste left to pollute the island by the scientists and the pharmaceutical companies that supported them.

Her position inspired a spirited and contentious discussion. The central conflict appeared to be more than confusion about whether community or environmental concerns should take priority, or whether they might come together in some yet undiscovered way. Rather, it hit on issues that were real to the students in our classroom as well as out in the world. Students native to the Dominican Republic, in particular, as they were the largest single language minority group, were irritated by the well-meaning but ignorant advice of outsiders. Other students felt they were themselves examples of these "outsiders" and took the stance that even small actions taken by individuals as "stewards of the earth" were meaningful. Some of this latter group genuinely felt that the prioritizing of community concerns put the environment low on the Latino community's list of priorities, in the Dominican Republic and elsewhere. They admitted that this made them hesitant to take on issues of social justice associated with the environment, as it did not resonate with the communities with which they were working. Their counterparts, who, for the most part, were themselves from the communities under discussion, were more aware of the negative effects of "outsiders" on the environment and the community.

What we came to understand was that, in the conversation, the term "outsiders" floated between two definitions. As long as these two meanings were confounded, the discussion could not move forward. One of the meanings related to the societal context of pharmaceuticals, as they were construed by the community as takers of knowledge and leavers of waste ("outsiders who did not care"). The second meaning of "outsider" referred to the environmentalist who, while ignorant of the community's perspective on the pharmaceuticals, tried and failed to inspire communities to engage in environmentally-sound initiatives. In summary, outsiders were both part of the group that introduced waste into the local community in the first place, and naïve about the broader societal context. In this mix of meanings, abstraction and decontextualization were occurring in at least two ways. The first occurred as a result of viewing environmental issues absent from their historical and social context. Second, in the unfolding of the conversation, politeness was communicated by generalizations and positionalities ("insiders" and "outsiders") while at the same time having perceived representatives of each among those who were speaking.

All of our students were bilingual, and the majority had had experiences in Latin America or in a Latino community within the U.S., and yet they were still grounded in different ethnocultural backgrounds. We were faced with progressive whites and students of color in our classroom feeling the need to move from the mentality of being an "insider" or an "outsider" to a more transformational way of thinking about the issues. This was neither easy nor straightforward. Some of the conversation was uncomfortable and even confrontational between students. In the face of this discomfort, we felt we needed to ensure that there were venues for informal as well as formal instructional conversations.

In this regard, much of this discussion, both formal and informal, was handled by Maria, for at least two reasons: (1) she had more experience in leading discussions concerning diversity; Elaine willingly deferred to her expertise in this area, choosing to listen and learn, and (2) students gravitated to Maria outside of class-time to describe their dissatisfactions, likely both because they knew her better than they did Elaine, and because they had her pegged as the "language and diversity" person, and Elaine as the "science" person. Both the relational and the perceived unequal understandings of the instructors might have been operating. Unfortunately, these perceptions of professorial expertise served to reinforce cultural images of scientists as not concerned with or naïve about issues of diversity and the uses of power, thus working *against* our desire to help students see science as thoroughly embedded in culture.

ETHNOMEDICINE AND WESTERN SCIENCE: MEDIATING LANGUAGE
AND CULTURE

The Ethnomedical Community Study also allowed us to explore the relation-ships (or lack thereof) between traditional healing language and the language of Western science. This assignment was designed to help students bring knowledge from the home domain into the classroom, examine the legitimation of know-ledge, and see how cultural and popular knowledge could be valued as real and useful. One of the first examples of this arose during an in-class discussion where one of our quiet Latinas, Juana, who had not yet spoken in the public forum of the classroom, told us about her Grandmother's deep knowledge of healing. We take the following excerpt from her written assignment, which repeats what she orally shared in the classroom:

> Being that [my grandmother] was born and raised in Mexico, she is very knowledgeable about traditional and natural ways of healing.... She always dis-couraged her children and grandchildren to stay away from fast food and to lead a healthier lifestyle.... She frowned upon being reliant on over the counter drugs, such as aspirin, cough syrup, etc. In her opinion, there are people who live simple, and who use what is available to cure illnesses and ailments.... Nearly all the remedies we used were from vegetables, fruits and herbs. Even better, according to my grandmother, we grew them in our own backyards.

This student positioned her Mexican grandmother as an expert in traditional and natural ways of healing. Juana's grandmother stood out from and even criti-cized family members not born in Mexico and who ignored traditional ways and knowledge. Juana noticed that what her grandmother believed was reflected in the way she lived her life—she ate healthy, she did not rely on processed or fast food, and she grew her own vegetables, fruits, and herbs. Her family life was organized to depend less on the market than the typical US resident. As Juana states, the intergenerational knowledge was filtered to her generation as she grew up with the home remedies (vegetables, fruits, and herbs with healing qualities) and her grandmother's linguistic mediating forms that "discouraged" less health-ful lifestyles. The activity of gardening was embedded in a functional purpose (to grow that which will be used to cure what ails us).

This assignment served to affirm Juana's cultural ways and cultural know-ledge. Through her sharing in class, she and other students were able to consider how scientific knowledge can be distant and alienating, and how bringing in home knowledge might help students give meaning to the kind of science that usually occurs in classrooms.

Setting up the assignment the way we did led students to entertain the possi-bility of reconsidering their definitions of science, as they were required to describe traditional, scientifically-based and new-age remedies. Several students, in class discussions and in their written assignments traveled ambivalently between these worlds. Some wrote about the "scientific studies" that identified the chemical components of traditional plant remedies. In doing so, they expanded their perspective of traditional medicine, specifically concerning their efficacy. One student (Anne), for example, chose to write in a hybrid way by including the scientific name—*Marinda citrofolia*—as well as the "common Hawaiian name" for the Noni plant. She reported that the Noni plant had been subjected to "studies [that] have shown that the fruit juice contains several healing attributes including, antibacterial, anti-inflammatory, analgesic, anticongestive, and cancer-inhibiting compounds."

Students wrote about how their interview subjects also created hybrids of the traditional and Western healing approaches. Juana's "grandmother and aunts have tried science and natural remedies [and] used to try all types of herbs and other plants that were either prescribed to them by the local pharmacy or *botánica.*'" Another student (Serena) reported on the complexity involved in her father's knowledge, as it included traditional, personal, scientific, and geographical understandings. Serena wrote:

> The *nopal* is known in the US as the prickly pear; in Mexico the *nopal* is used as a healing plant and as food. The *nopal* contains pectin and mucilage that are beneficial to the digestive system, and that are good for the liver and diabetes.... My father told me that the *nopal* offers iron, vitamin C, protein, sodium, fiber [and] is very nutritious and it has many other nutrients that are beneficial to the health. My father also believed that the best thing about the *nopal* is that it is found all over Mexico, and is easy to grow.

This juxtaposition of traditional and scientific knowledge led Elaine to a realization concerning how scientific language disguises sources of medications. The ingredients of over-the-counter remedies associated with Western medicine are listed explicitly on the bottle as isolated chemical ingredients. These lists do not teach us where these ingredients are from, or how they work. In contrast, the ingredients of traditional remedies are not explicated in this way, except when they are transformed by the application of Western scientific approaches that identify particular molecules and their role in treatment. As noted above, many of the students' papers indicated that people utilize both traditional and Western medicine, and use the Western scientific names and roles of the constituents of traditional plant remedies. In general, however, the plants or parts of the plants (e.g., leaves) are used directly—their ingredients are not isolated and extracted, nor, as in many cases of Western medicines, synthesized.

In addition to the linguistic habit of itemizing each ingredient, Western medical treatments require "testing" of the scientific kind. As Melissa, who is a middle-school science teacher, and apparently comfortable with science, wrote:

> Similarities between over-the-counter drugs and the recipes that were discussed in the interview are difficult to uncover. The over-the-counter drugs are sited by their chemical names whereas the chemicals in the plant are not. It is difficult to know if these chemicals are the same or similar without chemical investigation.

What sense can we make of these juxtapositions in terms of students' understanding and applying our goal of creating connections among science, language, and cultural knowledge? We mentioned earlier that we view science as a culture with particular ways of using language to express knowledge. We see, in the student excerpts, scientific language and ways of healing side-by-side with those of traditional cultures. In order to compare or combine these two or more categories of knowledge, students needed to utilize the language of each, in close proximity. While it's possible that the privileging of the linguistic scientific discursive practices in the students' written work were used to justify "folk" scientific understandings, it could also be viewed as students engaging in the blurring of the typical separation between "home knowledge" and "science class knowledge." Could this be a way to help students think about language acting to further divide the haves and the have-nots through the specific rules of scientific language, rules that work to maintain objectivity and generalizability, as opposed to the personal, culturally-embedded language of traditional healing that focuses on the whole system in which the person exists (what they eat, where they get food from, etc.) rather than just the ailment?

We would like to suggest that these examples illustrate that carefully created assignments like the Ethnomedical Community Study can create spaces for students to discuss issues of concern to them that would not come up in traditional teaching methods within science. The first example illustrates the different network of relationships of meaning making that come with students, in this case around the environment and minority communities. In the second example, we focused on how the language of different types of knowledge can compete for public legitimacy when the conversation is opened to considering alternative knowledge about healing. These both illustrate the opportunity for students to explore issues of social justice and the environment. They also point to the necessity of teachers to be prepared for the tensions competing culturally-influenced ways of perceiving and conceptualizing create when they are brought into the classroom.

This assignment could also help students think about the contribution of the theory of funds of knowledge (Moll, Amanti, Neff, & Gonzalez, 1992) as a pedagogical strategy for mediating student learning, and to explore how to incorporate community knowledge in the content of science. Through this assignment, we were also able to explore how using different ways of thinking about scientific inquiry could counteract some of our students' dislike for science. Some of our students spoke about their fears with the subject and how the Ethnomedical Community Study had gotten them in touch with the knowledge of their own families or people from different backgrounds with such knowledge.

Upon reflection, we realized that there could be clearer connections made between the students' discussion of pharmaceutical companies and environmentalism, the social divides between "insiders and outsiders," and the uses of scientific knowledge to mystify and disregard traditional knowledge and cultural experience. However, the gap between students' understandings of science as an objective, fact-driven endeavor and science as a human-created culture was a difficult one to bridge. While the assignment did open spaces for new ways of thinking or for the exploration of topics that are not normally taken up in science, and we learned about their importance in teaching and thinking about science education with diverse student populations, we remain with questions about whether more explicit instruction, or maybe more role-switching between the language and the science educator, would have served to help construct such a bridge in those instances.

Proposition B: The Knowledge We (Instructors and Students) Bring to the Conversation About Science and Language(s) in Schools Are Important to Understanding How to Mediate Learning

We were struck by the complex ways in which we, as teaching partners, contributed to the construction of classroom science understandings. This is distinctly different from showing our understanding as individual experts (a traditional way of teaching—one expert with a group of non-experts). We were "not autonomous minds meeting in a rational parliament of equal individuals but, instead, a richer and more complex notion of learning-in-community, often among unequal participants with a role assigned to power relationships and differences" (Lemke, 2001, 298). We were, however, consciously attempting to break down hierarchies and balance the playing field, for each other and for students. Because our relationships with each other and with the students were intentionally flexible, we were able to explore the links between developing scientific understanding in particular areas and the language issues that arose therein.

COMPLEXITIES OF TRANSLATION

Some issues were anticipated, as the nature of the course was to focus on ELLs. The role of translation, multiple literacies, and their roles in content learning, code-switching and language variation, language policy, and language planning, with respect to program types, all came up through the curriculum planning and instructional stages of our co-teaching. For Maria, these were the key concepts related to the role of language and the relationship between language and cognition. To illustrate their importance, we highlight translation as a mediating force in the process of cognitive development and scientific understanding of second language learners. We illustrate this through our interactions during a planning session and its spillage into the classroom.

The concept of "buoyancy" and its translation provides the context for this illustration. Maria had written a paper on the relationship of science and language as it related to the concept of buoyancy, which captured some of her previous teaching on the topic (Torres-Guzmán, 1997). There were some differences this time. First, Maria's attempt to find websites for student use on the concept of buoyancy in Spanish caused her to reflect on her depth of understanding of the concept and issues of terminology and meaning. She initially searched the Internet using the terms *flotar* and *flotación,* which were direct translations of "floating" and "flotation," the phenomena explained by the scientific explanation named, in English, "buoyancy." When this failed to provide viable sites, she used an indirect search strategy by using the term *ciencias escolares* (school science in Spanish). The search led us to the term of *hidrostática* (www.ciencianet.com), where we found good material concerning teaching buoyancy with school age students.

The connection between hydrostatics and buoyancy made Maria wonder more about the concept of buoyancy. In class, Elaine connected gravity and buoyancy as force and counterforce. Maria took this explanation as a trigger to think through the conditions under which buoyancy occurs. It led her to thinking about other terms in Spanish (ex. *flotabilidad*) and to conversations with people outside the class. As a result, she gained understanding of the condition of fluidity, in liquids or air, in relation to buoyancy. This was a shift in her scientific as well as linguistic knowledge, which made visible the dynamic nature between language and cognition. Later she was able to share this conceptual and linguistic growth with students, at which time, some students talked about how they had wrestled with the translation themselves.

The translation of buoyancy was a rich event as it also served to ground Elaine's understanding of issues related to bilingual learners. While Maria read on *hidrostática* (www.ciencianet.com) and simultaneously translated from the Spanish website into English for Elaine, she hesitated when she came to the word *inclinar*. It was close to the English word "incline," but she intuitively knew this was not the word she would normally use in English. Soon enough she came to the appropriate translation as "tilt" or "tip," but the moment of mental searching for the appropriate translation, similar to an Internet search wait time, was constructed by Maria as possibly appearing to the outsider, Elaine, as not being competent. It illustrated the need to distinguish between behavior and meaning, as "being stuck" did not mean that Maria did not know the word. This led to a discussion of how many times when the second language learner does not respond immediately, the "being stuck" behavior is interpreted as "not knowing." Instead, students may be actually engaged in the complex cognitive operation of figuring out the best word choice in the context of the specific translation event, an even more complex process than one would expect to occur in an Advanced English class, when a student is refining a composition and searching for the right word. Or, they may be at the "periodic pause at crossroads to decide which path to travel next" as in Lemke's (1999, 87) metaphor for the engagement of researchers in search for new theoretical paths. Elaine later reflected on how the experience had made her more aware of the need for teaching in more than one language and why the native language was important for student understanding. She was making the connection between language and cognition, and the role of the native language for understanding in science. This issue was explored further in our planning, our discussions of readings, our journals, and conversations with students.

Our discussion of this experience also permitted Elaine to make connections with and explore the issue of cognates, the Latin base of scientific terminology, and the subtleties of semiotics in translations. In turn, when Elaine voiced her discovery, Maria felt it was also necessary to call attention to what is called "false friends," the common misuse of cognates, in translations. The aspects of our planning around translations were later brought into the classroom. The exploration of issues associated with language and concept in relation to cognates continued even in our interactive journals after the class had finished. Elaine wrote:

> This connects to the idea [Maria] brought up of Spanish being close to Latin—this is something that might give Spanish speakers an advantage in learning science vocabulary. On the other hand [as you taught me], the translations are not always direct or obvious. Is it likely that this would happen more with "conceptual" words, like buoyancy—something that explains a complex natural phenomenon—than with "names," e.g., of bones, of species, etc.? This may be something that would lead teachers away from conceptual learning and more toward vocabulary learning; something that is a disturbing habit in all science education, and something that is encouraged by high-stakes exams. Also, it is "easier" (although much less pleasant) to teach memorization of vocabulary than to teach for understanding.

The incident of translation during the computer search also created a rich context for us to discuss other pedagogical questions related to bilingual students. For example, Elaine shared a prior experience with a Latina student that had left her a bit puzzled. She had asked the Latina student in class if she would be able to translate from Spanish to English and had been perplexed by the student's negative response. Elaine was not sure what the response meant, and why the student would say no if she knew two languages, which seemed (to Elaine) something of which to be immensely proud. Elaine was able to ascertain with Maria the appropriateness and possible meanings of such a response. The student may have been communicating to Elaine that she did not have a mastery of the Spanish academic terminology. The result was a suggestion that we design an instructional activity similar to what we had experienced together around translation because this would help teachers understand experientially some of the issues around bilingualism. We soon realized that this was not an appropriate activity for our student population since they were all bilingual. This helped us highlight what teachers in classrooms need to consider—who their children are and what they bring into the classroom.

In addition to our multiple discussions about translations, these issues showed up frequently in class. The material for the first day's activity on buoyancy, which Maria had prepared (Torres-Guzmán, 1997), for example, included the translation of red snapper as *besugo*. A Columbian student blurted out that she had never heard that word; she attributed not knowing the word to not knowing the fish, red snapper, before she arrived in the U.S. Another student added to the discussion that the word *chillo* was used for red snapper in Puerto Rico. This was an in-class illustration of the role of language variation as bilingual teachers experience it in their classroom and it is illustrative of the translation between Spanish-speaking groups and how the lingua franca, English, sometimes can support understanding the meanings of words in Spanish.

Our experiences led both of us to a rationale for teachers who work with English language to learn about language variations, translations, and to the fluidity in languages and categorization. We experienced the conversation around the issues of translation as an important learning tool. Elaine had raised a question about how misunderstandings in science could occur given the back and forth of translations as she observed and thought about the possible explanations of Maria's experience. Maria, on the other hand, was raising questions about the multiple purposes of translation in learning because one experience of translation had led her to explore a concept more deeply while other experiences of translation led to exploring the arbitrariness of language, both through language variation and scientific terminology. Moreover, translation raised, for both of us, issues related to methodologies of teaching science that extended vocabulary or that focus on memorization, and the engagement or non-engagement of students given what the population thought interesting with respect to the act of translating—if viewed as a mere act of label exchange or as a complex cognitive operation that mediated learning. It further raised the question of how well teachers, in general, understood the complexity of translation.

Proposition C: By Integrating, and Paying Attention to Language in the Context of Learning Science, We Provide Participants (the Instructors and Students) Opportunities to Shift in Their Understandings of the Relationship of Culture and Language to Science and Nature

In our work with each other and our students (all of whom spoke both Spanish and English, many with Spanish as their first language); we expressed our understandings of the world from the perspective of how we defined ourselves culturally and linguistically. For the science educator, this meant being especially conscious of utilizing mechanistic and reductionist views of how living things work, with a reliance on, for example, chemically based explanations for curative powers of plants. This Western-scientific perspective was initially in tension with her desire to fully value, on its own terms, the knowledge that each student, culture, and language brings to understanding the natural world. She struggled with the impetus to explain traditional knowledge from a Western scientific perspective: Would this help students connect to science and learn its explanations? Or would it be seen as a denigration of traditional knowledge? Her response to this concern is currently found in viewing language as a means of expressing, describing, and utilizing the inextricable relationships of humans with nature (Balée, 2001). In this, Elaine felt that she was enriching her knowledge base in ways that would enrich her abilities to value and learn from students' traditional knowledge about the natural world.

For the bilingual educator, the significance of the relationship between language and cognition, and science was magnified and deepened. For example, her experiences with various indigenous populations, including Maori, Quechua, and Mayan, had provided her with a basis for connecting the role of science in different cultures but through this co-teaching experience she was able to understand the science, through the lens of different cultures, as explanations of the origins of humans. There was another level of understanding of the relationship between science, human origins, identity, and language as found in the case of the Mapuches of Chile. The identity of the Mapuche is related to the literal meaning of their name, people of the land, which is connected indirectly to how they conceptualize their relationship to the world. She could then better understand the rationale for a "Maori biology" which had initially seemed to not fit with her Western view of science. She understood why religious, spiritual, and traditional folklore explanations were included in such a course. In this way, she discovered potential sources of scientific explanations in traditional legends, folktales, and mythology, such as the Popol Vu and other indigenous legends, and tales available in Spanish literature.

In the Ethnomedical Community Study, as well as other assignments, students indicated recognition of the etymological connections between language, culture, and science. One student's cross-cultural knowledge was demonstrated when she noted with interest that "in my interview the malady was referred to as 'nerves' while the western sources of medicine such as the pharmacy, vitamin store and Internet used the more technical terms of 'stress' and 'anxiety'" (Kathy).

Much more common, however, were tight links between culture and nature, in the form of human health—links that could open up visions of science to include rich traditional folk knowledge. Several students chose diabetes as their malady to explore, as it "is a health issue that plagues many people, particularly people of color" (Serena). These students noted that diabetes had been an issue in their families, and thus there existed an accessible store of knowledge concerning its prevention and treatment. The subject was discussed in class, and one of the students, Serena, brought in the observation that many Mexicans develop diabetes when they encounter a different set of foods in the U.S. than those they had grown up with in Mexico (Serena researched the *nopal*, a plant that her father told her was used to fight diabetes). The knowledge that students unearthed through their interviews with traditional healers demonstrated the complex connections endemic to traditional knowledge. The interview subjects' knowledge was deeply rooted in understanding of the plants' many uses as medicines and food; knowledge of the geographical sources of the plants; and commentary on the economics of health and healing through traditional methods compared to Western methods. In addition, students *themselves* developed further understandings of their home cultures, as Juana's example below demonstrates:

> Grandma can be evasive about her recipes (all of them, whether cooking or medicine). She shares some recipes with my mother, although Mom is considered a non-believer in her eyes, and "this is why she is always sick." However, I (a believer) have never been privy to any of Grandma's recipes because I am still an outsider to this *dama de casa* club. I never obtained exact recipes. Grandma was content to leave me with only a vague impression that the leaves should be boiled in water and sugar added to taste, although I am certain that the teas I drink have several ingredients...

The deep knowledge bases and perspectives that students wrote about could be utilized in science classrooms to both validate traditional ways of knowing, and to critique the comparatively simplistic, although apparently mysterious, ways of Western science and medicine. In more folk traditional situations, the plants could be studied in-depth, in terms of their uses, and the recipes that best fulfill these uses. "Hands-on" classroom activities could be created such that they attend to the very plants that students find out about from their families and communities, and Western scientific knowledge about health and about plants could be fruitfully employed to widen students' understandings. And maybe Juana could recreate, on her own, the recipes to which her Grandmother will not yet allow her access.

Beyond the fascinating details of students' ethnomedicine papers, there lies an important pedagogical point: requiring that students choose a malady to research led *Elaine* to a new understanding of the place of diabetes and other health issues in some cultures of the U.S. As professors who are always exhorting our students to learn about and from *their* students, this situation provided us with good reason to keep promoting this approach. In terms of students' learning, they explored rich sets of data that they would not likely have encountered if we had been more specific about what plants and/or maladies they should study. These data sets could provide extensive material for classroom study that is rooted in students' interests and in the knowledge of their communities.

Discussion

We found, through teaching and doing inquiry together, that we had underestimated the possibilities provided by teaching science education from a perspective based in language uses, language learning, and the rootedness of language in the natural world (Abram, 1996; Wollock, 2001). While in our teaching we created assignments such that they represented and grew out of our collaboration, in our analysis we speculated about the influences between the act of team teaching and the evidence of the team teaching found in the students' work. We believe that the individual assignments (in particular, the Ethnomedical Community Study) and the course as a whole—including our relationships and interactions with each other and the students—took on their unique form *due to* the multiple resources we offered each other and students in our collaboration and our focus on the needs of ELLs.

We found that our experiences with language illustrated the development of the inter-relational qualities of our co-teaching. For example, Elaine had initially assumed a role of "apprentice" on the language issues. In one session, she had questions prepared to ask about how Maria thought about some of the issues. The eventual encounter with "ethnomedicine" opened a whole new dimension of our thinking. How we discovered and used language as symbolic representations of the world and how we used science talk were ways through which we expanded our own and students' ways of seeing. We saw how where the knowledge and language of the students are heterogeneous, teaching/learning can take on a stimulating intellectual quality. We believe this diversity in thinking can bring the excitement of discovery of language and the concepts embedded in the language to the learning of science.

In considering language development, bilingualism, and the learning of language through science, we found that our awareness of systems of power—systems that are at work in the classroom as well as in the larger world, their parallels, and connections—also emerged. For example, we were fascinated by the vital links between biodiversity and linguistic diversity and provided two key readings on the topic—from Skutnabb-Kangas (2000) and Nazarea (1998)—that focused on the inextricable connections between local natural systems, culture, and language creation and maintenance. We explored the relationship between linguistic diversity and biodiversity, the ethnocultural community, power relationships within the context of ethnomedicine, the sociohistorical issues of oppression within ethnobotany, and related questions of pedagogy. Our explorations, as most research, led to even more questions, such as: How can we support students in maintaining the knowledge of their heritage in ways that are respectful and nurturing, even in English only classrooms? How can we help teachers see support as other than "exoticizing" and patronizing? In what ways can we work against the reductionism and decontextualization of scientific knowledge, and of language, while doing justice to students' need to succeed in mandated curricula and exams? Does the perspective on how the healing powers of traditional remedies are represented enrich our knowledge? Or does it prioritize Western over traditional and engender loss of the traditional? Is there another way to value and learn from the traditional, so we're not comparing and judging?

Despite the difficulties and awkwardness of some interactions in and outside the classroom—for example, we made assumptions about each other, stumbled on each others interpretations, and continuously tried to accommodate to each other's needs—the process of discovery, in relation to each other as well as to the intellectual connections we made, was exciting and powerful. We learned about creating a classroom culture permitting us, and our students, to explore issues related to science education and language education that we would not have otherwise. By voicing with each other what we felt about interactions with students and each other in the classroom, we broke a silence. By breaking this silence, we created a new space for creating new knowledge. As Maria wrote:

> It was like transformed energy ... the word that came to mind was synergy. There was more excitement in our work than if it had been done individually.... Your presence gave me the opportunity to go deeper into my knowing and to broaden my horizon.

In teaching the course with Elaine, Maria felt she had deepened her scientific understanding and her understanding of how language carried with it meaning and how that meaning brought the language to life. We also expanded our understanding of the relationship between conceptual and linguistic development, and we began to develop an understanding of our relationship to each other, as it is manifested in the field of science.

Conclusion

Our data illustrate the different entry points and pathways to learning. In the same context (the course) Maria and Elaine connected to different aspects of what was going on and developed different learnings. What is even more powerful is the understanding that our language of science was developed in social interaction, through the collaboration, whether in the activities of team teaching, of reflecting on others' practices as a way of looking at one's own, or in dialogue with each other. Through both our team teaching and our teacher research experiences, we found that we each brought particular perspectives and knowledges to bear on the process. We used these differences to feed our team teaching—our differences fueled our discussions and inspired teaching structures (e.g., assignments) and situations that would not have occurred if we both had the same point of view. Thus we come together in concert in our interest in the uses of language in the creation of scientific knowledge.

Our study provides conceptualizations for the teaching and learning of pedagogical practices that, based in our understandings of science and language education, should create contexts within which children, bilingual and monolingual, can engage in powerful science learning and language learning in principled and fully integrated ways (Torres-Guzmán, 1997). This effort involves thoughtful critique of scientific inquiry as presented in mainstream education documents and venues, as well as deep discussion of the role of Western science in our global society. Team teaching is vital in making the connections that allow teacher education students to engage in these complex discussions; however, team teaching is not easy, nor is it well supported. Nevertheless, through our teaching and our analysis, we have come to believe that the challenges endemic in team teaching lead to worthwhile results, and as such, should be pursued in policy, in practice, and in research.

Targeting Enduring Understandings

1   What are the connections you see between the discussion in Torres-Guzmán's and Howe's chapter and the four elements of culturally-responsive instruction, as described by Villegas and Lucas?
2   How would you describe the "work" that language is doing in this multi-cultural science context?

Deepening the Reflection

1   In what ways could Torres-Guzmán and Howe, in their roles as team-teachers, have used their positions to further reveal and complicate the apparent boundaries between "science" and "culture"? If you were a student in that classroom, what discussions would you have liked to have had to enhance your interest and understanding in the science-culture relationship?
2   The *Ethnomedicine Community Study* was an attempt to support students in maintaining the knowledge of their heritage in ways that were respectful and

nurturing. What possibilities and issues does this assignment bring up? Can you envision yourself giving your students such an assignment? Why or why not?

3   In what ways can we work against the reductionism and decontextualization of scientific knowledge, while doing justice to students' need to succeed in mandated curricula and exams? How have you and do you experience the pressure of high-stakes standardized testing? What do you think it is doing to science teaching and learning? How will you continue to manage this pressure as a teacher?

Encouraging Engagement

1   Torres-Guzmán and Howes used the *Ethnomedicine Community Study* to raise awareness about traditional and community-based knowledge concerning health and the natural world. Design another assignment that you could use in your classroom to enhance students' knowledge of the everyday scientific understandings that exist in their families and communities. Where could you place this assignment in the scope and sequence of your science curriculum?

2   Pair up with a student in a "diversity"-oriented course such as Multicultural Education or ESL Education. Like Torres-Guzmán and Howes, have a conversation about a science lesson you are planning, sharing your respective insights about its science, culture, and language components and opportunities. What do you learn from your dialogue about the relationship between science, culture, and language? How do you now understand your own readiness to infuse cultural and linguistic resources into your science teaching? What more knowledge and skills do you need? How can you continue this learning?

Note

1   *Botánica is an establishment where herbs, candles, and other artifacts associated with folk healing are sold. The individuals attending are also consultants on remedies associated with different maladies.*

References

Abram, D. (1996). *The spell of the sensuous: Perception and language in a more-than-human world.* New York: Pantheon Books.

American Association for the Advancement of Science (AAAS). (1989). *Science for all Americans.* New York: Oxford University Press.

Balée, W. L. (2001). Environment, culture, and Sirionó plant names. In L. Maffi (Ed.), *On biocultural diversity: Linking language, knowledge, and the environment* (pp. 298–310). Washington, DC: Smithsonian Institution Press.

Cochran-Smith, M. (2003). Learning and unlearning: The education of teacher educators. *Teaching and Teacher Education, 19,* 5–28.

Cochran-Smith, M., & Lytle, S. L. (1993). *Inside/outside: Teacher research and knowledge.* New York: Teachers College Press.

Crawford, T., Kelly, G. J., & Brown, C. (2000). Ways of knowing beyond facts and laws of science: An ethnographic investigation of student engagement in scientific practices. *Journal of Research in Science Teaching, 37*, 237–258.

Gallard, A. J. (2003). *Creating a multicultural learning environment in the science classroom.* NARST research matters to the science teacher. http://www.enc.org/focus/multi/document.shtm?input=ACQ-111327–1327.

Gallas, K. (1995). *Talking their way into science: hearing children's questions and theories, responding with curricula.* New York: Teachers College Press.

Grosjean, F. (1989). Neurolinguists, beware! The bilingual is not two monolinguals in one person. *Brain and Language, 36*(3), 3–15.

Halliday, M. A. K. (1964). Comparison and translation. In M. A. K. Halliday, M. McIntosh, & P. Strevens (Eds.), *The linguistic sciences and language teaching.* London: Longman.

Halliday, M. A. K., & Martin J. R. (1993). *Writing science: Literacy and discursive power.* New York: Routledge.

Howes, E. V. (2002). Learning to teach science for all in the elementary grades: What do preservice teachers bring? *Journal of Research in Science Teaching, 39*, 845–869.

Kelly, G., & Breton, T. (2000, April). *Framing science as disciplinary inquiry in bilingual classrooms.* Paper presented at the Annual Meeting of the American Educational Research Association, New Orleans.

Kluth, P., & Straut, D. (2003). Do as we say and as we do: Teaching and modeling collaborative practice in the university classroom. *Journal of Teacher Education, 54*, 228–240.

Lee, O. (2001). Promoting scientific inquiry with elementary students from diverse cultures and languages. In G. Griffin (Ed.), *The education of teachers: Ninety-eighth yearbook of the National Society for the Study of Education* (pp. 23–69). Chicago, IL: University of Chicago Press.

Lee, O., & Fradd, S. H. (1998). Science for all, including students from non-English language backgrounds. *Educational Researcher, 27*, 12–21.

Lemke, J. L. (1990). *Talking science: Language, learning, and values.* Norwood, NJ: Ablex.

Lemke, J. L. (1999). Meaning-making in the conversation: Head spinning, heart winning, and everything in between. *Human Development, 42*, 87–91.

Lemke, J. L. (2001). Articulating communities: Sociocultural perspectives in science education. *Journal of Research on Teaching, 38*, 296–316.

Maffi, L. (Ed.). (2001). *On biocultural diversity: Linking language, knowledge, and the environment.* Washington, DC: Smithsonian Institution Press.

Moll, L. C. Amanti, K., Neff, D., & Gonzalez, N. (1992). Funds of knowledge for teaching: Using a qualitative approach to connect homes and classrooms. *Theory into Practice, 31*, 132–141.

National Research Council (NRC). (2000). *Inquiry and the national science standards.* Washington, DC: National Academy Press.

Nazarea, V. D. (1998). *Cultural memory and biodiversity.* Phoenix, AZ: University of Arizona Press.

Rodriguez, A. (1997). The dangerous discourse of invisibility: A critique of the National Research Council's National Science Education Standards. *Journal of Research in Science Teaching, 34*, 19–37.

Rodriguez, A. (1998). Strategies for counterresistance: Toward sociotransformative constructivism and learning to teach science for diversity and for understanding. *Journal of Research in Science Teaching, 35*, 589–622.

Rosebery, A. S., Warren, B., & Conant, F. R. (1991). *Appropriating scientific discourse: Findings from minority language classrooms.* Cambridge, MA: TERC.

Rothschild, D. (Ed.). (1997). *Protecting what's ours: Indigenous peoples and biodiversity.* Oakland, CA: South and Meso American Indian Rights Center (SAIIC).

Skutnabb-Kangas, T. (2000). Connections between biodiversity and linguistic and cultural diversity. In *Linguistic genocide in education—or, worldwide diversity and human rights?* (pp. 63–99). Mahwah, NJ: Lawrence Erlbaum Associates.

Stoddart, T., Pinal, A., Latzke, M., & Canaday, D. (2002). Integrating inquiry science and language development for English language learners. *Journal of Research in Science Teaching, 39,* 664–687.

Torres-Guzmán, M. (1997). *An integrated biliteracy approach.* Bethesda, MD: TG & Associates.

van Dijk, T. A. (1980). *Macrostructures: An interdisciplinary study of global structures in discourse, interaction, and cognition.* Hillsdale, NJ: Lawrence Erlbaum Associates.

Vygotsky, L. S. (1962). *Thought and language* (E. Hanfmann & G. Vakar, Eds. & Trans.). Cambridge, MA: MIT Press. (Original work published 1934).

Wallace, C. S. (2004). Framing new research in science literacy and language use: Authenticity, multiple discourses, and the "third space." *Science Education, 88,* 901–914.

Warren, B., & Rosebery, A. S. (2002). *Teaching science to at-risk students: Teacher research communities as a context for professional development and school reform.* Final Report, Executive Summary, Project 4.1. Cambridge, MA: TERC.

Wertsch, J. V. (1991). *Voices of the mind: A sociocultural approach to mediated action.* Cambridge, MA: Harvard University Press.

Wollock, J. (2001). Linguistic diversity and language diversity: Some implications for the language sciences. In L. Maffi (Ed.), *On biocultural diversity: Linking language, knowledge, and the environment* (pp. 248–262). Washington, DC: Smithsonian Institution Press.

Yore, L. D., Hand, B. M., & Florence, M. K. (2004). Scientists' views of science, models of writing, and science writing practices. *Journal of Research in Science Teaching, 41,* 338–369.

Conclusion

Having arrived at this, the end of the volume, we hope that the authors' cumulative message is clear: science education is in a state of crisis. To return to Gates' remarks from our Introduction, the system that has taken the failure of science education to meet the needs of all students as business-as-usual must be confronted and transformed. Like many crises, the magnitude of concern and promptness of action over this state of affairs will depend a great deal on the public's perception of the problem and its will to address it.

The disparity between dominant and non-dominant populations with respect to participation and outcomes in science classrooms and professions does not go completely unacknowledged. Certainly efforts to begin to address the disparity have been laudable. But the fact remains that there is something inherently wrong when, in a democratic system, the community that produces the scientific policies and practices that influence the way we live in and understand our world is not reflective of the diversity in the population whom those policies and practices ultimately affect.

To be clear, we are not saying that students from non-dominant backgrounds lack the opportunity to receive science instruction; rather, we are saying that the forms of science instruction in which they participate often lack the means to advance learning and incite long-term interest. There is little material and human support to build a space in which these students can make meaningful connections between the ways of knowing and talking they encounter in their homes and communities and those they encounter in school science.

To use a familiar metaphor, school science is a dinner table at which, ostensibly, through the call for "Science for All," a place has been set for everyone. In reality, however, the fare is meager for and unfamiliar to many of the invitees and, even worse, other dinner guests and their discussions are alienating and exclusionary. Those who feel uncomfortable will either leave the dinner early or make sure they never return. In order to make the science dinner table a more welcoming setting, we as science education professionals must transform how we understand the purposes and processes of the meal. We must create a place for all around the table, not merely by setting down placards that announce our awareness of the presence of our guests, but by honoring their presence through consideration of their tastes and conversation about their talents. In the end, we may find that the enjoyment of our own meal has been enhanced and that, at the next meal, the list of satisfied invitees and future hosts has grown beyond our expectations. Our hope is that you consider the culturally-responsive framework we utilized here as a way to expand the science dinner menu. Socioculturally-conscious, diversity-affirming, change-oriented, constructivist approaches offer alternative ways of meal planning that stimulate the appetite for a wider variety of dishes whose appeal may be more far-reaching than the standard science *du jour.*

Our use of this science dinner table metaphor, though playful, is meant to ease entry into the serious tension between the ideas of recognition and redistribution in efforts for social justice (Fraser & Honneth, 2003). Discourse around equity issues in science education often takes a thin, recognition-only approach to social justice where the goal is understood as "a difference-friendly world, where assimilation to majority or dominant cultural norms is no longer the price for equal respect" (Fraser & Honneth, 2003, p. 7). This requires "upwardly revaluing disrespected identities and the cultural products of maligned groups; recognizing and positively valorizing cultural diversity" (p. 13). By merely calling everyone to the table, as in the movement for "Science for All," we extend this thin kind of recognition. What is needed, however, to achieve a more substantial kind of recognition, a thick approach to social justice, is the accompanying goal of redistribution. This approach focuses on "injustices it defines as socio-economic and presumes to be rooted in the economic structure of society" (p. 13). This requires redistributing income and/or wealth, reorganizing the division of labor ... or transforming other basic economic structures" (p. 13). When we not merely call everyone to the table, but, going further, ensure that everyone is well-served and well-fed, we redistribute these important socio-economic resources. It is through such acts of redistribution, when we share the tools and positions of power, that we demonstrate a genuine recognition for all whom we've invited to the table. Science education, we firmly believe, is not only a place where these thick interwoven processes of redistribution and recognition can happen, it is a place where they must happen.

The chapters you have read present three, not mutually exclusive, avenues through which redistribution and recognition, to address the science education crisis, can occur: The Language of Science Schooling, Science Learning Funds of Knowledge, and the Development of Science Learner Identities. Students of science need access to and explicit instruction in, through transformed curriculum and instruction, the academic language by which mastery of conceptual understanding is attained and measured. Students of science also need transformed curriculum and instruction that encourages deep and enduring connections between their everyday science experiences and the more specialized ways of knowing and talking in classroom science settings. And students of science need transformed curriculum and instruction that engages them agentively and generatively in building meaningful science-learning identities. If we understand the message of "Science for All" in a thick sense, it is our responsibility to change our materials and practices with these student needs in mind.

In closing, we urge you to consider what the teaching and research initiatives highlighted in this volume's chapters and end-of-chapter activities mean for your science teacher identity. Where does your professional journey intersect with those of the authors, or with those of the teachers, students, and families featured in their chapters? As the editors, we shared a sense of excitement that our paths have led us to this very moment where the world seems poised to recognize the crisis in science education and to realize the potential for and promise of change. What excites you? Where might your path and ours converge? Let us go out into the world together, and with the students and families it is our privilege as educators to serve, let us build it anew. At the table of change, there is always a place for us all.

Reference

Fraser, N., & Honneth, A. (2003). *Redistribution or recognition?: A political-philosophical exchange.* London: Verso.

Afterword

Sonia Nieto, Series Editor

As this text has made abundantly clear, teachers today face a far different landscape than did their peers half a century ago. For one, the children in our classrooms are different from the children in schools of several decades ago. To be sure, our nation has always been characterized by tremendous diversity of race, language, national origin, social class, and so on, but this diversity is more evident today than at any time in our history. Not only is the diversity obvious in our large urban schools, but it is also increasingly visible in suburban and rural schools. Yet the teaching force is overwhelmingly middle-class, White, and monolingual English speaking. Most teachers, including those who teach science, often have had little experience with diversity, either personally or professionally. Add to this the tremendous differences in achievement among students of different backgrounds and we begin to see the challenges faced by science teachers today.

The chapters in this book make it clear that science teaching, too, is different from what it was when many of us went to school. The rapidly changing nature of science means that teachers need to be flexible, keeping abreast of daily changes in order to reflect these changes in their curriculum. The pedagogy of science is also changing, as teachers try to heed calls to not just prepare students with current and reputable information, but also to encourage them to be curious, questioning, and creative. The enormous ethical challenges posed by technological advances in biology, ecology, and other fields also make this a difficult but exciting time in which to teach science. This is the field in which you, the teachers of today and tomorrow, will make your mark.

The editors and authors of *The Work of Language in Multicultural Classrooms: Talking Science, Writing Science* have taken it upon themselves to address these two crucial issues in education today, that is, the growing diversity in our nation's schools, and the increasing need for citizens with strong skills in technology and science. But rather than simply write a book about diversity and science teaching that includes a few "ethnic titbits" for science teachers to use in their practice, editors Katherine Richardson Bruna and Kimberley Gomez, along with their chapter authors, have addressed these needs through a social justice perspective. They have asked you to consider questions of great import: How can we best teach the language of science so that all students, especially those who have been educationally marginalized, can learn to "master the code"? How can we draw on students' abilities, identities, and experiences to make science more appealing and significant in their particular contexts? How can science teaching take into account the sociopolitical context in which education takes place? What does it mean to be culturally responsive within the science classroom? These and other questions undergird the chapters in this book because the intention of the authors is not to tinker with science education, but rather to transform it in fundamental ways. Instead of viewing science as a mere technical skill or set of procedures, the authors of this text have challenged the neutral and "culture-less" space so often inhabited by science. Their reasoning is that if students' linguistic, cultural, and experiential resources are respected and incorporated into the science education they receive, this approach will diminish the tremendous marginalization of students whose cultural, racial, and linguistic backgrounds differ from the so-called "mainstream."

This is indeed a noble aim. This text, I believe, will push the field of science education to consider how diversity can be an asset—rather than a deficit, as it is so often presumed to be—in the science classroom. As preservice and practicing teachers of science, through reading this text, you have been privy to classroom conversations among students and their teachers. You have witnessed how teachers struggle to teach science in ways that make it exciting, meaningful, and respectful to their students, and you have seen how they sometimes fail to do so. You have read about the challenges of using pedagogy that connects with students' lives, and you have been challenged to think about how all of these studies and stories might affect your own teaching. In a word, you have benefited from the insights and experiences, and the successes and mishaps, of the teachers and researchers who have shared their knowledge and their journeys with you. It is my hope that all of these will have an impact on your own teaching so that science can become, in the words of Richardson Bruna and Gomez, "a site of cultural and linguistic responsiveness."

Contributors

Doris Ash is an Assistant Professor within the Education Department at the University of California Santa Cruz.

Patricia Baquedano-López is an Associate Professor of Language and Literacy, Society and Culture within the Graduate School of Education at the University of California, Berkeley.

María Estela Brisk is a Professor and Chair of the Department of Teacher Education within the Lynch School of Education at Boston College.

Katherine Richardson Bruna is an Assistant Professor of Multicultural and International Curriculum Studies within the College of Human Sciences at Iowa State University.

Tamara Ciesla is a Ph.D. student and Research Assistant within the College of Education at the University of Illinois at Chicago.

KimMarie Cole is an Associate Professor of English at SUNY Fredonia.

Rhiannon Crain is a Graduate Student within the Education Department at the University of California Santa Cruz.

Noel Enyedy is an Associate Professor at the Graduate School of Education and Information Studies at UCLA. He is also the Director of Research at the University Elementary School.

Eugene E. Garcia is Vice President of Education Partnerships at Arizona State University.

Jennifer S. Goldberg is an Assistant Professor within the Graduate School of Education and Applied Professions at Fairfield University in Fairfield, Connecticut.

Kimberley Gomez is an Assistant Professor in the Literacy, Language and Culture Program within the Department of Curriculum and Instruction at the University of Illinois at Chicago.

Louis Gomez is the Aon Professor of Learning Sciences and Computer Science within the School of Education and Social Policy at Northwestern University.

Holly Hansen-Thomas is an Assistant Professor of TESOL and Literacy Education within the School of Education at Binghamton University—SUNY.

Phillip Herman is a research Assistant Professor within the School of Education and Social Policy at Northwestern University.

Elaine V. Howes is a Professor in the Science Education Program within the Department of Secondary Education at the University of South Florida.

Julian Jefferies is a Ph.D. candidate within the Lynch School of Education in the Department of Teacher Education at Boston College.

Stacy Kaczmarek is a Graduate Student within the Lynch School of Education in the Department of Teacher Education at Boston College.

Shlomy Kattan is a Ph.D. candidate in Language and Literacy, Society and Culture within the Graduate School of Education at the University of California, Berkeley.

Yu-Min Ku is an Assistant Professor within the Graduate Institute of Learning and Instruction at National Central University in Jhongli City, Taiwan.

Joel Kuipers is a Professor within the Department of Anthropology at The George Washington University.

Okhee Lee is a Professor in the Department of Teaching and Learning at the University of Miami.

Lindsey A. Massoud is a Research Assistant within the Department of Anthropology at The George Washington University.

Christine C. Pappas is a Professor within the College of Education at the University of Illinois at Chicago.

John M. Reveles is an Assistant Professor at the Michael D. Eisner College of Education in Northridge, California.

Jennifer Zoltners Sherer is a Research Associate in the Learning, Research and Development Center at The University of Pittsburgh.

Jorge L. Solís is a Ph.D. candidate in Language and Literacy, Society and Culture within the Graduate School of Education at the University of California, Berkeley.

Lisa Patel Stevens is an Assistant Professor within the Lynch School of Education in the Department of Teacher Education at Boston College.

Kip Tellez is an Associate Professor within the Education Department at the University of California, Santa Cruz.

María E. Torres-Guzmán is an Associate Professor in the Program in Bilingual/Bicultural Education International and Transcultural Studies within Teachers College at Columbia University.

Eli Tucker-Raymond is a Ph.D. student and Research Assistant within the College of Education at the University of Illinois at Chicago.

Maria Varelas is a Professor within the College of Education at the University of Illinois at Chicago.

Gail Brendel Viechnicki is an Assistant Research Professor in the Department of Anthropology at The George Washington University.

Kate Muir Welsh is an Assistant Professor within the College of Education at the University of Wyoming.

Jolene White is a Research Associate at The University of Illinois at Chicago.

Adam Williams is a Research Associate within the School of Education and Social Policy at Northwestern University.

Laura J. Wright is a Research Assistant within the Department of Anthropology at The George Washington University.

Monica S. Yoo is a Ph.D. student at the University of California, Berkeley.

Author Index

Subject Index